UNMANNED AERIAL VEHICLES

LICENSE, DISCLAIMER OF LIABILITY, AND LIMITED WARRANTY

By purchasing or using this book and its companion files (the "Work"), you agree that this license grants permission to use the contents contained herein, but does not give you the right of ownership to any of the textual content in the book or ownership to any of the information, files, or products contained in it. *This license does not permit uploading of the Work onto the Internet or on a network (of any kind) without the written consent of the Publisher.* Duplication or dissemination of any text, code, simulations, images, etc. contained herein is limited to and subject to licensing terms for the respective products, and permission must be obtained from the Publisher or the owner of the content, etc., in order to reproduce or network any portion of the textual material (in any media) that is contained in the Work.

Mercury Learning and Information ("MLI" or "the Publisher") and anyone involved in the creation, writing, production, accompanying algorithms, code, or computer programs ("the software"), and any accompanying Web site or software of the Work, cannot and do not warrant the performance or results that might be obtained by using the contents of the Work. The author, developers, and the Publisher have used their best efforts to insure the accuracy and functionality of the textual material and/or programs contained in this package; we, however, make no warranty of any kind, express or implied, regarding the performance of these contents or programs. The Work is sold "as is" without warranty (except for defective materials used in manufacturing the book or due to faulty workmanship).

The author, developers, and the publisher of any accompanying content, and anyone involved in the composition, production, and manufacturing of this work will not be liable for damages of any kind arising out of the use of (or the inability to use) the algorithms, source code, computer programs, or textual material contained in this publication. This includes, but is not limited to, loss of revenue or profit, or other incidental, physical, or consequential damages arising out of the use of this Work.

The sole remedy in the event of a claim of any kind is expressly limited to replacement of the book and only at the discretion of the Publisher. The use of "implied warranty" and certain "exclusions" vary from state to state, and might not apply to the purchaser of this product.

Companion files of color plates and figures from the book also available for downloading from the publisher by writing to info@merclearning.com.

Unmanned Aerial Vehicles

An Introduction

P. K. GARG

MERCURY LEARNING AND INFORMATION
Dulles, Virginia
Boston, Massachusetts
New Delhi

Copyright ©2021 by Mercury Learning and Information LLC. All rights reserved.

This publication, portions of it, or any accompanying software may not be reproduced in any way, stored in a retrieval system of any type, or transmitted by any means, media, electronic display or mechanical display, including, but not limited to, photocopy, recording, Internet postings, or scanning, without prior permission in writing from the publisher.

Publisher: David Pallai
Mercury Learning and Information
22841 Quicksilver Drive
Dulles, VA 20166
info@merclearning.com
www.merclearning.com
1-800-232-0223

P.K. Garg. *Unmanned Aerial Vehicles.*
ISBN: 978-1-68392-709-9

The publisher recognizes and respects all marks used by companies, manufacturers, and developers as a means to distinguish their products. All brand names and product names mentioned in this book are trademarks or service marks of their respective companies. Any omission or misuse (of any kind) of service marks or trademarks, etc. is not an attempt to infringe on the property of others.

Library of Congress Control Number: 2021940431

212223321   Printed on acid-free paper in the United States of America.

Our titles are available for adoption, license, or bulk purchase by institutions, corporations, etc. For additional information, please contact the Customer Service Dept. at 800-232-0223(toll free).

All of our titles are available in digital format at *academiccourseware.com* and other digital vendors. The sole obligation of Mercury Learning and Information to the purchaser is to replace the book, based on defective materials or faulty workmanship, but not based on the operation or functionality of the product. *Companion files of color plates and figures from the book also available for downloading from the publisher by writing to info@merclearning.com.*

Dedicated to the memory of my loving parents

CONTENTS

Preface xv

Chapter 1: **Overview of Unmanned Aerial Vehicles** 1

    1.1 Introduction 1
    1.2 Historical Developments 3
    1.3 Uses of UAVs 9
    1.4 Some Technical Terms 17
    1.5 Characteristics of UAVs 19
        1.5.1 Photogrammetry UAVs 22
        1.5.2 LiDAR UAVs 27
        1.5.3 A Comparison of Photogrammetry UAVs and LiDAR UAVs 31
    1.6 Working of a UAV 33
        1.6.1 Altitude 36
        1.6.2 Vertical and Oblique Photographs 36
        1.6.3 Positioning 37
    1.7 Advantages and Disadvantages of UAVs 37
        1.7.1 Advantages of UAV 38
        1.7.2 Disadvantages of UAV 40
    1.8 Summary 41
    Review Questions 42

Chapter 2:	**Various Components of UAV**	**45**
	2.1 Introduction	45
	2.2 Basic Aerodynamics	46
	2.2.1 Lift	47
	2.2.2 Drag	47
	2.2.3 Thrust	48
	2.3 Stability and Control of UAV	48
	2.3.1 Roll	49
	2.3.2 Pitch	50
	2.3.3 Yaw	50
	2.3.4 Throttle	51
	2.4 UAVs Classification	51
	2.4.1 According to Operation & Use	53
	2.4.2 According to Size	54
	2.4.3 According to Altitude	55
	2.4.4 According to Wings Type	58
	2.5 Components of a UAV	61
	2.5.1 Transmitter	63
	2.5.2 Receiver	63
	2.5.3 Body Frame	64
	2.5.4 Motors	67
	2.5.5 Electronic Speed Controller (ESC)	69
	2.5.6 Propellers	70
	2.5.7 Flight Controller	72
	2.5.8 The Payloads	74
	2.5.9 Gimbals and Tilt Control	75
	2.5.10 Sensors	75
	2.5.11 Global Positioning System (GPS)	80
	2.5.12 Remote Sensing Sensors	81
	2.5.13 Battery	91
	2.5.14 Ground Control Station (GCS)	95
	2.6 Summary	96
	Review Questions	97

Chapter 3:	**Autonomous UAVs**		**99**
	3.1	Introduction	99
	3.2	The Automatic and Autonomous UAVs	100
	3.3	Architecture of an Autonomous System	103
	3.4	The Concept of Autonomy	105
	3.5	Various Measures of UAV Autonomy	108
		3.5.1 Fully Autonomous Operations	108
		3.5.2 Semi-autonomous Operations	108
		3.5.3 Fully Human Operated	109
		3.5.4 Sheridan Scale for Autonomy	109
		3.5.5 UAV Rating and Evaluation	110
	3.6	Simulation of the Environment	112
	3.7	Path Planning	116
		3.7.1 Mission Planning	117
		3.7.2 Flight Planning	119
	3.8	Controlling the Mission Planning	120
	3.9	Controlling the Payloads	122
	3.10	Flight Safety Operation	123
		3.10.1 Geofencing	124
		3.10.2 Collision Avoidance System	126
		3.10.3 UAV Detection Technology	127
		3.10.4 Air Traffic Management	131
	3.11	Path Planning Algorithms	132
		3.11.1 Open-Source Software	136
		3.11.2 Commercial Software	139
		3.11.3 Simultaneous Localization and Mapping (SLAM) Algorithm	140
	3.12	Intelligent Flight Control Systems	141
	3.13	Innovative Technological UAVs/Drones	144
	3.14	Cost of UAVs/Drones	147
	3.15	Summary	148
	Review Questions		149

Chapter 4:	**Communication Infrastructure of UAVs**	**151**
	4.1 Introduction	151
	4.2 UAV Communication System	152
	4.3 Types of Communication	154
	4.4 Wireless Sensor Network (WSN) System	158
	4.5 Free Space Optical (FSO) Approach	162
	4.6 The FPV Approach	164
	4.7 Cybersecurity Approach	166
	4.8 Swarm Approach	167
	4.8.1 Infrastructure-based Swarm Architecture	169
	4.8.2 Flying Ad hoc Networks (FANETs)-based Swarm Architecture	170
	4.9 Internet of Things (IoT)	172
	4.10 Summary	177
	Review Questions	178
Chapter 5:	**UAV Data Collection and Processing Methods**	**179**
	5.1 Introduction	179
	5.2 Data Products	180
	5.2.1 Aerial Remote Sensing Data	181
	5.2.2 LiDAR Data	183
	5.2.3 Ground Control Points (GCPs)	184
	5.3 Data Processing Approaches	185
	5.3.1 Georeferencing Approach	185
	5.3.2 Terrain Modeling Approaches	187
	5.3.3 Image Processing Approaches	200
	5.3.4 Cloud Computing Approaches	208
	5.3.5 UAV Photogrammetry Software	209
	5.3.6 LiDAR UAVs Software	217
	5.4 Summary	223
	Review Questions	223

Chapter 6:	Regulatory Systems for UAVs	225
	6.1 Introduction	225
	6.2 Issues of Concern for UAVs	226
	6.2.1 Air Traffic Control	228
	6.2.2 Wi-Fi/Bluetooth Security Issues	229
	6.2.3 Privacy Issues	230
	6.2.4 Moral and Ethical Issues	231
	6.2.5 Safety Issues	231
	6.2.6 Legal Issues	232
	6.3 Aviation Regulations	233
	6.3.1 The FAA Regulations	235
	6.3.2 The European Union (EU) Regulations	236
	6.3.3 The International Civil Aviation Organization (ICAO)	240
	6.3.4 Australia	242
	6.3.5 Canada	243
	6.3.6 China	243
	6.3.7 Germany	244
	6.3.8 Hong Kong	244
	6.3.9 Israel	244
	6.3.10 Japan	245
	6.3.11 New Zealand	245
	6.3.12 Russia	246
	6.3.13 South Africa	246
	6.3.14 Ukraine	247
	6.3.15 India	247
	6.4 Summary	253
	Review Questions	254
Chapter 7:	Various Applications of UAVs	255
	7.1 Introduction	255
	7.2 Overview of UAVs/Drones Applications	256
	7.3 Various Applications	263

	7.3.1	Surveying and Mapping	263
	7.3.2	Industrial Applications	263
	7.3.3	Inspection and Monitoring	271
	7.3.4	Forestry	276
	7.3.5	Agriculture	280
	7.3.6	Hydromorphological Studies	294
	7.3.7	3D Modeling-Based Applications	294
	7.3.8	Wireless Network-Based Applications	295
	7.3.9	Smart Cities	297
	7.3.10	Traffic Monitoring and Management	298
	7.3.11	Swarm-based Applications	300
	7.3.12	Surveillance Operations	301
	7.3.13	Search and Rescue (SaR) Operations	304
	7.3.14	Logistics Operations	306
	7.3.15	Crowd-Sourcing	309
	7.3.16	Post-disaster Reconnaissance	311
	7.3.17	Security and Emergency Response	312
	7.3.18	Insurance Sector	312
	7.3.19	Health Sector	313
	7.3.20	Covid-19 Pandemic	315
	7.3.21	Entertainment and Media	319
	7.3.22	Military	320
	7.4	Summary	324
		Review Questions	325
Chapter 8:	**The Future of UAV Technology**		**327**
	8.1	Introduction	327
	8.2	Future Insights	328
	8.3	Key Challenges	330
		8.3.1 Topography and Environmental Aspects	331
		8.3.2 Legal Aspects	332
		8.3.3 Privacy Aspects	332
		8.3.4 Air Safety and Security Aspects	332

		8.3.5	Cyber-security Aspects	333
		8.3.6	Reliable Communication System	335
		8.3.7	Swarm-based Operations	337
		8.3.8	IoT-based Operations	339
		8.3.9	Big Data and Cloud Computing	340
		8.3.10	Energy Sources	342
		8.3.11	Drone Taxis	349
	8.4	Swot Analysis of UAV Technology		349
		8.4.1	Strength	349
		8.4.2	Weakness	350
		8.4.3	Opportunity	350
		8.4.4	Threats	351
	8.5	The UAVs Market Potential		352
		8.5.1	Based on Regions	354
		8.5.2	Based on Applications	356
		8.5.3	Based on Autonomy	360
		8.5.4	Based on Size	360
		8.5.5	Based on Hardware and Software	361
	8.6	Emerging Research Areas		362
	8.7	Summary		366
	Review Questions			367

References   369
Index   409

PREFACE

The technology and applications of Unmanned Aerial Vehicles (UAV) is growing globally at a fast pace. This book covers the theory, hardware, and software components of UAVs. It also explores the utilization of UAVs in different application areas, such as construction, oil and gas, mining, agriculture, forestry, search and rescue, surveillance, transportation, disaster, logistics, health, journalism, retail delivery, and many more.

The book is comprised of eight chapters which provide an overview of basic concepts, components of UAVs, various sensors used, architecture of autonomous UAVs, communication tools and devices to acquire real-time data from UAVs, the software needed to analyze the UAV data, required rules and regulations to fly UAVs, and various application areas. The final chapter highlights several areas which need further research in UAV hardware and software components as they become even more sophisticated. Review questions are included at the end of chapters to enhance comprehension of the subject matter. Generally, this book would be helpful to readers who want to understand the intricacy of UAVs and to appreciate their utility for the benefit of mankind and society.

Companion Files

Color plates and figures from this book may be obtained by writing to the publisher at *info@merclearning.com*.

P. K. Garg
June 2021

CHAPTER 1

OVERVIEW OF UNMANNED AERIAL VEHICLES

1.1 INTRODUCTION

The drone is more formally known as an Unmanned Aerial Vehicle (UAV) or an Unmanned Aircraft System (UAS) or a Remotely Piloted Aircraft (RPA) that can fly autonomously or guided remotely without a human in control (Glover, 2014). A drone is like a male honey bee. Several names are available in the literature, like UAV, Unmanned Air Vehicle, Unmanned Aircraft Vehicle, Unmanned Aerospace Vehicle, Uninhabited Aircraft Vehicle, Unmanned Airborne Vehicle, Unmanned Autonomous Vehicle, Upper Atmosphere Vehicle, Aerial Robots, etc. In this book, both terms, drone and UAV have been used interchangeably.

The UAV is defined as an aircraft that operates without an onboard pilot. Essentially, a UAV is a flying robot that can be remotely controlled or fly autonomously through software-controlled flight plans in their embedded systems (Austin, 2010). It can work in conjunction with several onboard sensors, including Inertia Measurement Unit (IMU) and Global Positioning System (GPS). The UAV is a lightweight flying device similar to an aircraft; however, different from an aircraft since it is not operated by a pilot onboard (Gupta *et al.*, 2018). Based on how its movement is controlled, it can fly with a human remotely controlling the UAV, or in the most advanced cases, fly itself without any human intervention.

The terms UAV and UAS are closely related; however, the distinction should be made. The UAS is a much broader term describing the entire system that enables UAV operations. The UAS terminology is frequently used by the Federal Aviation Administration (FAA) of the United States, the European Aviation Safety Agency, and the UAV Systems Association (Dawson, 2018). A UAS consists of a UAV platform, in addition to the ground control station (GCS), communication, data link, piloting, sensing, flight planning, and other critical components needed for the operation of a UAV. A simple generic UAS system is shown in Figure 1.1.

The UAV is a component of the UAS since it refers to only the vehicle/aircraft itself. The UAV-related topics cannot be covered completely without reference to the relevant UAS components. A UAS unit may consist of more than one UAV; for example, a US Air-force Predator unit contains four UAVs. Most UASs have been developed from the military world, but in the last decades, UASs have penetrated a variety of civilian applications (Boucher, 2015). This is mainly due to the lowering of price and the development of sensors, imaging technologies, and UAV components. The UAS is an emerging sector of the aerospace industry with great opportunity and market demands (≈70% of global growth and market share is in the US) that can be leveraged to high profitability in the near future (PwC, 2018).

The International Civil Aviation Organization (ICAO) employs the RPA system as these systems, based on cutting-edge developments in aerospace technologies, offer advancements for civil-commercial applications and improvements to the safety & efficiency of the entire civil aviation (ICAO, 2015). Many pilots prefer RPA because flying certain types of UAVs requires more skill. The RPA is essentially interchangeable with UAV. Taking control of an RPA requires more than simple handheld controls (Dawson, 2018).

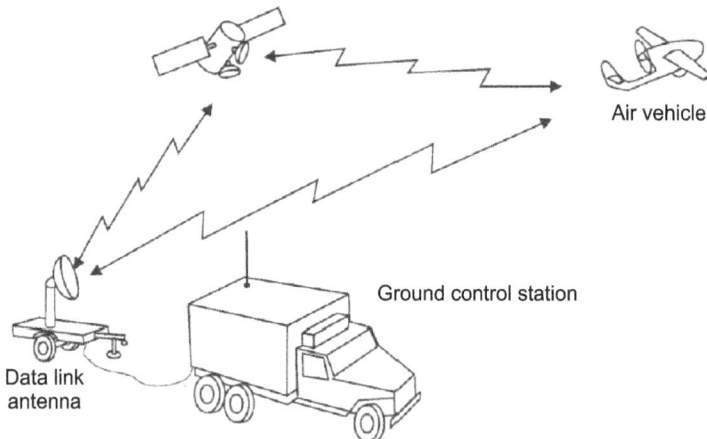

FIGURE 1.1 A typical UAS system.

The UAVs carry cameras, sensors, communications equipment, or other payloads. Remotely operated UAVs are actively controlled by a pilot on the ground, while autonomous UAVs perform functions without active human involvement. Varying degrees of automation are available in most modern UAVs (Singhal *et al.*, 2018). For example, a UAV may be remotely operated by a pilot but may use automation to maintain its position without commands from the pilot. The relative autonomy of a specific platform will also depend on the software used. The UAVs are opening up many new opportunities, from pilots for UAVs to electronics, sensors, and cameras. UAVs' increase in awareness and mission capabilities are driving innovations and new applications worldwide (White Paper, 2018).

1.2 HISTORICAL DEVELOPMENTS

Hot air balloons and explosives were first used for military reconnaissance in the French Revolutionary Wars to attack the enemies. The history of drones traces back to 1849 in Italy when Venice was fighting for its independence from Austria. Austrian soldiers attacked Venice with hot air, hydrogen- or helium-filled balloons equipped with bombs (Newcome, 2004). The kite was also used to collect information. In 1883, Douglas Archibald, an Englishman, attached an anemometer to a kite line and measured wind velocity at altitudes up to 1,200 ft. Mr. Archibald attached cameras to kites in 1887, providing one of the world's first reconnaissance UAVs. William Eddy took hundreds of photographs from kites during the Spanish-American War, which may have been one of the first uses of UAVs in combat (Dronethusiast, 2020).

During World War I, interest in UAVs was created in 1916 when Hewitt-Sperry Automatic Airplane was developed, and the Dayton-Wright Airplane Company invented an unmanned aerial torpedo in 1917 that carried explosives to a target at a preset time (Keane and Carr, 2013). This led to full-sized airplanes with primitive controls capable of stabilizing and navigating without a pilot onboard. The conversion of manned airplanes to target drones continued during the 1920s and 1930s (NIAS, 2018). During World War II, a large number of radio-controlled, radar-controlled, and television-controlled glide bombs were invented and used. During World War II, the first large-scale production of drones was made by Reginald Denny Industries (IWM, 2018). They manufactured nearly 15,000 drones for the US Army to be used for training anti-aircraft gunners. The Cold War also contributed to the evolution of drones. The most famous was the US Navy TDR-1, the US Army AZON, and the German Fritz X.

In 1917, Charles Kettering (of General Motors) developed a biplane UAV for the Army Signal Corps. It took him about three years to develop, which was called the Kettering Aerial Torpedo, but is better known as the "Kettering Bug" or just the "Bug." The Bug could fly nearly 40 miles at 55 mph and carry 180 lb of high explosives (Blom, 2011). The air vehicle was guided to the target by preset controls and had detachable wings released when it was over the target, allowing the fuselage to plunge to the ground like a bomb (Keane and Carr, 2013). In 1917, Lawrence Sperry developed a UAV, similar to Kettering, for the Navy, called the Sperry-Curtis Aerial Torpedo. It made several successful flights out of Sperry's Long Island airfield but was not used in the war.

The first used drone appeared in 1935 as a full-size retooling of the de Havilland DH82B "Queen Bee" biplane, which was fitted with a radio and servo-operated controls in the back sea (NIAS, 2018). The term *drone* dates to this initial use, a plane on the "Queen Bee" nomenclature. The plane, launched in 1941 (shown in Figure 1.2), could be conventionally piloted from the front seat, but generally, it flew unmanned, and was shot at by artillery gunners in training. The first target drone converted for a battlefield unmanned aerial photo reconnaissance mission was a version of the MQM-57 in 1955 (Dronethusiast, 2020).

FIGURE 1.2 Launch of a DH.82 Queen Bee (mother of drones) target drone (1941) (Dronethusiast, 2020).

Several pioneers developed important parts of the UAV system. For example, Archibald Montgomery Low (Professor Low of England), known as the "Father of Radio Guidance Systems," developed the first data link and solved interference problems caused by the UAV engine. On 3 September, 1924, he made the world's first successful radio-controlled flight. In 1933, the British flew three refurbished Fairey Queen biplanes by remote control from a ship (IWM, 2018). Two of them crashed, but the third one flew successfully, making Great Britain the first country to fully appreciate the value of UAVs, especially after they decided to use one as a target but could not shoot it down. In 1937, another Englishman, Reginald Leigh Denny, and two Americans, Walter Righter and Kenneth Case, developed a series of UAVs called RP-1, RP-2, RP-3, and RP-4. They formed a company in 1939 called the Radioplane Company, which later became part of the Northrop-Ventura Division. Radioplane Company built thousands of target drones during World War II. (One of their early assemblers was Norma Jean Daugherty, later known as Marilyn Monroe). Of course, the Germans used lethal UAVs (V-1s and V-223s) during the later years of the war, but it was not until the Vietnam War era that UAVs were successfully used for reconnaissance (Sholes, 2007).

The development of UAVs took place in the 1980s (Keane and Carr, 2013). The usefulness of UAS for reconnaissance was demonstrated in Vietnam. At the same time, early steps were being taken to use them in active combat at sea and on land. Military drones used solidified in 1982 when the Israeli Air Force used UAVs to wipe out the Syrian fleet with minimal loss of Israeli forces. The Israeli UAVs acted as decoys, jammed communication, and offered real-time video reconnaissance. The Predator RQ-1L UAS from General Atomics was the first UAS deployed in the Balkans war in 1995 (Blom, 2011). This trend continued in modern conflicts.

UAV systems have been used largely in military applications throughout history, as is true of many areas of technology, with civilian applications tending to follow once the development and testing had been accomplished in the military arena (NIAS, 2018). The UAV technology continued to be of interest to the military, but it was often costly and not completely reliable to put into operational use. After concerns about the shooting down of spy planes arose, military revisited the UAVs. Military use of drones soon expanded to play roles in dropping leaflets and acting as spying decoys. Since then, drones have continued to be used in the military, playing critical roles in intelligence, surveillance and force protection, artillery spotting, target following, battle damage assessment and reconnaissance, and weaponry (IWM, 2018). Recognizing the potential of non-military, non-consumer drone applications, the FAA issued the first commercial drone permits in

2006 (FAA, 2010). It opened up new possibilities for companies or professionals who wanted to use drones in business ventures.

The widespread use of the UAV began in 2006 when the US Customs and Border Protection Agency introduced UAVs to monitor the USA and Mexico borders. In late 2012, Chris Anderson, editor-in-chief of Wired magazine, started his own drones company, called 3D Robotics, Inc. (3DR), which started off specializing in hobbyist personal drones, and now marketing its UAVs to aerial photography, film companies, construction, utilities and telecom businesses, and public safety companies, among others (Keane and Carr, 2013). In late 2013, Amazon announced a plan to use commercial drones for goods delivery activities (Kimchi, 2015). In July 2016, Reno-based startup Flirtey successfully delivered a package to a resident in Nevada *via* a commercial drone. Other companies have since followed it. For example, in September 2016, Virginia Polytechnic Institute and State University began a test with Project Wing, a unit of Google-owned Alphabet Inc., to make goods deliveries. In December 2016, Amazon delivered its first Prime Air package in Cambridge, England. In March of 2017, it demonstrated a Prime Air drone delivery in California (Intelligence, 2018).

Table 1.1 presents a summary of the historical developments of UAVs since 1782.

TABLE 1.1 A Summary of the Evolution of UAVs.

Year	Developmental Activity
1782	The first unmanned vehicle flight took place by the Montgolfier brothers.
1849	On 22 August 1849, the Austrians used UAV for the first time and attacked Venice using unmanned Balloons loaded with explosives.
1800s	The first drone with a "camera" was deployed at the end of the 19th century.
1894	Unmanned Balloons used by Austrians.
1898	Nikola Tesla invented a small unmanned boat that changes direction on verbal command by using radio frequencies to switch motors on and off and presented it in an exhibition at Madison Square Garden.
1915	During the Battle of Neuve Chapelle, aerial imagery was used by the British Military to capture about 15,000 sky view maps of the German trench fortifications in the region.
1916	During World War I, the first UAV took flight in the United States; though the success of UAV in test flight was erratic the military stamped their potential in the combat.

Year	Developmental Activity
1917	French artillery officer, René Lorin proposed flying bombs using gyroscopic and barometric stabilization and control.
1918	Charles Kettering (USA) flies Liberty Eagle "Kettering Bug."
1920	Elmer Sperry perfects the gyroscope, and the first enabling technology makes flight control feasible.
1920	Manned quadcopters were first experimented, but their effectiveness was hampered by the technology available at the time.
1920	Etienne Oemichen gave a design that had four rotors and eight propellers; all driven by one motor and recorded over 1000 successful flights. The first recorded distance was 360m in 1924 for a helicopter.
1935	The first used drone appeared as a full-size retooling of the de Havilland DH82B "Queen Bee" biplane.
1940	Reginald Denny sold about 15,000 radio-controlled target drones to the US Military to train the anti-aircraft gunners for World War II.
1941	The first large-scale production of drones by US initiated by Reginald Denny.
1943	Advanced technologies provided control, guidance, and targeting. Gyroscope governed by magnetic compass controlled azimuth—aneroid barometer used for altitude control. Speed was determined by engine performance at max. power and propeller driven "air-log" governed the range.
1943	The German Military inaugurated the FX-1400, or "The Fritz X," a bomb with four small wings and a radio controller weighing 1362 kg. The first remotely controlled munitions were put into operational use and a great breakthrough for guided aerial weapons.
1956	Model A quadcopter, designed by George de Bothezat and Ivan Jerome, was the first successful quadcopter that used varying thrust of all four propellers in order to control pitch, roll, and yaw. However, it was difficult for the pilot to fly because of the workload of controlling all four propellers' thrust simultaneously.
1958	Curtis Wright company designed Curtis Wright VZ-7 as a US Army project that used variable thrust in the four propellers to control flight to simplify the pilot's workload.
1960s	UAVs took a new role during the Vietnam War, that is, stealth surveillance from the early use of target drones and remotely piloted combat vehicles.

(continued)

Year	Developmental Activity
1970s	During this time, Israel developed two un-piloted surveillance machines, MASTIFF UAV and the IAA Scout. Also, the success of the Fire Bee continued through the end of the Vietnam War. In the 1970s, the US set its sights on other kinds of UAVs while other countries began to develop their own advanced UAV systems.
1982	Military drone use solidified. Israel outwits Soviet anti-aircraft technology at the outbreak of hostilities with Syria, by revealing its location using a swarm of unmanned aircraft.
1985	Pioneer UAV program was introduced in the 1980s when a need for an on-call, inexpensive, unmanned, over the horizon targeting, reconnaissance and battle damage assessment (BDA) capability for local commanders was identified by the US military operations in Grenada, Lebanon, and Libya. The Navy started the expeditious acquisition of UAV systems for fleet operations using non-developmental technology in July 1985.
1986	Reconnaissance Drone was a joint US and Israeli Project which produced the "RQ2 Pioneer," a medium-sized reconnaissance drone.
1990	Miniature and Micro UAVs became part of research and soon came into action.
2000	Predator Drone was first deployed in Afghanistan.
2005	The first commercial multicopter was developed by Microdrones in Germany.
2006	The use of drones become widespread.
2010	The Parrot AR Drone, a smartphone-controlled quadcopter for consumers, was introduced at the Consumer Electronics Show in Las Vegas.
2012	Congress required the FAA to integrate small drones into the national airspace by 2015.
2013	The use of commercial drones for delivery activities is planned.
2014	The drone was used for delivery by Amazon for the first time in 2014. Also, the FAA grants an exemption to film and TV production companies for drone use.
2016	Amazon delivered its first Prime Air package in Cambridge, England.
Present	In the past 10 years, many small advanced quadcopters have entered the market, including the DJI Phantom and Parrot AR drone. This new breed of quadcopters is cheap, lightweight, and uses advanced electronics for flight control.

1.3 USES OF UAVs

UAVs were most often associated with the military applications in the recent past, where they were used initially for anti-aircraft target practice, intelligence gathering, and as weapons platforms. Still, these days, UAVs are also being employed for the large number of recreational and civilian applications (Sholes, 2007). Table 1.2 lists a few example areas where UAVs are being used for various military and civilian applications (Stoica, 2018). These tasks are performed by a number of different types of UAVs/Drones, from do-it-yourself drones to drones used for the purposes of domestic surveillance to combat drones used in military strikes. There are different models of UAVs; however, some of them are hand-launched and operated remotely or fly autonomously.

TABLE 1.2 A Few Military and Civilian Applications of UAVs/Drones.

Military Uses	Civilian Uses
Reconnaissance	Aerial photography
Surveillance and Wide Area Aerial Surveillance (WAAS)	Agriculture and forestry
Target Acquisition	Coastguard and conservation
Artillery fire correction	Cities
Battlefield damage assessment (BDA)	Transportation and traffic agencies
Land mine detection	Fire services and forestry
Elimination of unexploded Improvised Explosive Device and land mines	Rivers authorities and fisheries
Battlefield situation awareness and understanding	Gas and oil supply companies
Illumination of targets by laser designations	Information services
Psychological impact on militants	Local authorities/Police authorities
Combat roles	Meteorological services
Swarms and electronic intelligence	Survey organizations/Ordinance survey

Based on the types of the wing, UAVs are broadly classified as fixed-wing or rotary wing types (Cai *et al.*, 2011). The fixed-wing UAVs move horizontally and need a larger open space for take-off and landing. On the other hand, the rotary-wing UAVs move vertically, requiring only a small open space for landing. Table 1.3 presents the basic difference between these two types of UAVs.

TABLE 1.3 Basic Characteristics of Fixed-wing and Rotary-wing UAVs.

	Fixed-Wing	**Rotary Wing**
Mechanism	The lift generated using wings with forwarding airspeed	The lift generated using blades revolving around a rotor shaft
Advantages	Simpler structure, usually higher payload, higher speed	Can hover, able to move in any direction, vertical take-off and landing
Limitations	Need to maintain forward motion, need a runway for take-off and landing	Usually lower payload, lower speed, shorter range

The UAVs are the most predominant segment of the UAS market. The UAS market includes all unmanned vehicles, such as UAVs, Blimps, and Zeppelins (Freudenrichn, 2010). A Blimp (or a "pressure airship") is a powered, steerable, large-sized balloon with fins and an engine, lighter than air (Figure 1.3a). It has a long endurance but flying at low speeds. It does not have a rigid internal structure; if a Blimp deflates, it loses its shape. Blimps are best known today for their role as advertising and promotional vehicles. Their primary military use is for anti-submarine and reconnaissance roles. They are low-tech and relatively low-cost airships. The Zeppelins are rigid or semi-rigid airships (Figure 1.3b), originally manufactured by the Luftschiffsbau-Zeppelin. They have a rigid metal skeleton consisting of a cigar-shaped, trussed, and covered frame supported by internal gas cells. They are more suitable for long trips in a wider variety of weather conditions, making them expensive (Marsh, 2013). After World War I, Zeppelins were extensively used as bombers and scouts. They are equipped with powerful engines and are capable of lifting heavier loads. The application segments where UAV, Blimps, and Zeppelins may be used are presented in Table 1.4.

TABLE 1.4 Major Segments of UAV, Blimps, and Zeppelins.

Segments	Sub-Segments	Intended Uses		
		UAV	Blimps	Zeppelins
Civil	Natural disasters	Low	Medium	Medium
	Humanitarian relief	Low	Least	Least
	Environment	Medium	Medium	Medium
	Precision agriculture	Medium	Low	Low
	Cargo transport	Low	Medium	Medium
Commercial	Weather & storm tracking	Low	Medium to high	Medium
	Advertisement	Low	Medium to high	Medium
Military/ Security	Defense	Medium to high	Medium	Low
	Wireless communications	Medium to high	Medium	Medium

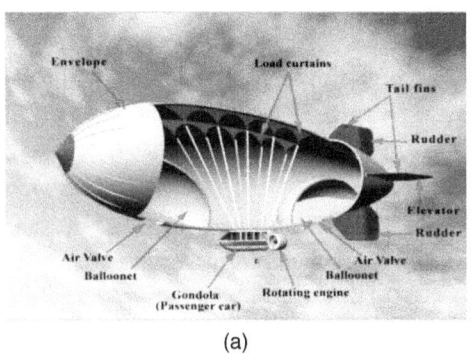 

(a)   (b)

FIGURE 1.3 (a) The Blimp, and (b) The Graf Zeppelin (Source: Encyclopedia Britannica, 1999).

The UAVs have applications in a large number of fields (Figure 1.4), such as aerospace, military, aerial surveying, law enforcement, search and rescue, mining, conservation, pollution monitoring, surveying, oil, gas & mineral exploration & production, disaster relief, archeology, cargo transport, agriculture, precision farming, construction, passenger transport, monitoring criminal & terrorist activities, motion picture film-making, hobby & recreational, journalism, and light show (Ayranci, 2017; Berie, and Buruda, 2018). The uses of UAVs are growing rapidly across many civilian applications, including real-time monitoring, providing wireless coverage, remote sensing, search and rescue, delivery of goods, security and surveillance, precision agriculture, and civil infrastructure inspection (Taladay, 2018). Coupled with the applications from other synergetic technologies, such as 3D modeling, Internet of things, artificial intelligence, and augmented reality (AR), as well as Virtual Reality, it has opened up new possibilities for organizations to leverage the use of UAV and its associated technologies across their operations (Dupont et al., 2016).

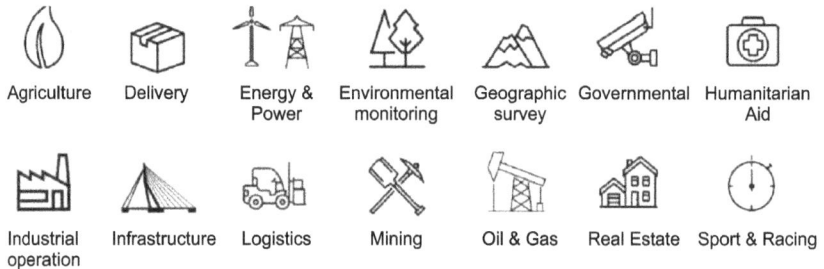

FIGURE 1.4 Various uses of UAVs.

Figure 1.5 shows the uses of UAVs to various maturity levels by different sectors of industry (White Paper, 2018). In the lower half of Figure 1.5, humanitarian relief, research & science, safety & security, insurance, and

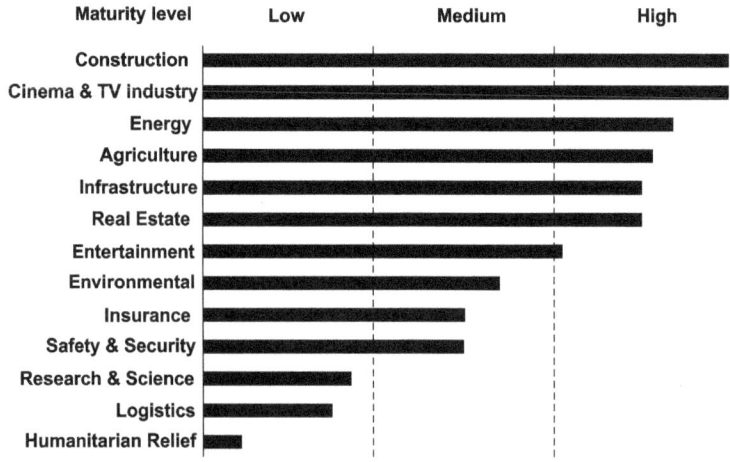

FIGURE 1.5 Uses of UAVs in industries to various maturity levels (White Paper, 2018).

environmental have lower maturity levels due to lack of funds. In contrast, logistics has lower maturity due to regulatory problems (Jha, 2016). The top half of Figure 1.5 shows the potential of full uses of UAVs in these industries. Applications in agricultural mapping usually happen in rural areas; therefore, regulatory hurdles for UAV operations are lower than in urban, densely populated areas.

In many business activities, drones can substitute traditional methods of operation. With less human operation and safety infrastructure, drones can reduce the time and costs of work. They can also enhance data analytics, which allows companies to better comprehend and predict operating performance. The drone market is growing steadily in the consumer, commercial, and military sectors. In a 2016 report, Goldman Sachs (2016) estimated that drone technologies would reach a total market size of US $100 billion between 2016 and 2020, though 70% of this figure would be linked to military, 17% to consumer, and 13% to commercial activities. The commercial business represents the fastest growth opportunity, projected to reach US $13 billion between 2016 and 2020.

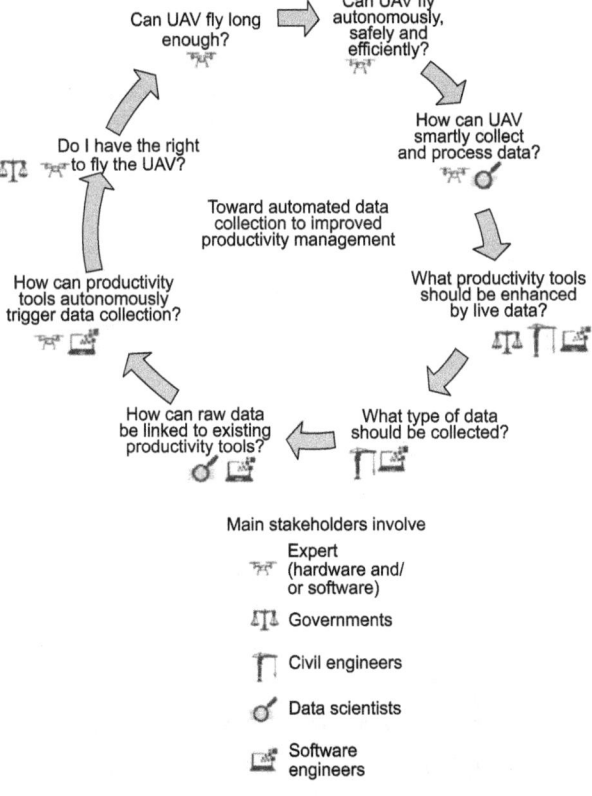

FIGURE 1.6 Cycle of a UAV (Dupont *et al.*, 2016).

The abilities of UAV to reduce the cost of compliance and cost of the technology while also enhancing the value of the information gathered through these systems have been the key drivers for the increased adoption of UAV worldwide. Throughout the world, there has been an increasing awareness about UAV in the industry and the government. The stakeholders using the UAV represent the government, UAV developers, and the UAV operators. As we see in Figure 1.6 attempts to describe the interaction of the stakeholders required to realize an automated data collection to improve productivity management through the deployment of UAVs.

Gartner (2018) presented the Hype Cycle for emerging technologies in 2017 (Figure 1.7) which provides a graphical representation of the maturity and adoption of technologies and applications and how they would be potentially relevant to solving real business problems and exploiting new opportunities. Gartner Hype Cycle methodology provides a view of how a technology or application will evolve over time, providing a sound source of insight to manage its deployment within the context of specific business goals. It is a unique graph because it garners insights from more than 2,000 technologies into a small set of 12 emerging technologies and trends. The commercial use of UAVs/drones is approaching the "plateau of productivity" in the Gartner Hype Cycle. This Hype Cycle specifically focuses on the set of technologies that are showing promise in delivering a high degree of competitive advantage over the next 5–10 years.

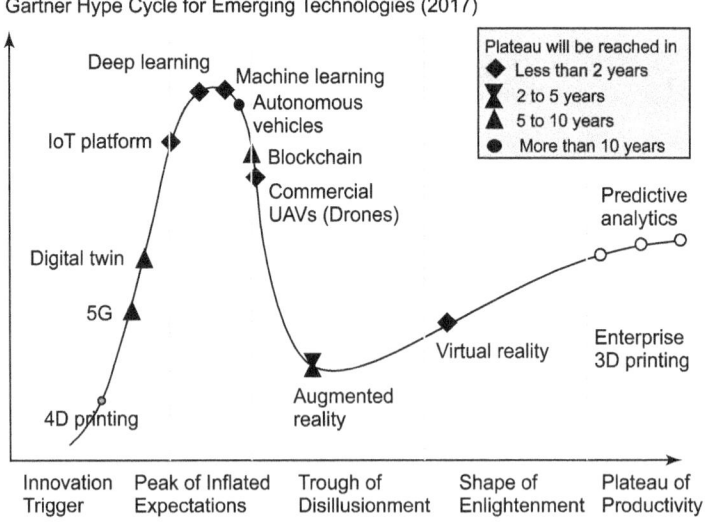

FIGURE 1.7 Hype Cycle for emerging technologies (Culus, 2018).

The UAVs allow users to increase the frequency of data collection and generate accurate and reproducible datasets. The recurring cost for these data sets is going to be very low as compared to other solutions. Thus, a well-designed setup would be required to use UAVs effectively and efficiently (Glover, 2014). All the components of this setup will have a significant role in the derived results. For example, a specific combination of sensors, software, and platform may guarantee accurate data in a particular application. The model in Figure 1.8 explains what it means to choose the right approach to include UAVs in several operations (White paper, 2018). It starts with the client helping users work through requirements and standards to narrow their way down to the required software and hardware. The UAVs application research carried out in many sectors have shown that the emerging interest in UAV-based technologies is related to the reduced size of the vehicles, affordability in terms of cost, low energy consumption, flexibility, minimizing risks, and the resulting high spatial resolution data (Puliti *et al.*, 2015).

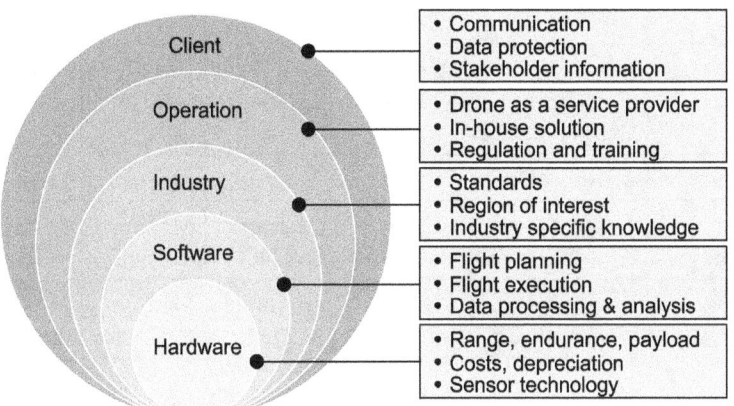

FIGURE 1.8 The onion model of UAV components (White Paper, 2018).

Smart UAVs are the next big revolution in UAV technology, promising to provide new opportunities in different applications, especially in civil infrastructure, to reduce risks and lower cost (Intelligence, 2018). They can be used to inspect nearly any structure, both indoors and outdoors. Still, they are most useful in inspecting structures that are difficult to reach by traditional means, including tall structures, such as buildings, towers. It can also include confined spaces in which space is limited for traditional inspections or structures that are over water, such as bridges or the undersides of oil rig platforms (Jha, 2016).

Globally, some countries have created a conducive environment for organizations to benefit from the applications of UAV and its associated technologies. Hence, these technologies are driving innovation in the UAV market. The UAV market is opening up many new opportunities, from UAV pilots to electronics and cameras. For example, vertical take-off and landing (VTOL) UAVs will be effective and useful in humanitarian aid missions. The relative market of UAVs in different segments in various parts of the world is presented in Table 1.5 (Luncitel, 2011).

TABLE 1.5 Relative Market of UAVs in Different Regions.

Segments	Sub-Segments	USA	Europe	Mid-East	Asia-Pacific	Others
Civil	Natural disasters/ Humanitarian relief	Medium	Low	Low	Low	Low
	Environment/ Weather & storm tracking	Medium	Low	Low	Low	Low
	Precision agriculture/ Cargo transport	Low	Low	Least	Least	Least
Commercial	Advertisement	Least	Least	Least	Least	Least
Military/ Security	Defense	High	Medium to High	Medium to High	Medium to High	Medium
	Wireless communications	High	Medium to High	Medium to High	Medium to High	Medium

As estimated by the US government, 110,000 drones are flying in US airspace, projected to more than quadruple by 2022. According to recent research PwC (2018b), the global UAV market is projected to grow to US $19.85 billion by the end of 2021, at a Compound Annual Growth Rate (CAGR) of nearly 13% over during 1997–2023. Thus, in the future, the drones will be used in a large range of applications, including construction, surveillance, search and rescue operations, logistics, traffic monitoring, exploring hazardous sites, military surveillance, weather and climatic mapping, photography and journalism, delivery services, and sustainable agriculture.

1.4 SOME TECHNICAL TERMS

Here some of the basic technical terms used for the operation of UAV have been explained.

Autopilot: A drone/UAV can conduct a flight without real-time human control, for example, following preset GPS coordinates.

Autonomous flight: The flight of a UAV is controlled by internal programming that guides it was to fly without sending radio signals by humans. For example, a UAV may use its onboard GPS system to fly from one predetermined point to another.

Airborne collision avoidance system: An aircraft system based on secondary surveillance radar transponder signals, which operates independently of ground-based equipment to provide advice to the pilot on potential conflicting aircraft that are equipped with Secondary Surveillance Radar transponders.

Waypoints: A waypoint is a set of coordinates that identifies specific points in physical space. Waypoints are used to create a flight path for UAVs. The UAVs with waypoint technology typically utilize GPS and the global navigation satellite system (GLONASS) to create waypoints.

Waypoints navigation: Waypoints allow a UAV/drone to fly on its own to its flying destination or points pre-planned and configured into the remote control navigational software. Waypoints are useful in designing various autonomous missions for UAVs. Mapping would be impossible without the possibility to define these physical locations.

Point of interest: The "point of interest," also known as "region of interest," designates a spot that a UAV must necessarily collect the data of that point.

Line of sight (LOS): Many small UAVs are LOS machines, meaning the operator controlling the device must be in direct sight of the aircraft so that radio signals can be transmitted back and forth. Many larger UAVs are not LOS aircraft because the radio signals that control them are bounced off of satellites or manned aircraft.

Visual line of sight (VLOS): The operation of an unmanned aircraft within the pilot's LOS at all times, without the aid of any device (binoculars) other than corrective lenses (glasses) defines the VLOS (Figure 1.9).

Beyond VLOS (BVLOS): The ability to operate an unmanned aircraft beyond the pilot's LOS is called BVLOS. Flying UAV beyond the VLOS requires a special permit from the regulating authority, for example, FAA in the USA (Figure 1.9).

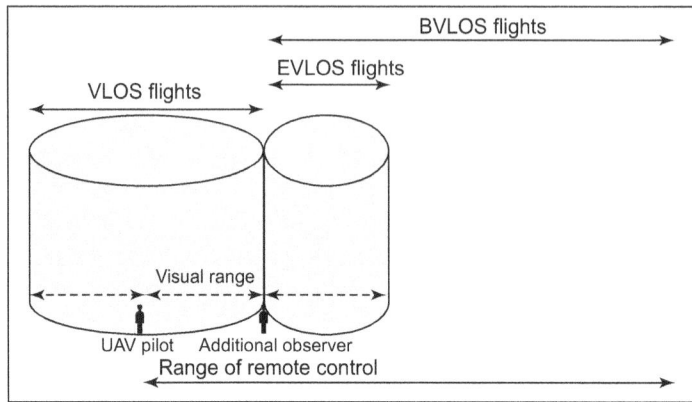

FIGURE 1.9 Distinction between UAV flight ranges-VLOS, EVLOS, and BVLOS.

Extended VLOS (EVLOS): It is a flight from 500 m range to a distance at which the UAV is still within the pilot's sight. It allows remote pilots to extend the proximity of UAV beyond their direct vision (Figure 1.9).

Geofencing: The use of GPS technology to create a virtual geographic boundary of an area, enabling software to trigger a response when a UAV/drone enters or flies within that particular area, is called geofencing.

First person view (FPV): In FPV, a UAV with a camera transmits video feed wirelessly to goggles, a headset, mobile, or any other display device. The user has a FPV of the UAV/drone flies environment and captures video or still images.

Air traffic control (ATC): A service operated by appropriate authority (such as the FAA in the US) to promote the safe, orderly, and expeditious flow of air traffic.

Ground control station (GCS): It contains software running on a computer that receives telemetry information from a UAV, and displays its progress and status, often including video and other sensor data. It can also be used to transmit in-flight commands to the UAV.

Inertial measurement unit (IMU): The IMU is an electronic device typically used to maneuver aircraft and UAVs (an attitude and heading reference system). It measures the aircraft's specific force, angular rate, and orientation using a combination of accelerometers, gyroscopes, and magnetometers.

Position hold: It is automatic position control of UAV in space using barometric or ground sensors. It is mainly used for inspections and flights to the area where it is not safe to land the UAV, for example, city, forest, etc.

Automatic take-off: It is done with pre-defined height. The runway alignment during automatic take-off is defined by a straight line joining two waypoints surveyed using the navigation system. Take-off sites must be defined as waypoints either inside or outside the mission area. For large sites, it may be important to have the take-off sites located in the middle.

Automatic landing: It is done with pre-defined speed. The automatic landing is initiated from a circuit after passing a waypoint defining where the vehicle exits the circuit and commences the landing approach. Landing sites must be defined as waypoints either inside or outside the mission area. For large sites, it may be important to have landing sites located in the middle.

Return to home: When a UAV/drone automatically returns to the point of take-off or when the operator triggers the function on the remote control to return home. The UAV also returns home if the battery power is low or the signal from the transmitter is lost.

Coming home: It is mainly used if the UAV is flown to longer distances and/or the flight area is not easy to access, for example, mines, open water.

No fly zone: Government regulations restrict areas where flying a UAV. Regions where a UAV could interfere with an airplane or record sensitive information make up most of these areas.

Fail safe: System that helps protect a UAV in case of some error. For example, if a UAV loses control of signals, a fail safe will have the UAV return to the point of take-off (return home).

1.5 CHARACTERISTICS OF UAVs

The UAVs are aeronautical platforms that operate without the use of onboard human operators (Garg, 2019). There are broadly two types of UAVs; fixed-wings and rotary-wings (Molina, 2014). The development of new sensors, microprocessors, and imaging systems led to the growth of UAV systems based on fixed and rotatory wings UAVs. Fixed-wing UAVs (Figure 1.10a) have a predetermined airfoil that makes flight possible by generating the lift caused by the UAV forward airspeed. Rotary-wing UAVs (Figure 1.10b) consists of a number of rotor blades that revolve around a fixed mast. The blades themselves are in constant motion, which produces the airflow required to generate the lift. Rotary-wings UAVs are also known as multicopters having more than two motors. Multicopter UAVs are classified depending on the rotor configurations as tricopters, quadcopters, hexacopters, and octocopters.

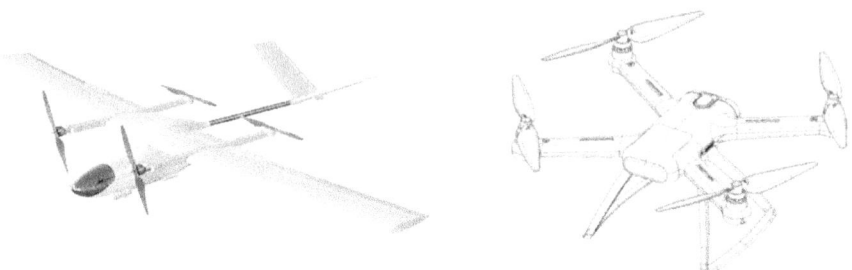

FIGURE 1.10 (a) Fixed-wing type UAV, and (b) Rotary wing UAV (Quadcopter).

The first commercial multicopter appeared in Germany in 2005, developed by Microdrones, and then the industrial sector of multicopters quickly expanded (Glover, 2014). Multicopters control the vehicle motion while varying the relative speed of each rotor to change the thrust and torque with fixed-pitch blades. The great advantage lies in the simplicity of the rotor mechanics required for the flight control, in contrast to conventional single- and double-rotor helicopters, which use complex blade rotations (Vergouw et al., 2016).

The concept of a quadcopter (rotary wing) vehicle is not new; manned quadcopters first experimented in the 1920s, but their effectiveness was hampered by the technology available at the time (Cai et al., 2011). A quadcopter has four propellers (Figure 1.10b) which are fixed and vertically oriented. Each propeller has a variable and independent speed, which allows a full range of movements. Propellers control the conventional helicopters with blades that dynamically pitch around the rotor hub. The components required for blade pitch are expensive; which is one of the reasons why quadcopters are becoming common in many applications (Šustek and Úøednîèek, 2018). Today, UAVs are executing search and rescue missions, tracking cattle rustlers, and monitoring wildfires or landslides with minimal cost and little risk of loss of life (Xu et al., 2014). Other applications of this technology include geomatics, precision agriculture, infrastructure monitoring, and logistics.

The UAVs are generally categorized with reference to several parameters, such as payloads, weight, size, function, payload, geographical range, flight endurance, and altitude (Sholes, 2007). More details are given in Chapter 2. The payload refers to the carrying capacity of the UAV, including the contents. Geographical range, flight endurance, and maximum ceiling are the maximum distance, time, and altitude, respectively, that UAVs can reach during an interrupted mission (Glover, 2014). Size and weight are important characteristics where the size of the UAV (such as nano, mini, regular, large) may vary from that of an insect to the size of a commercial airplane. In contrast, weight may vary from hundreds of grams to hundreds of kilograms (Vergouw et al., 2016). They are also recognized as per their range (such as very close range, close range, short-range, mid-range, etc.). In addition, today, UAVs are characterized as having different energy sources, such as battery cells, solar cells, and traditional airplane fuels. The UAVs are also classified based on autonomy, being fully autonomous to fully control by a remote pilot.

Two primary types of remote sensing systems can be employed in UAVs: active and passive remote sensing systems. The active remote sensing systems include laser altimeter, Light Detection and Ranging (LiDAR), radar, and ranging instrument, while passive remote sensing systems include accelerometer, hyperspectral sensor, and imaging sensor (Wingtra, 2019). In a passive remote sensing system, the sensor detects natural radiation that is emitted or reflected by the object. In contrast, active sensors carry their own light source to illuminate the object and detect natural radiation emitted or reflected by the object. With the advancement of electronic technology in sensors, batteries, cameras, and GPS systems, quadcopters became widely employed over the past decade, both recreationally and commercially (Šustek and Ůøedníèek, 2018; Albeaino et al., 2019). A comparison of UAVs, airborne, and satellite systems is given in Table 1.6.

TABLE 1.6 A Comparison of UAV, Airborne and Satellite Systems (Singhal et al., 2018).

System	System Resolution	Degree of Availability	Operation Mode	Payload Capacity	Operating Cost
UAV	cm to m	High	Autonomous or remote control	Limited	Low
Helicopters	100 m	High	Human pilot	Limited	Medium
Airborne	Up to 50 m	Moderate	Human pilot	Much more than helicopters	High
Satellite	1 m-1 km	Poor	Autonomous	Limited	Very high

Based on the sensor systems, UAVs are classified as photogrammetry-based UAVs and LiDAR-based UAVs (Austin, 2010). The photogrammetry UAV and LiDAR UAV are actually quite different from each other, even if their three-dimensional (3D) outputs look similar. In survey missions, the choice between photogrammetry UAV and LiDAR UAV depends mainly on the application in hand (Nex and Remondino, 2014; Dering *et al.*, 2019). In addition, operational factors, such as cost and complexity, are considered. UAV photogrammetry indeed opens various new applications in the close range aerial domain, introducing a low-cost alternative to the classical manned aerial photogrammetry for large-scale topographic mapping or detailed 3D recording of ground information and being a valid complementary solution to terrestrial acquisitions (Figure 1.11). UAV images are also often used in combination with terrestrial surveying to complement data for 3D modeling and create orthoimages (Albeaino *et al.*, 2019).

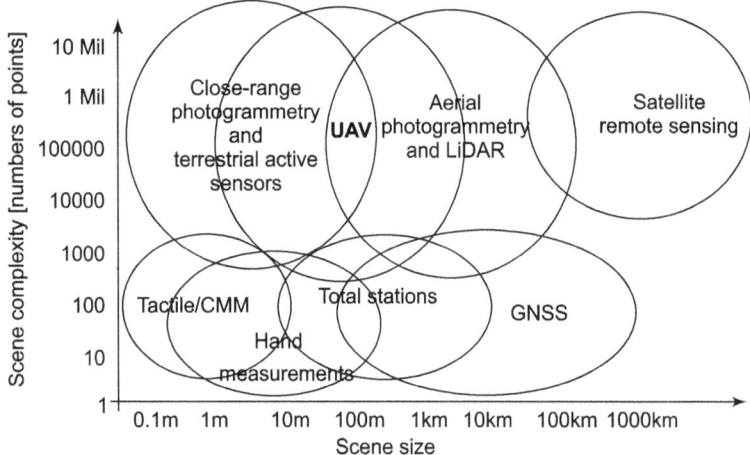

FIGURE 1.11 Available geomatics techniques, sensors, and platforms for 3D recording purposes, according to the scene's dimensions and complexity (Nex and Remondino, 2014).

The latest UAV developments can be explained with low-cost platforms combined with digital cameras and global navigation satellite system (GNSS), Inertial Navigation System (INS), necessary to navigate the platforms, predict the acquisition points, and possibly perform direct georeferencing. Although conventional airborne images still have some advantages, very high-resolution satellite images are filling up the gap between airborne and satellite mapping applications. UAV platforms are an important alternative and solution for studying the environment in 3D (Colomina and Molina, 2014).

1.5.1 Photogrammetry UAVs

UAV photogrammetry is described as "a photogrammetric measurement platform, which operates remotely controlled, semi-autonomously, or autonomously, without a pilot sitting in the vehicle." Photogrammetry UAV captures a large number of high-resolution photos over an area (Wingtra, 2019) with overlap such that the same point on the ground is visible in multiple photos. In a similar way that the human brain uses information from both eyes to have depth perception, photogrammetry uses these multiple images to generate a 3D model. The high-resolution 3D reconstruction contains elevation/height information and texture, shape, and color for every point, enabling easier interpretation of terrain. Photogrammetry UAVs are cost-effective and provide outstanding flexibility regarding where, when, and how 2D or 3D data are captured (Federman *et al.*, 2017).

A UAV coupled with a high-resolution camera can map a large area in a single flight, generating 2D and 3D data (Figure 1.12). Images can be collected from a UAV with a resolution of the order of a few centimeters. Figure 1.13 shows the appropriate camera platform for a photogrammetric survey which depends on both the areal extent and required sampling resolution (Dering *et al.*, 2019). The coverage per flight-hour for a photogrammetry-based UAV exceeds by more than 5x as compared to LiDAR-based UAVs. So, high-quality data can be captured in less time and at an overall lower cost (Buczkowski, 2018). After the acquisition, images can be used for stitching and mosaicking purposes and input in the photogrammetric process. For this purpose, camera calibration and image triangulation are initially performed to create a Digital Terrain Model (DTM) or a Digital Surface Model (DSM), or orthoimages (Colomina and Molina, 2014).

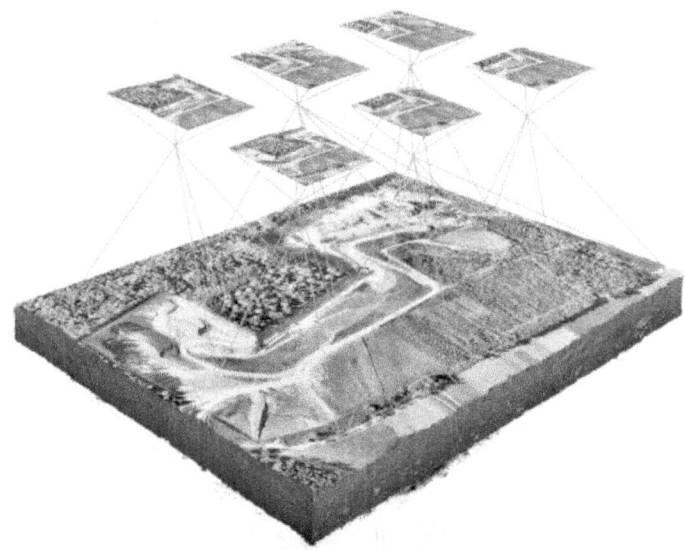

FIGURE 1.12 Photogrammetry UAV (Wingtra, 2019).

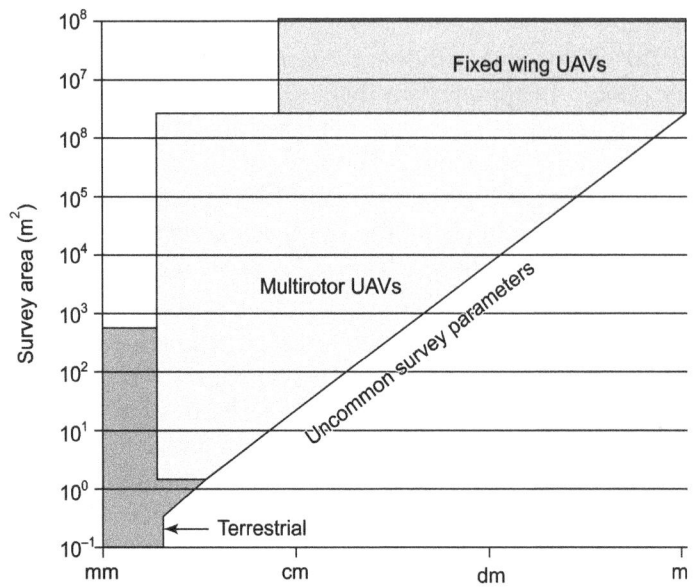

FIGURE 1.13 Sampling resolution with area for the photogrammetric survey (Dering et al., 2019).

The flight may be undertaken in manual, assisted, or autonomous mode, according to the mission specifications, platform type, and environmental conditions. The onboard GNSS/INS navigation device is used for the autonomous flight (take-off, navigation, and landing) and guide image acquisition. The navigation system, generally called autopilot, allows performing a flight according to the plan and communicates with the platform during the mission. Double-frequency positioning mode or the use of the real-time kinematic method would improve the accuracy of positioning to decimeter level. Still, they are expensive to commonly use on low-cost solutions (Kan et al., 2018). During the flight, the autonomous platform is normally observed with a GCS which shows real-time flight data, such as position, speed, altitude and distances, GNSS observations, battery or fuel status, rotor speed, etc. (Buczkowski, 2018). On the contrary, remotely controlled systems are piloted by the operator from the ground station.

The workflow of photogrammetry UAV, presented by Nex and Remondino (2014), is shown in Figure 1.14, where the input parameters are flight parameters, available devices, and additional parameters for workflow steps. The flight planning and data acquisition mission is normally planned beforehand using required software, starting from the knowledge of the area of interest, required ground sample distance, and the intrinsic parameters of the onboard digital camera. The desired image scale and camera focal length are generally fixed to derive the flying height of the UAV. The camera perspective centers (also called waypoints) are computed after deciding the longitudinal and lateral overlap of the strips. All these parameters can vary according to the goal of the flight; for example, missions for a detailed 3D model generation usually require higher overlaps in images.

A typical photogrammetry UAV platform requires flight or mission planning and Ground Control Points (GCPs) measurements for georeferencing purposes (Federman *et al.*, 2017). Mission planning and flight altitude are flexible and controlled based on the real needs of users. The flight paths are designed to meet the objectives of the flight mission. Here, mission planning may also have features that contain all safety formalities and applications. Additional information, such as completing a flight log after each flight, can also be attached to this category. The logbook contains information about the flight pattern, time-of-flight, battery life, wind speeds, temperature, etc. (Shakhatreh *et al.*, 2018).

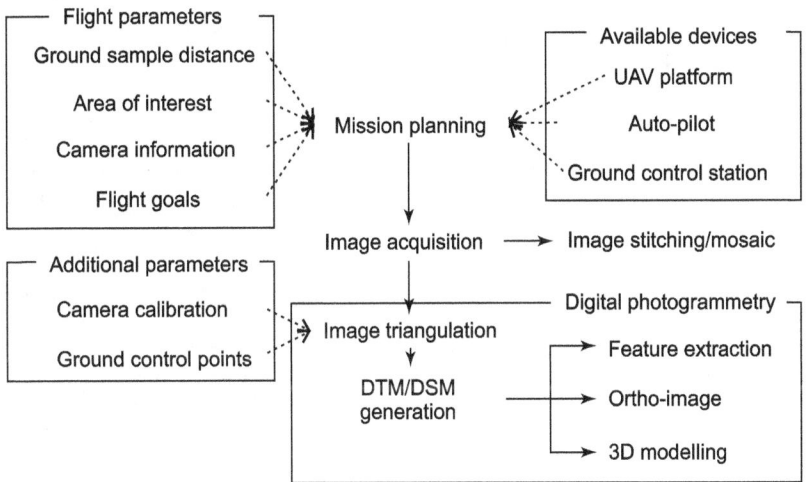

FIGURE 1.14 Typical acquisition and processing of UAV images (Nex and Remondino, 2014).

Camera calibration and image orientation tasks require the extraction of common features visible in as many images as possible (tie points) followed by a bundle adjustment, that is, a non-linear optimization procedure (Colomina and Molina, 2014). The collected GNSS/INS data can help for automated tie point extraction and allow direct georeferencing of the captured images. Some efficient commercial software that could be used includes, for example, PhotoModeler Scanner, Eos Inc; PhotoScan, Agisoft. Once a set of images has been oriented, 3D reconstruction and modeling steps are surface measurement, orthophoto creation, and feature extraction.

Table 1.7 presents some of the popular software which can be used for photogrammetry UAVs flight planning mission. A good UAV system with professional mission planning and post-processing workflow helps ensure that quality data is captured to get accurate results. Good overlapped images increase the accuracy and provide better error correction as compared to complete reliance on the direct georeferencing method used in the LiDAR survey. Photogrammetry UAV generates not only accurate 3D models but also full-color, high-resolution information for every point on that model (Dering et al., 2019). This provides visual context for 3D data and makes interpretation and analysis of the results much easier than LiDAR point cloud data.

TABLE 1.7 Selected Flight Planning Software/Apps for UAV Photogrammetry (Dering et al., 2019).

Software	Operating System(s)	Cloud Processing Capabilities	Web Source	Applications
MapPilot	iOS only	Yes	https://support.dronesmadeeasy.com	Flat or hilly landscapes (e.g., coastal platforms, stockpiles, etc.)
FlyLitchi	iOS Android	No	https://flylitchi.com/	Complex surveys, such as steep cliffs, pit walls, or tall structures (e.g., towers)
DJI GS Pro	iOS only	No	https://www.dji.com/ground-station-pro	Flat or hilly terrain (e.g., coastal platforms, stockpiles, etc.) and tall structures
UGCS Pro	Android Windows	No	https://www.ugcs.com	All
Pix4D Capture	iOS Android	Yes	https://pix4d.com/product/pix4dcapture	Flat or hilly terrain and tower-shaped structures
DroneDeploy	iOS Android	Yes	https://www.dronedeploy.com	Flat or hilly terrain (e.g., coastal platforms, stockpiles, etc.)

Photogrammetry UAV has reached its saturation in terms of cost/accuracy/speed offered. It is preferred in (i) mapping of bare earth sites, (ii) areas not obstructed by trees, buildings, or other objects, (iii) only DSM is to be created, and (iv) when cost consideration is the governing factor.

1.5.2 LiDAR UAVs

The LiDAR technology has been around for many decades, but it is used only recently on UAVs. A LiDAR sensor emits laser pulses and records the time-of-flight and intensity of the reflected energy to be returned to an onboard sensor (Wingtra, 2019). A LiDAR uses oscillating mirrors to send out laser pulses in many directions to generate a "sheet" of light as the UAV moves forward (Figure 1.15). Measuring the timing and intensity of the returning pulses can provide measurements of the ground points. This range is converted to the coordinates of the points (x, y, and z) on the ground using onboard GPS and IMU (Garg, 2019).

FIGURE 1.15 The LiDAR data collection (Wingtra, 2019).

The UAVs which carry these sensors because of their weight and power requirements tend to be significantly larger (typically multicopters). All of these high-end systems carry out direct georeferencing of the raw data. The final output is a large number of cloud points with their known 3D coordinates, which could be used to create DTM or DSM. Laser UAVs can collect data at point densities between 60 and 1500 points/m^2 (Puliti *et al.*,

2015). A high-end LiDAR UAV system can generate results with horizontal (x, y) and elevation (z) accuracies in the range of 1–2 cm (Buczkowski, 2018). To get higher vertical accuracy of 1 cm (0.4 in), the cost of the employed system will go high with an increased level of operational complexity compared to photogrammetry UAV. Recently, efforts to make LiDAR scanners smaller, lighter, and cheaper to be employed in LiDAR UAVs (Barnes, 2018). For example, the Velodyne LiDAR Puck has less than 1 kg weight and is about 100 mm in diameter with a laser pulse range of 100 m.

The LiDAR instrument mounted on UAV collects signals at certain intervals. The instrument sends continuous pulses and can fire a large number of pulses in a second, thus collecting huge point cloud data in a very short time (Dering *et al.*, 2019). The first pulse is used to survey the top of objects, while the last pulse is used to survey the ground below the object. Intermediate pulses convey information about the vertical structure of the object.

In some specific situations, a terrain model below vegetation may be needed as an output. While photogrammetry UAV can effectively create 3D models of the ground in areas with sparse vegetation, the LiDAR UAV is very useful when mapping the areas beneath the dense vegetation, as shown in Figure 1.16 (Dering *et al.*, 2019). This is because LiDAR light pulses can filter through small openings between the leaves and reach the ground below. Due to these advantages, LiDAR UAVs have been successfully used to collect data under trees/forest cover (Corrigan, 2016, Cao *et al.*, 2019). However, the LiDAR systems are expensive for surveying and take a longer time to cover large areas. Some common LiDAR systems and UAV used are given in Table 1.8.

The use of airborne LiDAR has a number of advantages which include:

- Rapid data acquisition is possible.
- The density of points collected enables the determination of man-made objects and infrastructure, such as buildings, bridges, and roads.
- Due to the large number of points collected, a sufficient number of laser pulses will penetrate the gaps between the tree canopy and return back to the sensor, enabling the determination of terrain under the tree canopy.
- Determination of heights of infrastructure, such as power lines or tree canopy, is easy due to a large number of points with recorded coordinates of features. Such information can be very useful for forest management (i.e., tree height, forest density, wood volume, and biomass estimation).
- The generation of digital elevation models is easy and accurate, which is required for many applications.
- LiDAR can be used both as an airborne sensor and as a terrestrial sensor.

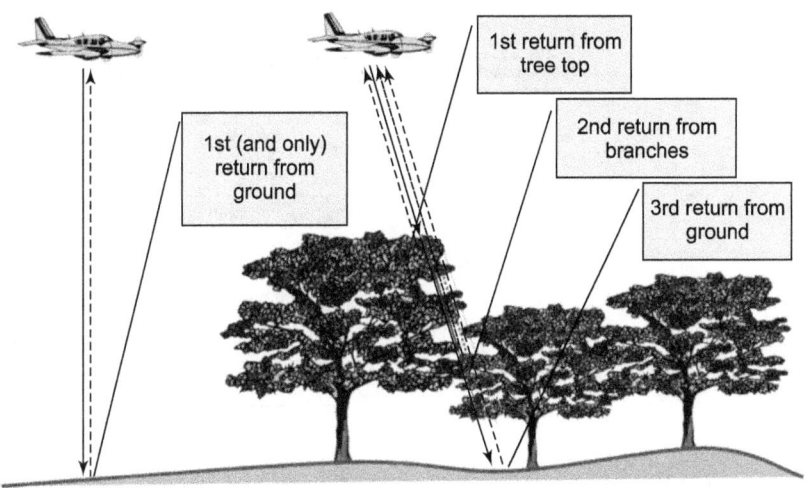

FIGURE 1.16 The LiDAR multiple returns (Ortiz, 2018).

TABLE 1.8 Common LiDAR UAVs and Sensors Used.

S.No.	LiDAR UAVs	LiDAR Sensors
1	DJI M600 Pro LiDAR quadcopter	LeddarTech Vu8
2	Draganflyer Commander	LeddarOne
3	Riegl RiCopter LiDAR UAV quadcopter	Velodyne HDL-32E
4	Harris H4 Hybrid HE UAV quadcopter	Velodyne Puck VLP-16
5	VulcanUAV Harrier Industrial multirotor	Velodyne Puck Lite
6	VelosUAV helicopter	Velodyne Puck Hi-Res
7	Robota Eclipse fixed-wing drone	VUX-1 UAV
8	DJI Matrice 200 series quadcopter	Routescene—UAV LiDAR Pod
9	OnyxStar Xena drone	YellowScan—3 UAV
10	OnyxStar Fox-C8 HD quadcopter	YellowScan Vx Long Range
11	GeoDrone X4L LiDAR quadcopter	YellowScan Mapper
12	Tron F9 VTOL fixed-wing LiDAR drone	YellowScan Surveyor
13	Boreal long range fixed-wing drone	Leica Geodetics
14	Vapor 55 UAV helicopter	Leica Geo-MMS SAASM

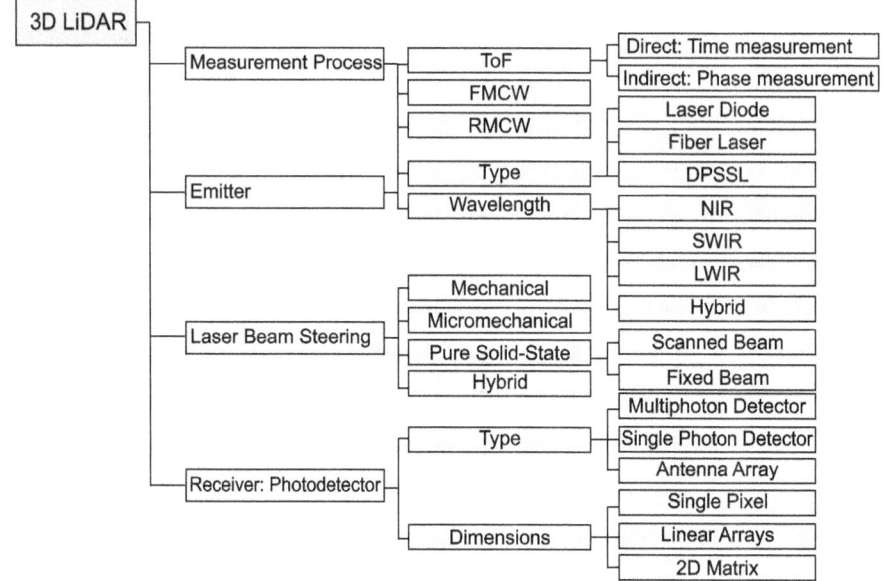

FIGURE 1.17 Four fundamental technologies for the development of LiDAR systems (Wijeyasinghe and Ghaffarzadeh, 2020).

Wijeyasinghe and Ghaffarzadeh (2020) have identified four fundamental technology choices (given in Figure 1.17 and explained below) that will strongly impact LiDAR product performance and business growth. The global market for 3D LiDAR in autonomous vehicles is expected to grow to US $5.4 billion by 2030.

- *Measurement process:* The measurement process determines how the light signal is generated and how it is measured. Most LiDAR companies develop conventional time-of-flight technology, but there is a trend toward emerging frequency-modulated continuous wave and random-modulated continuous wave LiDAR technology.
- *Laser beam:* The short-wave infrared (SWIR) LiDAR is considered safer for human vision than near-infrared (NIR) LiDAR operating at shorter wavelengths. Most LiDAR companies develop conventional NIR technology, but there is a trend toward emerging SWIR LiDAR because it can operate safely at higher power and detect distant objects.
- *Beam steering mechanism:* The laser light must be diffused or scanned by beam steering technology to illuminate the scene. Conventional

automotive LiDAR uses a rotating mechanical assembly to scan laser beams by 360°, which enables the vehicle to see around it using a single LiDAR mounted on the roof. They are large, expensive, and prone to mechanical failure. The technology trend is toward small, cheap, and stable module designs. Emerging beam steering options include mechanical systems with limited motion, micro-mechanical systems based on micro-electro-mechanical systems mirrors, and pure solid-state systems. A solid-state LiDAR sensor is a compact and robust device, which is easily embedded into vehicles. For example, 3D flash LiDAR is a type of solid-state LiDAR that usually illuminates the entire scene at once, such as a camera. This enables fast imaging without distortion; however, the typical object detection range is shorter than scanned LiDAR due to laser power limitations, but the technology readiness is too low.

- *Photodetector:* The light-absorbing semi-conductor material must match the laser wavelength. Single photon-sensitive detectors are an emerging technology trend, enabling LiDARs to register signals quickly and detect weak signals from distant objects.

1.5.3 A Comparison of Photogrammetry UAVs and LiDAR UAVs

Most of the current photogrammetry UAVs are used for land surveying, which can import geographic information by putting together many aerial photographs. Photogrammetry UAV is good for surveying the sites without any obstacles, but most of the time, lands are not free from obstacles, so laser-based UAV surveying becomes the best choice (Chatzikyriakou and Rickerby, 2019). Photogrammetry UAVs do not provide optimum data when flying through forested areas, as the land features beneath the trees cannot be photographed, while LiDAR surveying gets land information beneath the trees (Cao *et al.*, 2019). In addition, LiDAR systems can take images at night and also work inside the tunnels. Moreover, laser-based surveying is considered to be more accurate than photogrammetry-based (Barnes, 2018).

LiDAR UAVs are preferred (Corrigan, 2019) where (i) mapping below tree canopy or mining sites or bathymetric survey required, (ii) other kinds of obstructions present to sun-light, (iii) need the vertical structure of the object, that is, by using waveform or multiple returns, (iv) mapping the objects, like transmission line, pipelines, roads in a corridor, (v) survey to be done at night or low light condition, (vi) higher accuracy across terrain required, (vii) not possible to establish the GCPs, and (viii) derive the benefits of both LiDAR and photo camera.

Both photogrammetry and LiDAR UAVs can provide a high level of 3D model accuracy, especially compared with terrestrial sampling methods (Corrigan, 2016). In order to generate the best results, a LiDAR system requires all of its components to work perfectly in synchronization. Small gaps or errors in sensor measurements can lead to significant errors in outputs. The GCPs, which are useful in photogrammetry to correct errors, are difficult to implement with the LiDAR data (Buczkowski, 2018, Chatzikyriakou and Rickerby, 2019). Most of the time, the only solution for erroneous LiDAR data would be to repeat the flights. LiDAR UAV projects require an expert who understands the details of each sub-system and can recognize consistent and accurate data. In contrast, photogrammetry-based workflows are more flexible.

Figure 1.18 illustrates a comparison of UAV-based photogrammetry to other surveying methods in terms of total coverage area and survey error. Arrows demonstrate the UAV photogrammetry region's existing and potentially future expansion through camera hardware, imaging methodologies, and UAV control advancements. It is seen that centimeter-level DTMs developed from UAV-based photogrammetry are a competitive alternative to LiDAR systems. Table 1.9 presents a comparison of UAV-based photogrammetry to LiDAR UAV. Considering the advantages offered and cost coming down, LiDAR UAVs might lead in the future (Dering *et al.*, 2019).

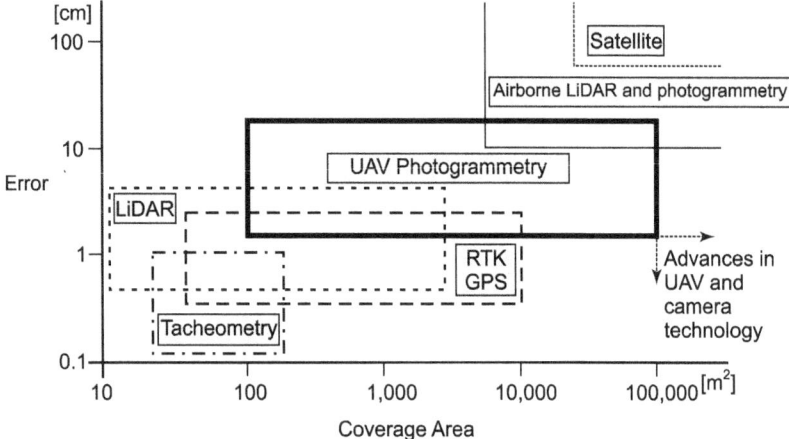

FIGURE 1.18 The UAV application to surveying tasks (Siebert and Teizer, 2014).

TABLE 1.9 A Comparison of Photogrammetry and LiDAR UAVs (Wingtra, 2019).

Parameters	Photogrammetry	LiDAR
Cost	US $20000-30000 for a professional UAV + high-resolution camera system	US $50000 or more for just the sensor. Survey-quality complete systems between US $1,50,000–300,000 range
Operational complexity	No additional sensors required; indirect georeferencing requires longer processing but is resistant to potential workflow errors	LiDAR uses direct georeferencing, which means that multiple components and sensors must work perfectly together in order to gather useful data
Outputs	2D orthomosaic maps, 3D models, point clouds, surface models with visual information as part of the 3D model	3D point clouds, intensity maps with multiple returns, and full-waveform information for classification
Accuracy	1 cm horizontal, 2–3 cm elevation (vertical) over hard surfaces	1–2 cm elevation (vertical) over soft and hard surfaces
Best for	Mapping, surveys, mining, broad-coverage combined with high horizontal and vertical accuracy	Terrain models below dense vegetation, forestry, 3D modeling of power lines or cables, 3D modeling of complex structures, etc.

1.6   WORKING OF A UAV

The most common UAV/drones have two units; receiver and transmitter. The receiver unit consists of the UAV/drone, while the transmitter unit is normally a ground-based controller system to send the commands (by a human) to the UAV/drone. Flight of the UAV may operate with various degrees of autonomy, either under remote control by a human operator or autonomously by an onboard computer (Dering *et al.*, 2019). The UAVs can also be flown autonomously, where modern flight controllers can use software to mark GPS waypoints so that the vehicle will fly to and land or move to a set altitude. This kind of autonomy is becoming increasingly common and contributes to many civilian applications. Although it is possible for the flight controller to control the UAV autonomously, it is generally good to have a remote-controlled transmitter so that the movement of the UAV can be controlled if something goes wrong or the UAV becomes out of control.

A handheld transmitter is adequate enough for larger aerial vehicles, and it is better to have a base station to help with all of the controls. Remote GCS also controls the UAVs, also referred to as a ground cockpit. Transmitters have many channels, and the number of channels of a transmitter is related to the number of separate signals it can send. Usually, a seven-channel radio is adequate enough to operate the UAVs in the beginning. A good transmitter must include a three-position switch or variable knob, as it would be required by most autopilots to switch between various flight modes (Gupta et al., 2018).

The UAV has a flight controller placed at the center and responsible for controlling the spin rate of the motors. Sticks on the controller allow movements in different directions, while trim buttons allow balancing the UAV. The UAV cannot function accurately and efficiently without using external sensors added with the flight controller. So for a stable and precise flight, different sensors, like accelerometer, gyroscope, and IMU, are used to interface with the flight controllers (Gupta et al., 2018). To fly UAV better, some additional sensors, like a barometer, distance measuring sensors, magnetometer, could also be used. The UAVs are also equipped with a different state-of-the-art technology equipment, such as infrared cameras, GPS, and laser devices (Dering et al., 2019). All these sensors and navigational systems are present at the nose of the UAV. The rest of the body is full of UAV technology systems. Computers can also be used to receive live video footage from the onboard camera and to display data being collected by sensors (Šustek and Úøedníèek, 2018).

A typical unmanned aircraft is made of light composite materials to reduce weight and increase maneuverability. The engineering materials used to build the UAV/drone are very lightweight and highly complex composites designed to absorb the vibrations, which decrease the noise produced. The composite material strength allows military drones to cruise at extremely high altitudes. To make the UAV agile, the acceleration and de-acceleration of the UAV must be quick, which means that it should pitch forward in the beginning and when it is time to stop, the UAV should pitch in the opposite direction quickly in an aggressive way. It is necessary to decrease the stopping distance. For the UAV to be agile, it must turn quickly with a small turning radius.

UAVs use a wide range of cameras for their missions. Most cameras used for UAV mapping are lightweight and can be programmed to shoot images at regular intervals or controlled remotely. The cameras required to carry out good mapping work are not the same as those used for professional video or photography work, but they are lightweight, high-resolution cameras designed to be used with UAVs. Some specialized devices that can be mounted on a UAV include LiDAR sensors, infrared cameras equipped for thermal imaging, and air-sampling sensors. Cameras can be mounted to the UAV in various ways. A motorized gimbal holds the camera and provides image stabilization, which can help compensate for turbulence and ultimately produce clear images. Gimbals are also used for changing the angle of the camera from vertical (straight down) to oblique. As UAV mapping is generally performed vertical or at some angle, gimbals may be relatively simple as compared to those used by filmmakers. The internal GPS functionality of some models can program the camera to take pictures at a certain interval or to take a picture based on distance or upon encountering a certain waypoint (Dering *et al.*, 2019).

Fixed-wing UAVs consist of a much simpler structure in comparison to rotary-wing UAVs. The simpler structure provides a less complicated maintenance and repair process, thus allowing the users more operational time at a lower cost. The only disadvantage to a fixed-wing system is the need for a runway or launcher for take-off and landing; however, "vertical take-off and landing" and "short take-off and landing" systems are very popular as they do not require a runway (Gupta *et al.*, 2018). Fixed-wing aircraft also require moving air over their wings to generate enough lift force to stay in a constant forward motion. It means they cannot stay stationary the same way as a rotary-wing UAV can. Therefore, fixed-wing UAVs are not best suited for stationary applications, like inspection work.

The biggest advantage of rotary-wing UAVs is that they can take-off and land vertically. This allows the users to operate within smaller vicinity with no requirement of a landing/take-off area. In addition, their capacity to hover and perform agile maneuvering makes them well suited to applications, like inspections where precision maneuvering for extended periods of time is required (Taladay, 2018).

There are various parameters that are to be considered for flight preparation, as explained below. Preflight checks are very important and help the operator to fly the UAV safely and provide professional results.

1.6.1 Altitude

Altitude is an important consideration when flying a UAV, both for practical purposes and in the interest of flying safely and legally. Legality is an important consideration when deciding an operating altitude. Caution should always be taken when flying UAV at higher altitudes, even if local regulations allow higher-altitude flight, to ensure that flights do not get in the way of manned aircraft. In many countries, it is illegal to fly above 500 feet (or 150 m).

Although higher altitude results in lower resolution, it allows the UAV to cover a larger area. Lower altitude photography gives better image quality but increases the time required to map a certain area. Suppose UAV flies at 100 m height and the area has a maximum elevation of 20 m, so from this height, UAV would cover 80% of the area in one photo compared to the area of one photo with the flat ground shown in Figure 1.19. Therefore, it is important to check the elevation range of the area being covered by the UAV to have proper coverage.

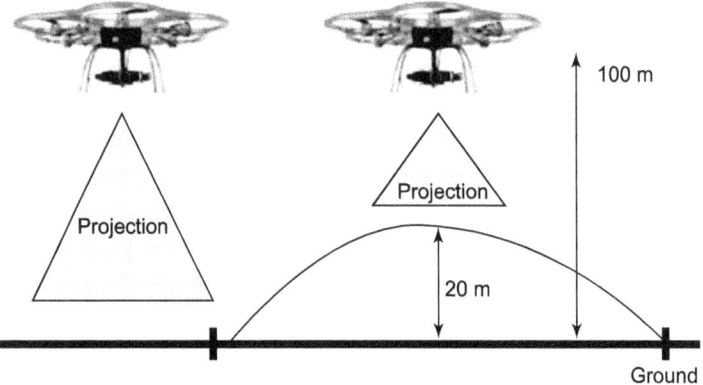

FIGURE 1.19 Terrain area covered by UAV.

1.6.2 Vertical and Oblique Photographs

The two aerial views most commonly used in UAV mapping are vertical (overhead) and oblique photographs (Newcome, 2004). Vertical photographs are taken directly above the object, with the camera looking vertically down. Oblique photographs are taken at a high or low angle to the object below. They can collect the information about the landscape that vertical photos cannot collect. During each flight, the angle of the camera remains fixed; else, this will make the resulting images considerably more difficult to process (Kaminsky et al., 2009). Photos taken from these two views can be combined using photogrammetry software, such as Agisoft PhotoScan or Pix4D, which gives users the ability to view and analyze multiple perspectives in a single computer-generated model.

The 3D models can be created using overlap from vertical photographs. Such 3D models are very useful for many applications, such as post-disaster damage assessment, urban modeling, smart city planning, and creating more accurate flood simulations. The UAV vertical images can be used for the creation of DTM or DSM but cannot capture the details of facades. For façade mapping, oblique images are becoming more popular in the UAV photogrammetric process. Oblique images are easy to capture because it only involves adjusting the built-in camera in the system to the desired angle using the remote control or adjusting the camera before each flight (Wu *et al.*, 2018).

1.6.3 Positioning

Highly accurate UAV navigation is very important when flying for applications, such as creating 3D maps, surveying landscapes, and search & rescue missions. As the UAV is switched on, it searches and detects GNSS satellites. High-end GNSS systems use a satellite constellation, a group of GPS satellites working together, giving coordinates of the point of observation (Garg, 2019). Many of the latest UAVs have dual GNSS, such as GPS and GLONASS, to provide better accuracy. The UAVs can fly in both GNSS and non-satellite modes (Jha, 2016). For example, DJI drones can fly in P-mode (GPS & GLONASS) or Attitude mode, which do not use satellite navigation (Kan *et al.*, 2018).

1.7 ADVANTAGES AND DISADVANTAGES OF UAVs

The choice of potential unmanned platforms for low-altitude earth observation is large and continues to grow. In recent years, UAV technologies are evolving rapidly and becoming appropriate for multiple applications. Various reasons make the technologies relatively preferred. Tethered systems navigated manually, such as Kites and Blimps, are ideal for the precise coverage of small sites that requires only a few images (Gupta *et al.*, 2018). However, these systems are hardly used for systematic surveys of larger areas where regular overlaps along evenly spaced flight lines are preferable. The GPS and INS technologies have led to the availability of a range of auto-piloted systems, such as satellites, planes, drones, and multicopters that can autonomously follow the prescribed flight lines (Kan *et al.*, 2018).

It is difficult to generalize the advantages or disadvantages of a particular platform because it's possible applications and working conditions vary greatly around the world (Jha, 2016). However, an attempt has been made to present the performance evaluation of satellites, manned aircraft, and UAVs in Table 1.10.

TABLE 1.10 Performance Evaluation of Satellites, Manned Aircrafts, and UAVs.

S.No.	Attributes	Performance		
		Satellites	Manned Aircrafts	UAVs
1	Endurance	High	Medium	Low
2	Payload capacity	Medium	High	Low
3	Cost	Medium	Low	Medium
4	Cloud free images	Low	Medium	High
5	Coverage	Low	Medium	High
6	Maneuverability	Low	Medium	High
7	Deployability	Low	Medium	High
8	Autonomy requirement	Medium	Low	Medium

1.7.1 Advantages of UAV

Any application that could utilize a highly mobile data collection or communication platform can utilize UAV as a primary data collection platform. UAVs have several advantages in military/army over manned aircraft. They provide a persistent presence over a specific area to capture still and video imagery, which provides military intelligence, surveillance, reconnaissance, and weapon delivery capabilities. In politically sensitive areas, UAVs have obvious advantages where the deployment of human soldiers would create too much controversy (Andrews, 2020). Humans on the ground (walking or using ground-based vehicles for data collection) can be challenged by hilly and difficult terrains, physical obstacles, or dangerous site conditions. The UAVs provide a safer method in high-risk situations where the presence of a pilot-driven aircraft is risky. They allow flight in difficult and inaccessible areas without having a risk of causalities since human life is not endangered during the process (Iizuka *et al.*, 2018).

The UAVs are also advantageous in many civilian applications, primarily for accelerating accessibility to remote and dangerous sites, sensor mobility, monitoring, and speed of data collection. The UAVs are important in a monitoring forest fire, especially in areas with rough topography where ground movement is not possible. Hence, UAVs are considered a highly preferred technology for risk management. In particular, UAVs have emerged as an essential data collection tool for applications involving natural hazards (e.g., earthquakes and landslides) where site accessibility is challenging post-event, and the need to collect recent data is urgent. As a result of these challenges, UAVs offer an economic advantage, including reducing costs associated with personnel, travel, and site logistics. Compared to a traditional inspection of structures, UAVs are cheaper, faster, and safer than traditional methods (Gupta et al., 2018).

The demand for UAVs to transport materials and medicines is increasing. As an example, the potential benefits of UAVs in the health sector are extensive. The UAVs can fly over vast distances, enabling the delivery of life-saving medical supplies to those in hard-to-reach communities. The technology could also be used to deliver non-emergency commodities for villages where visits of health workers are not regular, and delivery of commodities is infrequent; thus it greatly reduces the response time. The UAVs could bring meaningful benefits to public health through improved supply chain performance.

The UAV also offers the possibility of making a non-destructive and accurate measurement of many attributes. It enables work measurement and work certification remotely and accurately (e.g., site audits, civil surveys, and measurement) (Taladay, 2018). It can improve measurement accuracy through enhanced visualization of terrain for site surveys and measurement, construction progress, and 3D mapping (Jha, 2016).

UAV-based photogrammetry provides advantages over other remote sensing satellite platforms (Molina, 2014). The UAVs can be a smart and cost-effective complement to traditional photogrammetric data. Satellite images are limited by their revisit time, cloud coverage, and spatial resolution. Advances in UAV platform design have contributed to the innovation of small UAV that fly at low altitudes to provide very fine spatial resolution images (Iizuka et al., 2018). The UAVs can be deployed on demand, and flight parameters can be adjusted to acquire the images at desired high resolution. Hence, this technology provides an efficient method of spatial data collection for multiple applications, such as transportation, urban planning, crop species classification, and yield estimation that require detailed information (Taladay, 2018). The HALE UAV provides a cost-effective and persistent capability to efficiently collect and disseminate high-quality data across large areas than the ground-based systems.

The other advantage of UAV relates to low energy consumption (Banu et al., 2016). Small UAVs use either fuel (gas) or battery as a source of energy. Although the batteries or fuel need to be charged frequently, but the energy required for a mission is small. Therefore, its low energy requirement contributes to the wide acceptance of the technology. The UAVs, being smaller than manned aircraft, can be more easily and cost-effectively stored and transported.

The benefits of the technology are many as UAVs support operations in-

(i) *Safety:* by allowing users to acquire data from the air instead of entering difficult/unstable terrain.

(ii) *Costs:* that are related to reducing operational expenses for data collection, particularly if the area is large.

(iii) *Flexibility:* that is associated with operating missions on schedule or emergency or on demand.

(iv) *Speed:* that enables the collection of real-time data where infrastructure allows.

(v) *Increased productivity:* due to savings in time and a high degree of automation.

(vi) *Frequency:* that opens opportunities for monitoring and tracking of work in progress, and

(vii) *Accuracy:* by generating accurate and reproducible datasets (e.g., 3D models).

1.7.2 Disadvantages of UAV

The disadvantages of UAVs include low maneuverability, low operational speed, vulnerability to attack, cyber or communication link attacks or lost data links, limited area coverage, and the requirement of a sophisticated analysis system to handle Big Data. The UAVs could be used for lethal purposes, criminal activities, like drug smuggling, and terrorist activities when hijacked by terrorist networks under third-party control situations. Another issue relates to safety; when these are not operated by properly trained operators, the chances of UAVs failing are more, leading to fatal accidents (Iizuka et al., 2018). Therefore, the safety, security, and privacy concerns need to be taken into consideration in technology development, market development, and their use by the public and private parties. In addition, regulations must clarify the limits to use UAVs and ensure that such devices are flown under safe conditions. Some of the advantages and disadvantages of UAVs are presented in Table 1.11.

TABLE 1.11 Advantages and Disadvantages of UAV Types (Heutger and Kückelhaus, 2014).

Types	Advantages	Disadvantages	Photo
Visual Fixed-Wing	Long range High flight time Large coverage area Object resolution (cm/pixel)	Horizontal take-off, requiring substantial space (or support, e.g., catapult) Inferior maneuverability compared to VTOL	
Unmanned Helicopter	VTOL Maneuverability High payloads possible	Expensive Comparably high maintenance requirements	
Multicopter	Inexpensive Easy to launch Low weight The very small take-off area Object resolution (mm/pixel)	Limited payloads Susceptible to wind due to low weight Less flight time Small coverage area	

1.8 SUMMARY

UAVs have been widely adopted in the military world, and the success of these military applications is increasingly the driving force to use UAVs in civilian applications. They have recently received a lot of attention, being fairly economical platforms, having navigation/control devices and imaging sensors for quick digital data collection and production. The UAVs provide a real-time capability for fast data acquisition, transmission, and processing. In these conditions, automated and reliable software is required to process the data quickly. Both photogrammetry and LiDAR UAVs can provide images/3D data with high accuracy; each type of data has its own merits and demerits (Corrigan, 2016). The difference between photogrammetry and LiDAR UAV can easily be understood while considering operational and logistical factors.

The biggest advantage of UAV systems is the ability to quickly deliver high temporal and spatial resolution information and allow a rapid response in a number of critical situations where immediate access to this type of 3D geo-information may be difficult. The UAVs can be used in high-risk situations and inaccessible areas, although they still have some limitations, particularly the payload, battery life, and stability. Compared to traditional airborne platforms, UAVs reduce operational costs, providing high accuracy data. The UAVs offer three technical advantages (Andrews, 2020): (i) minimize the risk to the life of the pilot; (ii) not subject to natural human limits (e.g., fatigue); and (iii) data acquisition is generally cheaper than that of manned aircraft.

Besides very dense point cloud generation, the high-resolution UAV images can be used for texture mapping purposes on existing 3D data, orthophoto production, map generation, or 3D building modeling. In addition to large applications in military, UAVs are used for various other applications, such as disaster and healthcare, journalism and photography, early warning system and emergency services, marketing and business, activity monitoring, gaming and sporting, farming and agriculture, military and spy, and many other fields. On the other hand, UAVs might become the most dangerous devices on the earth if the regulations are not followed strictly (Gupta *et al.*, 2018). The same UAV that is used for the delivery of food and products could also be used to transfer bombs and other illegal materials.

The UAVs have been the subject of concerted research over the past few years due to their autonomy, flexibility, and broad application domains. It is predicted that the market for commercial end-use of UAVs might supersede the military market by 2021, cumulatively hitting approximately USD 900 million, while the global market size will touch USD 21.47 billion (Economic Times, 2018). With the implementation of proper monitoring, powerful law and security, and the pros of this technology, UAVs are expected to work for the benefit of society. As the cost will come down, UAVs will become widely available to the general public and used for wider applications. It is expected that in the future, the UAV technology will make the UAVs further smaller, lighter, cheaper, and much more efficient.

REVIEW QUESTIONS

Q.1. Define the terms: UAV, UAS, RPA, and Drone. What is the basic difference between these?

Q.2. Write the historical developments of unmanned aerial vehicles.

Q.3. Discuss the contribution of government and industry to the development of various UAVs.

Q.4. What are the major segments where UAVs, Blimps, and Zeppelins are being used? What is their relative market potential? Also, discuss the market potential of UAVs region-wise.

Q.5. Draw a diagram to show maturity levels of UAVs applications in various sectors.

Q.6. What do you understand by the Hype Cycle for the advancement of technology? Discuss it with reference to emerging technologies.

Q.7. Describe the terms: Waypoints, Point of interest, Line of sight, Position hold, Return to home, and No fly zone. Explain the difference between VLOS, BVOLS, and EVLOS.

Q.8. Discuss the general characteristics of a UAV.

Q.9. Differentiate the working of a Photogrammetry UAV and LiDAR UAV.

Q.10. Discuss the advantages and disadvantages of Photogrammetry UAVs and LiDAR UAVs.

Q.11. Enumerate the advantages and disadvantages of UAVs over satellite images.

CHAPTER 2

VARIOUS COMPONENTS OF UAV

2.1 INTRODUCTION

The UAVs are easy to deploy, as they have flexibility in performing difficult tasks and supporting high-resolution imagery even of remote areas. Understanding the aerodynamics of these UAVs is important to ensure flight safety and to improve task efficiency (Marques, 2013). The architecture of a typical UAV consists of several components having a mainframe, control system, GPS, sensors, monitoring system, data processing system, and the landing system (Jha, 2016). The internal system provides a wide range of functions ranging from navigation to data transfer to the ground. All these elements combined have to pilot the UAV without any human intervention. The UAVs are often accompanied by ground control stations consisting of laptop computers and other components that are required to be carried easily with the air vehicles. All parts and components of a UAV are vital to a smooth and safe flight.

The UAVs are classified based on the wing types, altitude range, endurance and weight, and support a wide range of applications, including military and commercial applications. The flight time of a UAV depends on several factors, such as energy source (e.g., battery, fuel, etc.), type, weight, speed, and trajectory of the UAV (Austin, 2010). The size, type, and configuration of UAVs could be different, depending on the specific application. The UAVs are fitted with high-resolution cameras and sensors and can navigate around the area through software (Gupta *et al.*, 2013).

The choice of selecting a UAV critically depends on many factors. Depending upon the goals and applications, an appropriate type of UAV can be used that would meet various requirements for the desired output, considering the nature of terrain, time & fund available, and the existing flying regulations. In fact, several factors, such as the UAVs' capabilities and their flying altitudes, must be taken into account to properly use them for any application. To date, some of the inhibiting factors for using UAVs in many civilian applications include the cost of acquiring these devices, building the required applications, and operating systems. Many organizations are developing UAVs or its components in order to reduce the cost of related services. The UAV market is growing, and UAVs are being used in new activities and solving new problems every day.

2.2 BASIC AERODYNAMICS

Increased demand for flight vehicles in various fields requires improved aeronautical techniques to ensure flight safety and improve task efficiency. To understand the aerodynamics of a drone/UAV, there is a need to understand some basic terminologies used for a flight, such as drag, thrust, lift, pitch, roll, and yaw (Marques, 2013). These terms define the position and orientation of the UAVs. So far, the aerodynamics of mini-UAVs is hardly understood as remote-controlled toy planes were developed with this knowledge. Aerodynamics is more important in fixed-wing UAVs than in rotary-wing aircraft (Ahmad, 2015). The blended-wing UAVs offer advantages in aerodynamic and efficiency over the conventional wing designs as they also simplify the control surfaces.

The available computational aerodynamics tools, including advanced computational fluid dynamics methods, are incapable of predicting the low Reynolds number ($< 2.5 \times 10^5$) airflow phenomena on wings. Low Reynolds numbers are the result of the low speeds in combination with the small wing chords, while the "low aspect ratio" < 4 is the result of a combination of different factors: high wing loadings due to high aircraft weight stall speed requirements and limits on the wingspan (Austin, 2010). In the last decade, a trend is observed toward smaller and mini-UAVs for information gathering and reconnaissance missions. Miniaturization of electronics and major leaps in battery technology have pushed the development of mini-UAVs toward practical applications, e.g., assisting fire-fighter crews, power line inspections, and filming sports events (Demir et al., 2015).

The primary forces that act on any air vehicle are thrust, lift, drag, and gravity (or weight), as shown in Figure 2.1. Weight is the force of gravity that acts downward toward the center of the earth. Lift, drag, and rotational moments are computed from dynamic pressure, wing area, and dimensionless coefficients (Marques, 2013). The expressions for these quantities are fundamental for aerodynamic equations that govern the performance of an air vehicle.

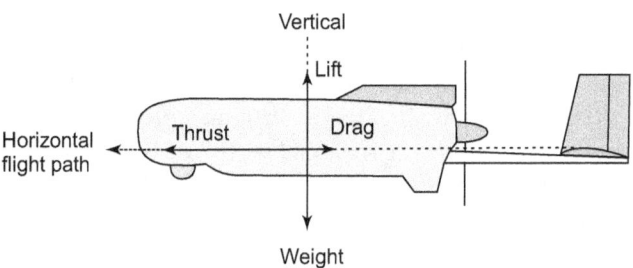

FIGURE 2.1 Primary forces acting on an air vehicle.

2.2.1 Lift

Lift is the force that acts at a right angle to the direction of motion through the air and is responsible for lifting the UAV up in the air acting against the weight of the UAV (Ahmad, 2015). It is created by differences in air pressure above and below the wings of aircraft. The propellers on the multicopter generate a lift force using Newton's third law of motion and Bernoulli's principle (Caughey, 2011). The lift force must be greater than the weight of the UAV/drone for a successful flight.

Wings generate lift due to two effects. As for all wings, the lift is generated by the air circulation around the wing, also referred to as linear lift. Its nature is linear with respect to the angle of attack (Jha, 2016). In contrast, the non-linear lift is generated due to the vortex lift which is produced on the leading edge or at the side edges of the wing. The leading edge vortex is generated when the wing is highly swept and has a sharp leading edge. The side edge vortex is present on all wings. In the case of "low aspect ratio" < 4, the vortex will have its effect on the wake and the wings (Austin, 2010).

2.2.2 Drag

The mechanical force applied by any fluid to oppose the motion of any object is called drag. Drag is the force that acts opposite to the direction of motion (Ahmad, 2015). It is caused by friction and difference in air pressure, which tends to slow down the flying aircraft. In the case of drones/UAVs passing through the air, "aerodynamic drag" is generated. The difference in velocity between the UAV and the air is responsible for generating the aerodynamic drag. The performance estimation of an aircraft largely depends on the correct calculation of the drag.

2.2.3 Thrust

Thrust is the force generated by the propellers of the multirotor to work against the aerodynamic drag. It is the force that moves the aircraft in the direction of motion (Jha, 2016). Engines produce thrust with the help of a propeller, jet engine, or rocket. The thrust is not the only force responsible for lifting the UAV up, but also it is the force that helps to overcome the drag resistance.

2.3　STABILITY AND CONTROL OF UAV

All UAVs must have some minimum level of stability and be controllable. Application of the fundamental principles of physics and aerodynamics provides an understanding of the design parameters, which have an influence on whether the air vehicle maintains stable flight when experiencing small disturbances or deviates uncontrollably from its intended flight path (David, 2011).

An object is said to be stable when a small external force is applied to the object, and the object comes back to its equilibrium position. An air vehicle must be stable if it has to remain in flight. In static stability, the forces acting on the air vehicle (thrust, weight, and aerodynamic forces) are in directions that would tend to restore the air vehicle to its original equilibrium position if it has been disturbed by wind or other force. If the air vehicle is not statically stable, the smallest disturbance will cause deviations from the original flight plans (Ahmad, 2015). In addition, it may overshoot, turn around, go in the opposite direction, overshoot again, and eventually oscillate to destruction. In this case, the aircraft would be statically stable but dynamically unstable.

It is well known that UAVs have many stability problems, such as shaking, out of control, and even crash, and if endowed with manipulators, its stability will further be affected by manipulator movements and environment contacts (Austin, 2010). Dynamic stability of UAVs, especially micro-UAVs, is necessary and important to ensure that the whole system is safe and reliable (David, 2011). If the oscillations are damped and eventually die out, the air vehicle is dynamically stable. An aircraft has three angular degrees of freedom (pitch, roll, and yaw), and equilibrium about each axis must be maintained. The pitch axis is most critical, and stability about it is called longitudinal stability. Some instability can be tolerated about the roll and yaw axis, which are combined in most analysis, and called lateral stability (Marques, 2013).

Control refers to the ability to change the state of motion of an air vehicle by some means. Control is accomplished through control surfaces actuated by control loops implemented in autopilots using basic sensors on board the UAVs and through operator's intervention (David, 2011). In order to take care of the inherent neutral stability of the drone/UAV, it is necessary to use sensors to control its behavior. A simplified functional block diagram of a closed-loop automatic control system is shown in Figure 2.2.

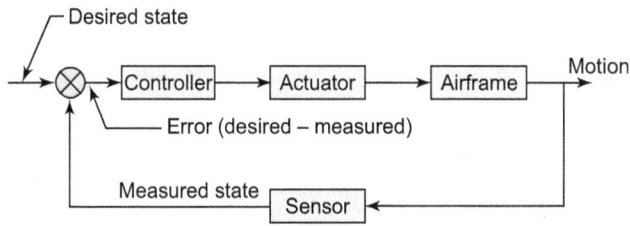

FIGURE 2.2 Block diagram of a control loop.

2.3.1 Roll

The rotation of the aircraft along the longitudinal axis is called "roll." The longitudinal axis is an axis along lengthwise of the air vehicle, usually passing through its center of gravity, as shown in Figure 2.3. For side to side motion (roll), the speed of the right motors is controlled relative to the speed of the left motors. In order to "roll" to the right, the speed of the left motors of the air vehicle is increased relative to the speed of the right motors (David, 2011). This rolls down the air vehicle to the right side in a sideways swaying movement as the torque is generated. In a similar manner, the speed of the right motors of the air vehicle is increased relative to the speed of the left motors to "roll" left. Thus, the torque on either side along the longitudinal axis continues to act on the same pair of rotors and then translates to the acceleration to move the air vehicle to left/right (Jha, 2016).

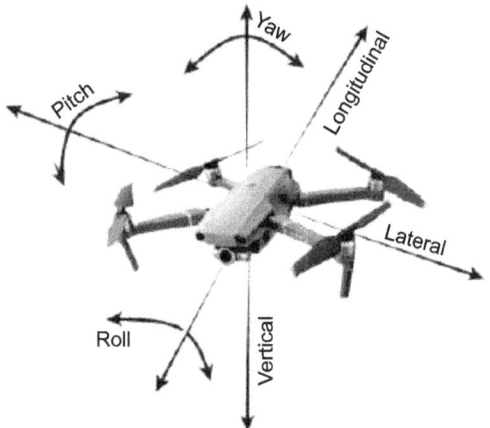

FIGURE 2.3 Representation of roll, pitch, and yaw.

2.3.2 Pitch

The rotation of the aircraft along the lateral axis is called "**p**itch" (Figure 2.3). The lateral axis is an axis that runs below the wing, from wingtip (left) to wingtip (right), passing through the center of gravity of air vehicle (Marques, 2013). To fly in the forward or the backward direction, the speed of rear motors is controlled relative to the front motors, and the term used for this motion is called "pitch." For the air vehicle to pitch forward, the torque is produced at the backward pair of rotors in order to generate more force (Jha, 2016). So, the speed of the rear motors will be increased as compared to motors at the front. Similarly, when the air vehicle has to pitch backward, the torque is generated at the forward side, thus increasing the speed of the front pair of motors as compared to rear motors.

2.3.3 Yaw

The rotation of the air vehicle (left or right) along the perpendicular axis is called "yaw." It is an axis perpendicular to the wings and body of the air vehicle (up-down), passing through the center of gravity of the air vehicle (Figure 2.3). The "yaw" is activated if the diametrically opposite pairs of motors increase their speed relative to the other pair (Gupta *et al.*, 2018). In such a case, the rotation speed of diametrically opposite pairs of motors is increased or decreased, varying the torque in the direction of rotation of that pair which causes the air vehicle to rotate in the direction of the increased torque (David, 2011). When a motor is turned clockwise, it generates counter-clockwise torque on the frame to equalize the torques.

2.3.4 Throttle

The throttle controls the speed of the motor, and adjusts the altitude, or height of the air vehicle (Marques, 2013). The "throttle" works the same way as any internal combustion engine throttle, by changing the amount of fuel and air that enters the combustion chamber of the engine. Thus, if the speed of all the rotors is increased at the same time, motors produce a strong downward force that helps the UAV/drone go higher.

2.4 UAVs CLASSIFICATION

The UAVs are classified into various types based on different criteria. The criteria for classifying UAVs include size and payload, flight endurance, flight range, altitude, and capabilities (Jha, 2016). Table 2.1 lists various types of drones/UAVs and their salient characteristics. These parameters of UAV are interrelated. In general, large UAVs carry more payloads, fly at a higher altitude, move a longer distance, and hence have better flight endurance and capability than small ones. Some commercial drone/UAV manufacturers include 3D Robotics, DJI, Elistair, Hubsan, Identified Technologies, Insitu, Measure, Parrot, Precision Hawk, and Yuneec.

The other types of UAVs are the groups of lightweight aerial systems, called blimps, such as balloons and airships, which are lighter than air and have long flight endurance, fly at low speeds, and generally are large-sized (Gupta *et al.*, 2013). In addition, there are flapping-wing UAVs, which have flexible and morphing small wings, similar to birds and flying insects (Dalamagkidis, 2014).

Broad classification of UAVs is given below:

- **Fixed-wing UAVs:** The UAVs have fixed wings that require a runway to take-off and land, or catapult launching. These generally have long endurance and can fly at high cruising speeds.
- **Rotary-wing UAVs:** They are also called multirotor or rotorcraft or VTOL UAVs. Rotary-wing UAVs have the advantages of hovering capability and high maneuverability, which is useful for many robotic missions, especially in civilian applications. They are typically more useful in situations where the terrain is difficult, or the area is unreachable otherwise.

TABLE 2.1 Various Types of Drones/UAVs and their Salient Characteristics.

Category	Weight of UAV (kg)	Operating Altitude (m)	The Radius of Mission (km)	Endurance (hr)	Typical Uses
Micro	< 2	250	< 5	1	Reconnaissance, inspection, surveillance
Mini	2–20	150–300	< 25	1–6	Surveillance, data gathering
Small	20–150	Up to 5000	50	8–24	Surveillance, data gathering
Tactical	150–600	Up to 10,000	300	> 24	Surveillance, data gathering
MALE	> 600	14,000	> 1,000	> 24	Surveillance, cargo transportation
HALE	> 600	Up to 65,000	> 1,000	> 24	Surveillance, data gathering, signal relay
Strike/Combat	> 600	Up to 65,000	> 1000	Days/weeks	Surveillance, data gathering, signal relay

- **Flapping-wing UAVs:** They have flexible and/or morphing small wings inspired by birds and flying insects. These are also available in hybrid configurations or convertible configurations, which can take-off vertically and tilt their rotors or body and fly such as airplanes, such as the Bell Eagle Eye UAV.
- **Nano-Air Vehicles:** They are extremely small and ultra-lightweight systems (e.g., DJI Tello, Eachine E10C, Holy Stone HS210) with a maximum wingspan length of 15 cm and a weight less than 20 g. These are used in swarms for ultra-short range surveillance.
- **Mini-UAVs (MUAVs):** These UAVs are between 2 and 20 kg (e.g., DJI Matrice 600 Pro, DJI Inspire 2) and are capable of being hand-launched with operating ranges up to about 25 km. These are used by mobile battle groups and diverse civilian purposes.
- **Micro-UAVs (MAVs):** These are less than 2 kg weight (e.g., DJI Spark, DJI Mavic, Parrot Bebop 2) and are used for operations in urban

environments, particularly within buildings where it is required to fly slowly and preferably to hover and to perch, (i.e., to be able to stop and to sit on a wall or post). The MAVs are generally expected to be launched by hand, and therefore winged versions have very low wing loadings, making them very vulnerable to atmospheric turbulence.
- **Close-range UAVs:** They are used by mobile army battle groups, other military/naval operations, and diverse civilian purposes. They usually operate at ranges of up to about 100 km and endurance time of 1–6 hours. They are usually used for reconnaissance and surveillance tasks.
- **Tactical UAVs (TUAVs):** They operate up to 300 km range. These are between 150 and 600 kg weight and operate within simpler systems by land and naval forces. Small and Tactical drones are of the size of a motorcycle/helicopter and are used for military purposes.
- **Medium Altitude Long Endurance (MALE):** They can fly between 5 and 15 km altitude with more than 24 hours of endurance. Their roles are similar to the High Altitude Long Endurance (HALE) systems but generally operate at somewhat ranges (>500 km) from the fixed bases.
- **High Altitude Long Endurance (HALE):** They can fly over 15 km altitude with more than 24 hours of endurance. The HALE UAVs carry out extremely long-range (trans-global) reconnaissance and surveillance and are increasingly used by Air Forces from fixed bases.

2.4.1 According to Operation & Use

Many types of UAVs exist today. These can be classified into five different types according to their operations (Dalamagkidis, 2014). These include (i) Remote human pilot (real-time control by remote pilot), (ii) Remote human operator (human provides the flight parameters to invoke the built-in functions for vehicle control), (iii) Semi-autonomous (human-controlled initiation and termination, autonomous mission execution), (iv) Autonomous (automated operation after human initiation) and (v) Swarm control (cooperative mission accomplishment *via* control among the vehicles). They are typically classified into six functional categories; target and decoy, reconnaissance, combat, logistics, personal/hobbyist, research & development, and civil & commercial. Civil and commercial drones are those dedicated to applications, such as infrastructure inspection, precision agriculture, traffic monitoring, law enforcement, emergencies, etc. According to uses, they can also be classified as:

- *Toy drones:* They are small (mini/nano) sized quadcopters, inexpensive and a good to start with most beginners.

- *Hobby drones:* They are mid-sized drones, used by novice flyers with some experience. Many have cameras attached.
- *Professional drones:* They are high-end copters with four or more propellers. Most have high definition (HD) cameras for aerial photography and offer more features along with the longest flight times.
- *Selfie drones:* They are small flying cameras that are easy to pack and carry. They enable to capture selfie shots from a unique perspective that can be shared on social media sites.
- *Racing drones:* They are small, fast, and agile quadcopters (multirotors) with an onboard camera that transmits real-time video back to the pilot for a cockpit perspective while flying through a course of gates and tunnels.
- *Foldable drones:* They are foldable drones with motor arms that can be folded and unfold. Because the motor arms can be folded, it reduces the risk of impact. They are designed to create compact space and are stored inside a travel case or backpack.
- *Inflatable drones*: The Diodon is the world's first drone with inflatable arms. The inflatable drones are easier to transport. They can be inflated fast, even with an air pump. They are waterproof and can safely land on water and even fly in the rains.

2.4.2 According to Size

The UAVs come in a wide variety of sizes. Many small UAVs/drones are now available for personal consumer use, offering HD video or still camera capabilities or to simply fly around. These UAVs/drones often weigh from less than a pound to 10 pounds (Ahmad, 2015). With increasing civilian applications, the class of mini/micro-UAVs is also emerging very fast (Dunbar *et al.*, 2015). A typical classification according to aircraft weight divides them into micro air vehicles (weighing less than 2 kg), mini-UAVs (2–20 kg), and heavier UAVs. Mini/micro type of UAVs has limitations with most computer embedded systems due to limited space, limited power resources, increasing computation requirements, the complexity of the applications, time to survey requirements, etc. (Jha, 2016).

Very small UAVs range from "micro" sized, which are about the size of a large insect with dimensions of the order of 30–50 cm. There are two major types of small UAVs; the first one uses flapping wings to fly such as an insect or a bird. The other uses a more or less conventional aircraft configuration, usually rotary-wing for the micro size range. The choice of flapping wings or rotary wings is often influenced by landing requirements and perch on small surfaces to allow surveillance to continue without spending the energy to hover. Another advantage of flapping wings is that it may look such as a bird or insect and be able to fly around very close to the objects of its surveillance or perch in plan view without giving a visual fact that it is not a bird but actually a sensor platform (Austin, 2010). However, for flapping wings, there are issues related to aerodynamics.

Some of the very small UAVs include (*i*) the Mosquito wing/fuselage, which is 35 cm (14.8 in.) long and 35 cm (14.8 in.) in total span, (*ii*) the Skate fuselage/wing, which has a wingspan of about 60 cm (24 in.) and length of 33 cm (13 in.). It folds in half along its centerline for transport and storage, and (*iii*) the CyberQuad mini, which has four ducted fans, each with a diameter of <20 cm (7.8 in.), mounted so that the total outside dimension that includes the fan shrouds is about 42 cm × 42 cm. This UAV resembles a flying "toy" called the "Parrot AR Drone." The sketches of some of these small UAVs are shown in Figure 2.4.

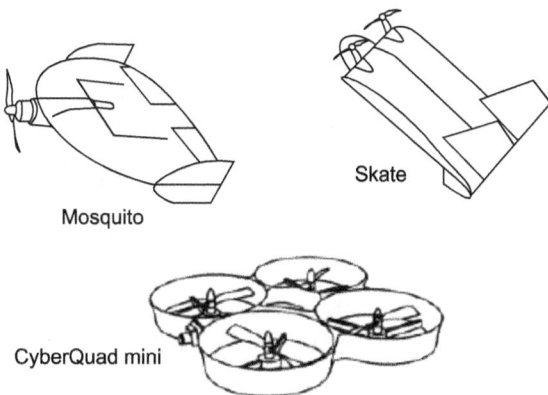

FIGURE 2.4 Very small UAVs (Austin, 2010).

Normally, larger UAVs in size are mostly used for military purposes, such as the Predator drone. The next in size are fixed wings UAVs which require short runways. These are generally used to cover large areas, such as surveying or combat wildlife poaching.

2.4.3 According to Altitude

In general, the use of UAVs as an aerial base station in communications networks can be categorized based on their altitudes as well as types. Figure 2.5 shows the classification based on low altitude platforms (LAPs) and high altitude platforms (HAPs) as well as based on wing types. This figure also presents the strengths and weaknesses of each classification.

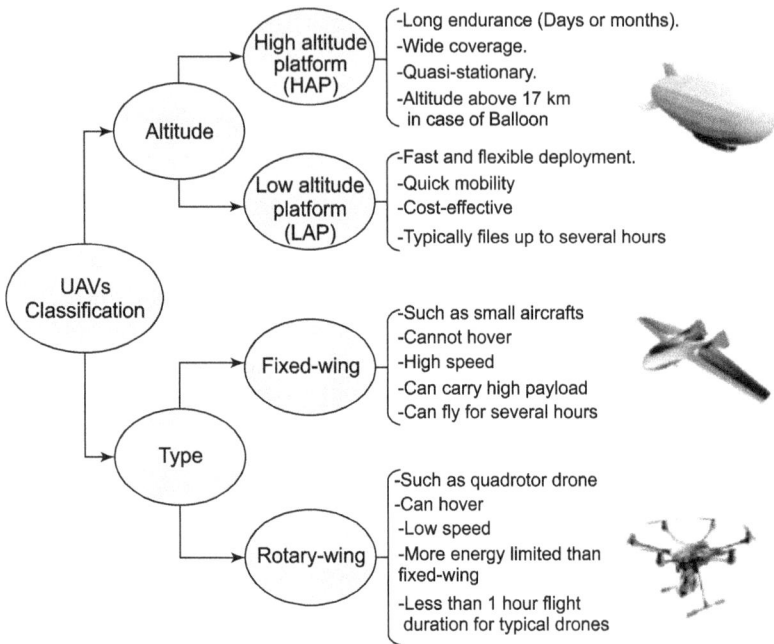

FIGURE 2.5 Categorization of UAVs (Mozaffari *et al.*, 2018).

The LAP is a quasi-stationary aerial communications platform. The LAP UAVs can quickly move and are flexible to fly at altitudes of tens of meters to a few kilometers but <10 km. Three main types of UAVs that fall under this category are; VTOL, aircrafts, and balloons (Dalamagkidis, 2014). On the other hand, HAPs operate at very high altitude above 10 km. Airships, aircrafts, and balloons are the main types of UAVs that fall under this category. Vehicles utilizing this platform are able to stay for a long time in the upper layers of the stratosphere (Jha, 2016). A comparison between the HAPs and LAPs is presented in Table 2.2. This table also summarizes the performance parameters, such as UAV endurance, maximum altitude, weight, payload, range, deployment time, fuel type, operational complexity, coverage range, and applications for each UAV type.

TABLE 2.2 HAP and LAP UAVs and their Performance Parameters (Shakhatreh et al., 2018).

Parameters	HAP			LAP		
	Airship	Aircraft	Balloon	VTOL	Aircraft	Balloon
Endurance	long endurance	15–30 hr JP-fuel >7 days Solar type	long endurance up to 100 days	Few hours	Few hours	Up to few days
Max. altitude (km)	Up to 25	15–20	17–23	Up to 4	Up to 5	Up to 1.5
Payload	Hundreds of kg	Up to 1700 kg	Tens of kg	Few kg	Few kg	Tens of kg
Flight range	Hundreds of km	1500 to 25000 km	Up to 17 million km	Tens of km	<200 km	Tethered balloon
Deployment time	Need runway	Need runway	Custom-built auto-launchers	Easy to deploy	Easy to launch by catapult	Easy to deploy 10–30 min
Fuel type	Helium balloon	JP-8 jet fuel Solar panels	Helium balloon Solar panels	Batteries Solar panels	Fuel injection engine	Helium
Operational complexity	Complex	Complex	Complex	Simple	Medium	Simple
Coverage area	Hundreds of km^2	Hundreds of km^2	Thousands of km^2	Tens of km^2	Hundreds of km^2	Several tens of km^2
UAV weight	Few hundreds of kg	Few thousands of kg	Tens of kg	Few of kg	Tens of kg	Tens of kg
Public safety	Considered safe	Considered safe	Need global regulations	Need safety regulations	Safe	Safe
Applications	Testing environmental effects	GIS imaging	Internet delivery	Internet delivery	Agriculture application	Aerial base station
Examples UAV	HiSentinel 80	Global Hawk	Project Loon balloon (Google)	LiDAR	EMT Luna X-2000	Desert Star 34 cm Helikite

Compared to HAPs, the deployment of LAPs can be done more rapidly, thus making them more appropriate for emergency situations. Unlike HAPs, LAPs can be used for data collection from ground sensors and readily recharged or replaced if needed. In contrast, HAPs have longer endurance, and they can be designed for long-term (e.g., up to few months) operations. Furthermore, HAPs are typically preferred for providing wide-scale wireless coverage for large geographic areas. However, HAPs are costly, and their deployment time is significantly longer than LAPs (Patel et al., 2017).

2.4.4 According to Wings Type

The UAVs can also be categorized based on wings type and the aerodynamic flight principle; fixed-wing, rotary-wing, flapping-wing, and VTOL. The salient features of some of the UAVs available in the global market are summarized in Table 2.3.

TABLE 2.3 Some Example of Commercial UAVs Available in the Global Market (Singhal et al., 2018).

Model	Flight Mode	Flight Time (min)	Speed (mph)	Weight (kg)	Camera
Autel Premium	X-Star Attitude GPS IOC	Up to 25	35	1.6	12 mp, Single shot
Blade BLH7480A	Bind-N-fly	5–10		95 g	EFC-721
DJI Phantom 1			22.36	1.2	4K
DJI Phantom Professional 3	GPS, GLONASS	23	35.7	1.28	1/2.3" complementary metal-oxide semiconductor (CMOS)
DJI Phantom 4		28	44.7	1.38	1/2.3" CMOS
DJI Inspire 2	GPS	27	13.42	3.44	Visible, multispectral
DJI Inspire Pro	GPS	15	11.18	3.5	Visible, multispectral
DJI Matrice 200/210/210 RTK	GPS (P and S mode)	27	51.4	3.80	Zenmuse series cameras

Model	Flight Mode	Flight Time (min)	Speed (mph)	Weight (kg)	Camera
Kaideng K70C	Headless mode	10	-	0.57	2.0 mp
Parrot ANAFI	GPS, GLONASS	25	34	0.32	1/2.4" CMOS
Parrot Bebop 2 Power	GPS, GLONASS	30	-	0.525	14 mp CMOS
JXD 509G	Headless mode Automatic landing	10	-	1.2	2mp, First Person View (FPV)
Quanum Nova	GPS	15	-	0.875	Go-pro
SYMA X8HG	Headless mode	7	-	1.5	8 mp, FPV
Yuneec Tornado H920	Waypoints, OrbitMe	24	24.85	4.9	16 mp
Typhoon K	GPS	25	4.5	1.7	12 mp
Typhoon H+	GPS	28	9	1.645	20 mp, CMOS
Walkera Scout X4	GPS	20	35	2.27	720 pixels HD video
Voyager 3	GPS/ GLONASS	25	-	3.650	1080 pixels, 60 fps
Kaideng K70C	Headless mode	10	-	0.579	2 mp

2.4.4.1 Fixed-wing UAVs

The fixed-wing UAVs benefit from a simpler structure, easy maintenance, efficient aerodynamics, and longer flight duration and range coverage. The ailerons, elevators, and rudder on the wing of the UAV control the movement of the UAV. They allow the UAV to turn around the pitch, roll, and yaw angles. The elevator controls the pitch, ailerons control the roll, and the rudder controls the yaw (Austin, 2010).

They are more suited to applications that demand longer flight endurance, such as monitoring large agriculture fields, forests, or linear infrastructure (roads or railways). Fixed-wing UAVs usually have more weights, move with high speed and carry more payloads (Figure 2.5), but they need to maintain a continuous forward motion to remain aloft, thus are not suitable for applications requiring stationary position (Jha, 2016). They require large space for take-off and landing, and as a result, they are also not suitable for areas crowded with tall objects, such as buildings, trees, and other infrastructures.

2.4.4.2 Rotary-wing UAVs

Rotary-wing UAVs (Figure 2.5) consist of a number of rotor blades that revolve around a fixed mast. The aerodynamics of rotor blades is similar to those of a fixed-wing; however, the air vehicle does not require forward movement to produce lift. The blades themselves are in constant motion, which produces the airflow required to generate the lift and help in hovering.

The rotary-wing UAV is controlled with the variation in thrust and torque from the rotors. For example, yaw movement uses the imbalance in the yaw axis forces produced by increasing/decreasing the speed of a pair of diagonal motors (Austin, 2010).

Multirotor UAVs are classified depending on the rotor configurations as helicopters, tricopters, quadcopters, hexacopters, and octocopters (Šustek and Úøedníèek, 2018). In terms of reliability, quadcopters have no possibility of landing if one motor fails, while a hexacopter can survive with limited yaw control, and octocopters can fly and land with one motor failure. The rotary-wing UAVs use batteries and have shorter flight times (Paneque *et al.*, 2014). These UAVs have the distinct advantage of needing small space for taking-off and landing compared to fixed-wing UAVs. As opposed to fixed-wing UAVs, they require a Gimbel system to support the camera.

Rotary-wing UAVs allow heavier payload, easier take-off and landing, better maneuvering than fixed-wing UAVs, and the possibility of hovering. The flight time of rotary-wing UAVs is <60 minutes with electric motors, while it is more than 1 hour in fixed-wing UAVs. Lesser payloads are carried by rotary-wing UAVs; however, fixed-wing UAVs can carry much higher weights. The rotary-wing UAVs usually have limited mobility and payload, but they are able to move in any direction as well as stay stationary in the air. They are used for applications, such as infrastructure inspection (bridges, dams, or windmills), monitoring of small fields, and spraying, and surveillance. The hybrid systems are also available in the market that combines rotary-wing systems for take-off and landing and fixed-wing systems for horizontal displacements. Table 2.4 presents the advantages and disadvantages of fixed wings and rotary wings UAVs, which may help in making the right selection of UAV for desired applications.

TABLE 2.4 Advantages and Disadvantages of Fixed Wings and Rotary Wings UAVs.

S.No.	Parameters	Advantages & Disadvantages of Using	
		Fixed-wing UAV	Rotary-wing UAV
1	Maneuverability	No	Yes
2	Price	No	Yes
3	Size/portability	No	Yes
4	Ease-of-use	No	Yes
5	Range	Yes	No
6	Suitability	Yes	No
7	Payload capacity	No	Yes
8	Safer recovery from motor power loss	Yes	No
9	Take-off/landing required	No	Yes
10	Efficiency for area mapping	No	Yes

The VTOL UAVs are generally quadcopters, and do not require take-off or landing runway (Watts *et al.*, 2012). The advantages of VTOL UAV are the portability of the platforms for remote area operations without the necessity of having a runway. These UAVs are very useful in situations where the limitation of terrain exists. Hovering capabilities, as well as maneuverability characteristics, provide a clear advantage of VTOL over other fixed-wing UAVs. In order for VTOL UAV to hover and make a transit flight, such as airplane, its propulsion system is very different than those of multicopter type or fixed-wing type UAVs (Yu *et al.*, 2016). However, high power requirements for hovering may limit the flight durations for VTOL.

The VTOL platforms are useful in rescue operations in a complex urban environment with narrow streets, and dense buildings, etc. Due to many advantages, the VTOL UAV is being used in commercial applications and is a topic for further research. The latest small drones, such as the DJI Mavic Air and DJI Spark, take VTOL to the next level and be launched from the palm of the hand.

2.5 COMPONENTS OF A UAV

The components of a UAV define its quality and type. The UAV hardware includes a frame, propellers, batteries, transmitters and receivers, Electronic Speed Controllers (ESCs), motors, and a flight controller interfaced with certain electronic devices, such as gyrometers, accelerometers, etc. To control the hardware, electronic modules, such as Arduino, Raspberry Pi, could also be used for self-programming (Dalamagkidis, 2014). The size and shape of the frame of a UAV completely depend on the needs and applications. The selection of motors, batteries, and ESC would depend on the weight that a UAV can carry.

Various components in a UAV system are payloads, control systems, data links, support elements, system users, etc. (Gupta *et al.*, 2018). The main components are shown in Figure 2.6 (Demir *et al.*, 2015). A UAV is a complex system composed of several components that work in coordination to obtain a highly valuable observation platform. Figure 2.7 depicts a schematic view of each hardware component of the UAV.

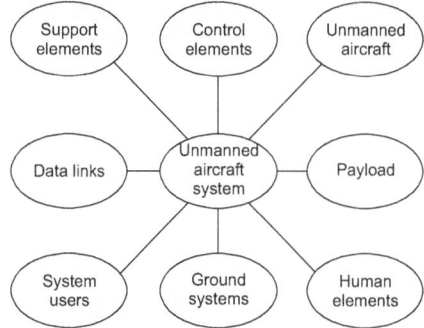

FIGURE 2.6 Main components of a UAV.

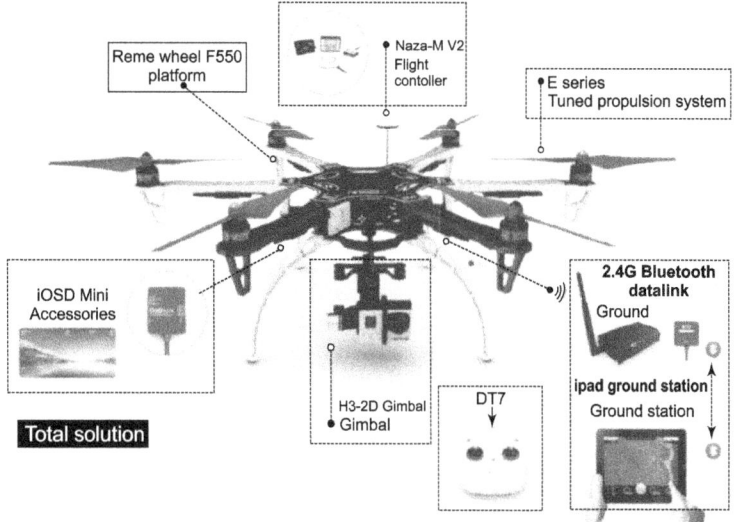

FIGURE 2.7 Pictorial view of each component of UAV (Uerkwitz *et al.*, 2016).

2.5.1 Transmitter

A transmitter or radio transmitter is an electronic device (Figure 2.8) that generates a radio frequency alternating current in electronics. When a connected antenna is excited by this alternating current, the antenna emits the radio waves. The number of channels in a transmitter determines the number of controls it has. For example, a four-channel radio will be able to control four different things. For a quadcopter, a minimum of four channels is required, not because of the four motors but because of the four different controls required to fly the quadcopter (Šustek and Úøedníèek, 2018). These controls are; throttle (how fast the motors are spinning), pitch (tilting the multicopter forward and backward), roll (tilting the multicopter to either side), and yaw (rotating the multicopter on its axis) (Gupta et al., 2018).

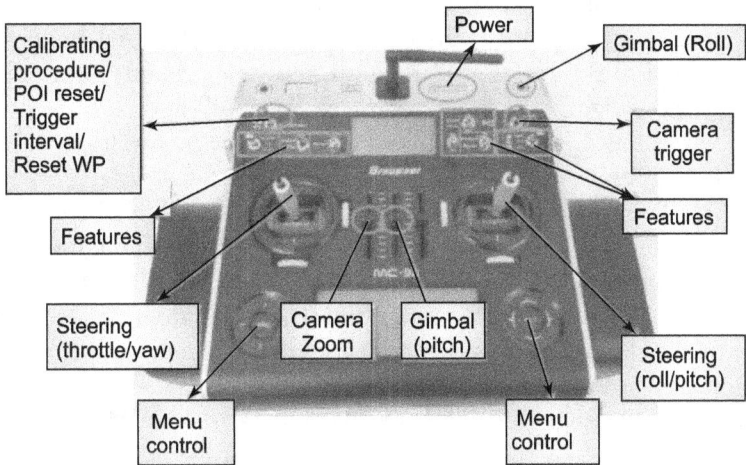

FIGURE 2.8 Transmitter and its components.

2.5.2 Receiver

The receiver is an electronic device used to receive radio waves and converts the information to a usable form (Gupta et al., 2018). There are four broad types of receivers.

1. *Pulse Width Modulation receivers*: use one servo wire for each channel. For four channels, four servo wires are used going to the channels' port on the receiver. These receivers are comparatively large because of so much wiring.

2. *Pulse Position Modulation receivers:* send multiple pulse width modulation signals through a single wire in succession. This receiver is preferred as it requires only a single wire and can carry all the channel signals required with a maximum of eight channels.
3. *Serial Bus receivers:* use just a single wire in a serial connection. They are much faster, and therefore are a preferred choice among UAV users.
4. *Radio receivers:* receive the control signals from the operator.

2.5.3 Body Frame

The frame is the skeleton and main element of the UAV, where all other components are fixed. For different tasks, different-sized frames are preferred, such as for aerial cinematography; for example, a big frame would be needed to lift the camera with tall landing gears and for high maneuverability, while a small frame is the best option for sports aerial cinematography. The design of the frame is a trade-off between strength (especially when additional weight, such as cameras are attached) and additional weight, which will require longer propellers and stronger motors to lift (González-Jorge et al., 2017).

The frame is an important part of the air vehicle because it needs to be light enough for the UAV to take-off but strong enough to provide support to components and not breaking in a minor crash. The UAVs have their own flight dynamics, and therefore the structural integrity needs to be met with. It should be a simple, light-weighted, aerodynamically efficient, and stable platform with limited space for avionics. Its shape must provide a smooth geometry having the least aerodynamic resistance (Marques, 2013).

It serves as the main support for other elements, such as motors, electronic subsystems, batteries, payload, and various other devices in such a way that they maintain stability during the flight and try to keep the UAV leveled. As mentioned before, there are two types of frames commonly used: fixed-wing and rotary-wing types; the most common types in rotary-wings include tricopter, quadcopter, hexacopter, and octocopter (Johnson, 2015), which are briefly explained below.

2.5.3.1 The Tricopter

This type of UAV has three arms, usually with a 120° angle between the arms (Figure 2.9a). This is, in theory, the lowest costing model because it needs fewer motors. Since this UAV is not symmetric, the rear motor is more complex than the other two in order to maintain a stable flight. All autopilots do not support this configuration.

2.5.3.2 The Quadcopter

This UAV with four arms is the most popular design (Figure 2.9b). The quatcopter is easiest to build and is quite versatile. All the flight controllers work with this design. But if one of the motors fails, then this type of UAV will most likely crash.

2.5.3.3 The Hexacopter

This type of UAV with six arms can also be used on a tricopter frame by doubling the motors on each frame (Figure 2.9c). The tricopter design is also called a Y6 configuration, giving the shape of the letter Y. If one of the motors on the hexacopter goes out, the other motors can usually make it to the ground more safely than a quadcopter. More motors will create more uplift thrust, allowing the UAV to lift heavier payloads but these models are usually more expensive.

2.5.3.4 The Octocopter

This UAV with eight arms can also be used on a quadcopter frame by doubling the motors on each frame (Figure 2.9d). The quadcopter is also called the X8 octocopter, as it gives the shape of the letter X. The octocopter has the same advantages and disadvantages as the hexacopter. Due to more motors, it is a more expensive model, but it can lift heavier payloads and recover much better from a failing motor than a hexacopter.

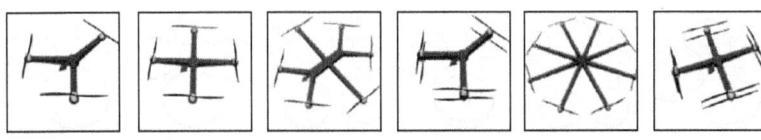

FIGURE 2.9 The Rotary-wing (*a*) Tricopter, (*b*) Quadcopter, (*c*) Hexacopter, and (*d*) Octacopter.

Fixed-wing systems exhibit greater flight duration; however, most of the market is occupied by rotor-wing systems as they carry a higher payload and show greater agility than fixed-wing systems. Figure 2.10 shows some fixed wings and rotary wings UAVs used in various agricultural applications.

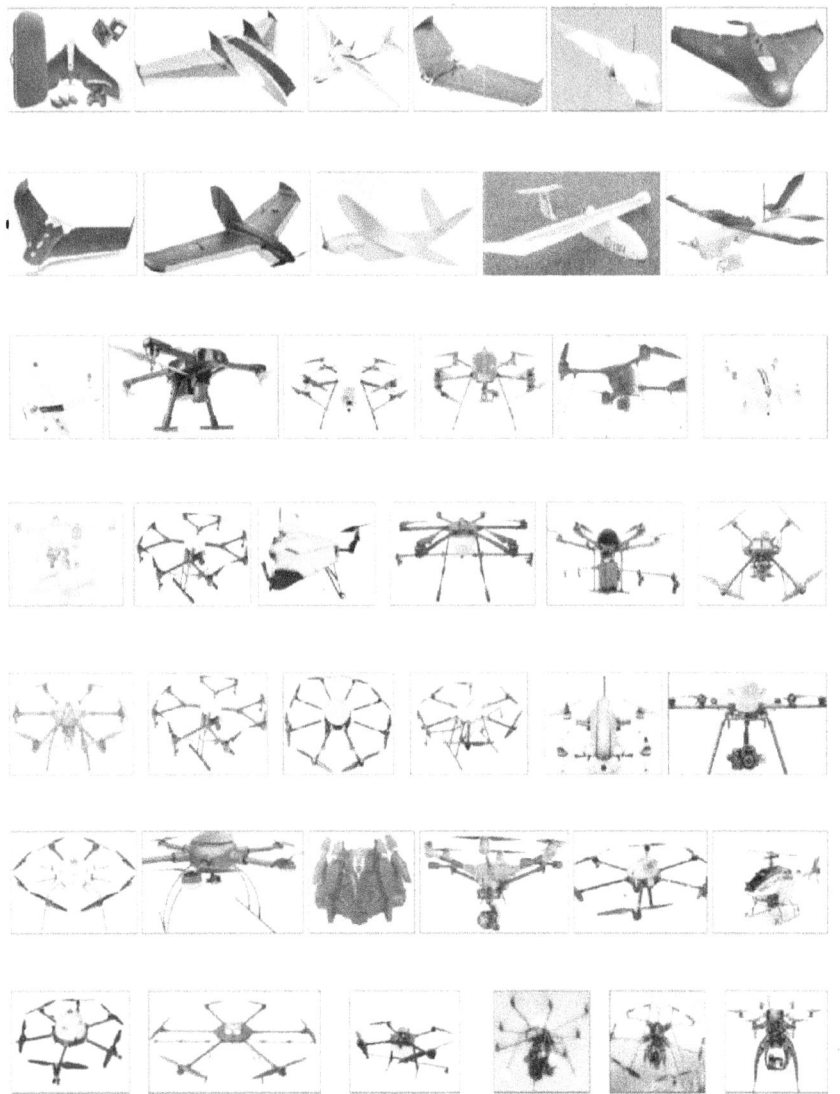

FIGURE 2.10 Some fixed-wing and rotary-wing UAVs used in agriculture studies (Shamshiri et al., 2018).

Among the rotary-wing types, the most common frame used is the quadcopter, but the hexacopter and octocopter can also be used (Cai et al., 2011). The concept of quadcopter UAV is not new; manned quadcopters first experimented within the 1920s, but their effectiveness was hampered by the technology available at that time (Bloom. 2011). In a quadcopter, four motors are placed at the edges of the arms of the frame. The direction of rotation of each motor is such that the torque generated by each of the motors cancels out. The adjacent motors rotate in the reverse direction while the opposing motors rotate in the same direction.

For the quadcopter to hover in place, it is necessary to ensure that all the motors rotate at the same rpm, which must be sufficient to generate a "lift" balancing its own weight. In addition, the torque due to all four motors must be canceled out so that the quadcopter does not yaw in a certain direction. In order to gain the altitude, all four motors of the quadcopter must increase the speed of rotation simultaneously, and similarly to descend down, the speed of all four motors must be decreased simultaneously. In terms of reliability, quadcopters have no possibility of landing if one motor fails, while a hexacopter can survive with limited yaw control, and octocopters can fly and land safely with one motor failure. However, with the advancement of electronic technology in sensors, batteries, cameras, and GPS systems, quadcopters have become widely employed over the past decade both recreationally and commercially (Šustek and Úøedníĉek, 2018).

A number of aerospace materials are used for fixed-wing and rotary-wing UAVs. The fiberglass (G10), carbon fiber, and aluminum materials are used for the frames, but the most important is how well the frame has been designed. Low-cost fixed-wing UAV, such as Parrot Disco uses a foam frame, while high-end UAVs use materials, such as carbon fiber. Rotary-wing UAVs use plastic, aluminum, and carbon fiber frames, depending on the quality of the aircraft. Low-cost models, such as DJI Phantom, use plastic, while, for example, the DJI Inspire or Microdrones use carbon fiber. In some models, there are frames that enclose the most important electronic systems in the interior, while in other models, the electronics are outside of the frame. The latter is more sensitive to inclement weather (mainly water) and impact (Yu *et al.*, 2016).

2.5.4 Motors

The motor is an equally important component, as the power system of a multicopter depends on the motor. The motor is required one per propeller. Although there are UAVs that use fueled motors, brushless direct current (DC) electric motors are the most commonly used in multicopter. Gasoline motors are too heavy for UAVs. Brushless DC motors appeared in the 1960s due to advances in solid-state electronics. These are electronically commutated systems (Figure 2.11a), and are powered by a DC electric power source using an integrated inverter, which produces an alternate current

(AC) electric signal to drive the motor (Gupta et al., 2018). The permanent magnet rotates, and the current-carrying conductors are fixed. The armature coils are switched by transistors at the correct rotor position in such a way that the armature field is in space quadrature with the rotor field poles.

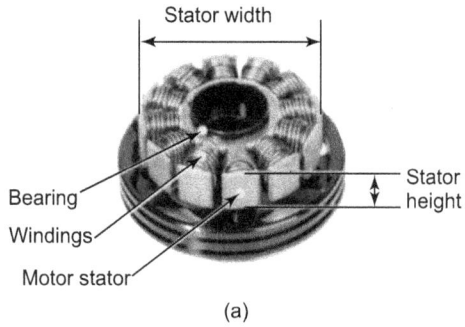

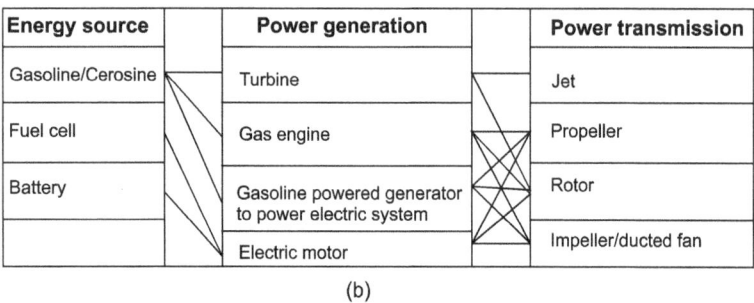

FIGURE 2.11 (*a*) A brushless DC motor, and (*b*) Link between energy source, power generation and power transmission (Wackwitz et al., 2016).

These motors have very high power density (in terms of size and light weight) and are incredibly efficient which can spin fast with precisely controlled speed. Moreover, they are less noisy and economical to manufacture. There is a wide variety of physical configuration, which depends on the stator windings. The windings can be configured as single-phase, two-phase, or three-phase motors. On the other hand, the most used brushless DC motors are three-phase (Šustek and Úøedníèek, 2018).

Brushless motors commute by software implementation using a microcontroller or in digital firmware using a field-programmable gate array and are used in two common electrical winding configurations. The delta configuration connects three windings to each other in a triangle-like circuit. The Y-shaped configuration connects all the windings to a central point, and power is applied to the remaining end of each winding. A motor with windings in delta configuration gives low torque at low speed but can give a higher top speed. The Y-configuration gives high torque at low speed. The reliability, easy control, and efficient power transmission of brushless motors have contributed greatly to many UAV developments in fixed-wing and rotary-wing systems. Figure 2.11b shows a connection between energy source, power generation, and power transmission.

With multirotors, it is important to maintain the thrust to weight ratio as two or more to ensure that double thrust is produced by the motor than the total weight of UAV, or in other words, UAV should be able to hover at half of the throttle. Motors are rated in "KV" units which equates to the number of revolutions per minute it can take when a voltage of 1 V is supplied to the motor with no load. For most multirotor UAVs, a low KV around 500–1000 is desired for stability. It is important to see the thrust rating that some manufacturers would provide. Usually, it is related to a given type of propeller option. This thrust is essential because if the thrust is 2.5 kg and the UAV also weighs 2.5 kg then it will have great difficulty in taking off and staying up. The propellers and motors are chosen so that the maximum thrust is obtained. A faster motor spin will give more flight power but requires more power from the battery resulting in a decreased flight time (Corrigan, 2020).

Motors must be matched to the propeller specifications that are being used. It is important that the maximum current ratings should not exceed by trying to get too much power out of them, forcing them to turn too slowly, or overheating them. The motors chosen will have a huge impact on the flight time and decide the quantum of heavy load a UAV can carry. It is highly suggested that the same type of motor be used for all of the rotors in a UAV so that they all do the same amount of work for the UAV. Even though the motors are the same make and model, their speeds may still vary, which is the function of the flight controller. The motors receive data for direction from the flight controller and the ESC to either hover or fly the UAV (Dalamagkidis, 2014).

2.5.5 Electronic Speed Controller (ESC)

One crucial element of the motor is the ESC, as shown in Figure 2.12(a). It provides electronically generated three-phase power to the motors and controls the rotation frequency of the motors. An ESC may have a complete programming facility, including battery mode and throttle range. The ESC helps the flight controller to be able to control each of the motors separately to produce the correct spin speed and direction so that the entire UAV can stay stabilized (Gupta et al., 2018).

The ESCs are typically rated according to the maximum current. They support nickel metal hydride, lithium polymer (LiPo), and Lithium Iron Phosphate (LFP) batteries with a range of input voltage (Vergouw et al., 2016). The rotation frequency is controlled using "pulse width modulation." Pulse width modulation signal from the receiver is converted to an appropriate level of electrical power that drives the brushless motor. The Battery Elimination Circuit (BEC) is a voltage regulator that converts the voltage from a battery to a lower voltage to control the supply. Usually, BEC is built into ESC, eliminating the need for another battery to power the 5V electronic devices. The universal BEC is used under the situation when ESC does not have built-in BEC. It is generally more reliable, more efficient, and able to provide more current than the BEC. The universal BEC is connected in a similar way as the ESC, that is, directly to the main battery of the UAV.

2.5.6 Propellers

A propeller is the most common propulsion device used in UAVs. A propeller is composed of several blades that rotate around an axis to produce thrust by pushing air, much such as a fan does. The performance of a propeller is determined by its geometry, and different propeller geometries are optimized to produce thrust efficiently at given rotational and forward flight speeds. The propellers are the components that are frequently damaged while working and need replacement. Propellers convert the rotational motion into thrust (González-Jorge et al., 2017).

There are five main variables that need to be considered when choosing a multirotor propeller; size, pitch, blade configuration, material (durability), and design (efficiency). Pitch is the distance traveled by UAV in one single prop rotation. A low pitch propeller provides perfect stability and vibration. More torque is produced by the propellers of the lower pitch and hence can operate on less current, whereas more air will get displaced for a high pitch propeller, and this will result in a turbulent motion. The variation in speed of the multicopter is easier in the case of small propellers, but in the case of large propellers, it takes a little time to change the speed. Also, the UAV with small props uses a high *rpm* motor, and the blades spin rapidly to make the UAV fly (Gupta et al., 2018). However, a motor having less KV is used for operating the larger propellers; otherwise, the motor may burn due to excessive heat produced. Due to this, the UAV would shake while hovering. Thus, a lower pitch propeller is suggested.

There are two types of propellers used: (*i*) plastic propellers, which are lightweight and inexpensive but have balancing issues and produce more sound and vibrations and (*ii*) carbon fiber propellers which are lighter in weight, more strong, more balanced, and produce lesser vibration and sound. But fiber propellers are expensive and produce lower thrust than plastic ones. Carbon fiber propellers are more rigid and more durable during small crashes. However, as they are more rigid, the motor bearings support higher impacts during crashes (Vergouw *et al.*, 2016).

Figure 2.12b shows blade propellers. A single-bladed propeller would be the most efficient, although it is not used because of the dynamic imbalance. Tricopters consist of three propellers, quadcopters four propellers, and hexacopters six propellers, and so on. Out of four propellers, two move in the clockwise direction in quadcopter while the remaining two moves in the anti-clockwise direction. Sometimes propellers with three or more blades are used because of space limitations or because propeller blades that are too large result in a higher speed close to supersonic. Two-blade propellers give more thrust than three- or four-blade propellers, which are more affected by turbulence. Ideally, the most efficient flight consists of the largest propeller diameter turning at the lowest velocity. Speed is thus a trade-off for efficiency. One example is racing drones, which commonly use three-blade propellers (Patel *et al.*, 2017).

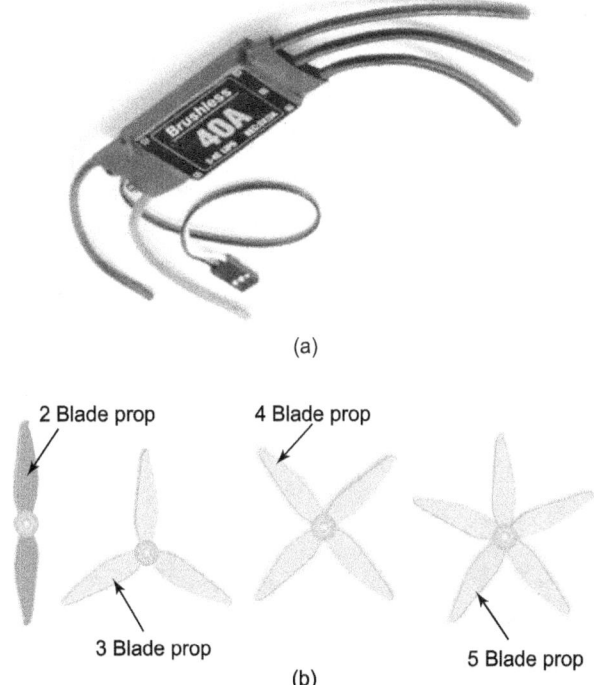

FIGURE 2.12 (*a*) Electronic speed controller and (*b*) Blade propellers used in co-opters.

Propellers can also be divided into fixed and adjustable pitch (Valvanis and Vachtsevanos, 2015). Civil UAV systems typically use fixed-pitch propellers. Adjustable pitch propellers are sometimes used in helicopter-type UAVs with a fueled motor. Fuel-injected propeller engines are used for fixed-wing UAV, and T-motors for VTOL and multirotor UAVs. In such a case, the working principle is based on the change of pitch of the propeller blades to absorb the selected motor power for any selected power. Increasing the blade pitch increases the drag while decreasing the blade pitch decreases the drag. Many UAV manufacturers recommend foldable propellers. They can minimize the damage during problematic landings and improve UAV transportability.

2.5.7 Flight Controller

The flight control system is the brain of the aerial vehicle, as it helps stabilize the motors and synchronize them so that even if the motors give a different output of thrust, the UAV can stay steady. The flight control system consists of hardware and software elements that will allow the aircraft to be controlled remotely, either directly by a pilot or autonomously by an onboard computer (González-Jorge *et al.*, 2017). The UAV flight dynamics are highly variable and non-linear, and so maintaining attitude and stability may require continuous computation and re-adjustment of the flight system. The hardware of the flight control system includes a main control module, signal conditioning and interface module, a data acquisition module, and a servo drive module (Huang *et al.*, 2007). It has a circuit board of varying complexity designed to collect aerodynamic information through a set of sensors (accelerometers, gyroscopes, magnetometers, pressure sensors, GPS, etc.) in order to automatically direct the flight of an aircraft along with its flight plan *via* several control surfaces present in the airframe (Figure 2.13). It reads the data provided by the sensors, processes it into useful information, and according to the type of control, either relays this information to the Ground Control Station (GCS) or feeds the actuator control units directly with the updated state. This operation is to be performed according to the development of the flight plan as well as the actual mission assigned to the UAV.

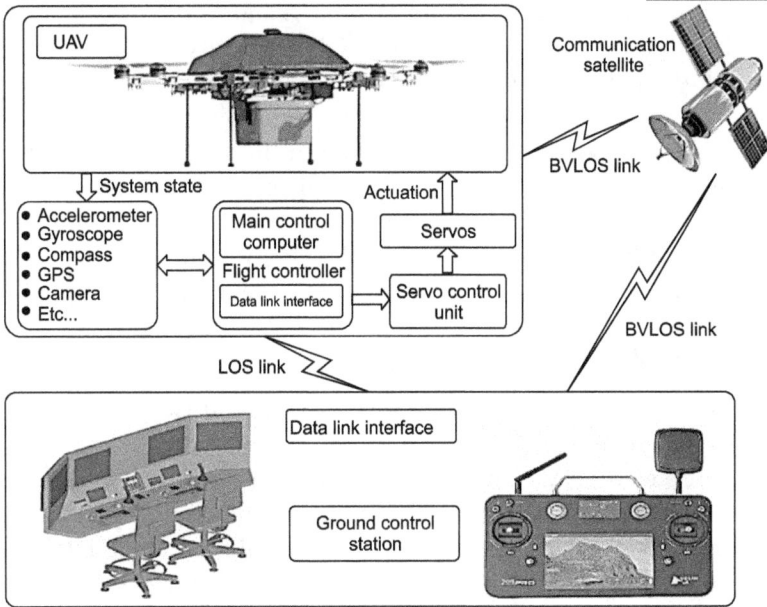

FIGURE 2.13 Flight controller and associated links (Altawy et al., 2016).

The flight controller can be programmed to take-off and fly to waypoints. For a small DIY quadcopter, instead of a flight controller, a gyro sensor may be used. It can also be self-programmed using an Arduino interfaced with the electronic components, such as gyrometer, accelerometer, etc., or else a variety of pre-programmed highly stable flight controllers are available; some of which are 3DR PixHawk, DJI A3, DJI NAZA-M V2, Lumenier LUX, Naza, KK, WKM, Ardupilot and Rabbit (Gupta et al., 2018).

The flight control system collects the flight state data measured by each sensor in real-time, and receives control commands and data transmitted by the radio monitoring and control terminal from the communication channel of the ground monitoring and control station (Caughey, 2011). After calculation, the control command is output to the executing mechanism to control the UAV of various flight modes and manage and control the mission equipment. It simultaneously transmits the status data of the UAV, the operational status parameters of the engine, onboard power supply system, and mission equipment to the onboard radio data terminal in real-time, and then sends it back to the ground monitoring station *via* the radio downlink channel. The sensors integrated on the aircraft frame produce electronic signals transmitted to the flight computer that determine how to move the actuators at each control surface to provide the expected response (Austin, 2010).

Autopilots are part of avionics and are key elements of UAVs. Geopositioning is provided by a GNSS receiver, which gives latitude, longitude, altitude, speed, track angle, and time. Attitude is sensed using an INS. The operator can use a computer, tablet, or mobile phone to provide GNSS waypoints that can be saved in the autopilot to navigate. There are many software options for mission planning. One free software example is the QGroundControl. It provides full flight control and UAV setup for PX4 or ArduPilot. It allows mission planning for autonomous flight, flight map display, video streaming, and support for managing multiple UAVs.

2.5.8 The Payloads

The payload is the weight a UAV can carry. It is usually counted outside of the weight of the UAV itself and includes extra cameras, sensors, or packages for delivery. The UAVs used in the military have much larger payloads than the civilian UAVs. The payload may include all the elements of a UAV that are not necessary for flight but are carried for completing the specific mission objectives. Two UAVs of the same mass, one with large rotors and one with small rotors, may carry different payloads, as the first one will create more air thrust with the same power, despite the airflow being slower than that generated by the smaller-rotored UAV (Vergouw *et al.*, 2016).

The payloads carried by UAVs include, such as cameras, sensors, radars, Light Detection and Rangings (LiDAR), communications equipment, weaponry, including distance sensors (ultrasonic, Laser), time-of-flight sensors, chemical sensors, and stabilization and orientation sensors, among others. These sensors are generally small, and lightweight with low power consumption and providing high accuracy. However, the UAV onboard devices and sensors vary depending on the missions. The greater the payload a UAV can carry, the more the flexibility to have additional technology and sensors on the UAV. The UAV devices and sensors are quite different in size and capability for different UAV types. For example, the sensor and device requirements for a HALE UAV will be quite different from a micro UAV requirement. The payload of a UAV should be as light as possible to have a long endurance (Patel *et al.*, 2017). A flight time is normally quoted with the model of UAV used; however, care should be taken to understand whether this flight time is with or without the payload. The flight time is normally reduced when the UAV is carrying extra weight, purely because of the additional power required to lift it.

2.5.9 Gimbals and Tilt Control

Aerial cameras need to be integrated with the UAV platform using gimbals. Gimbal technology is vital to capture quality aerial photos, film, or 3D images. A gimbal is a motorized platform (Figure 2.14) that uses several sensors, usually accelerometers, a magnetic compass, servos, and a microcontroller, to prevent spurious movement (Jha, 2016). They balance the vibration and spurious movements of UAVs, which otherwise could induce blurring in images. The gimbals with two axes are most common that compensate for movement in roll and pitch angles. The gimbal allows the camera to tilt while in flight, creating unique angles to capture oblique images. There are mostly three-axes stabilized gimbals with two working modes; First Person View (FPV) and non-FPV mode. Practically, all the latest UAVs/drones have integrated gimbals and cameras.

FIGURE 2.14 The Gimbal system (Gonz·lez-Jorge et al., 2017).

2.5.10 Sensors

A sensor is a device that detects and responds to some type of input from the physical environment. It converts physical data to electronic data. The specific input could be light, heat, motion, moisture, pressure, or any number of other environmental phenomena. UAVs can install the sensors, including accelerometer, gyro sensor, magnetometer, barometer, Inertial Measurement Unit (IMU), LiDAR, RADAR, laser ranger, ultrasonic, time-of-flight (ToF) sensor, thermal sensors, chemical sensors, and GPS (Colomina and Molina, 2014).

Figure 2.15 shows various sensors fitted in a modern UAV. The specific type of sensor that will be required by the UAV depends entirely upon the application and the specific data that will be collected during the UAV flight. Some of the most commonly used sensors that can be found in UAVs are explained below:

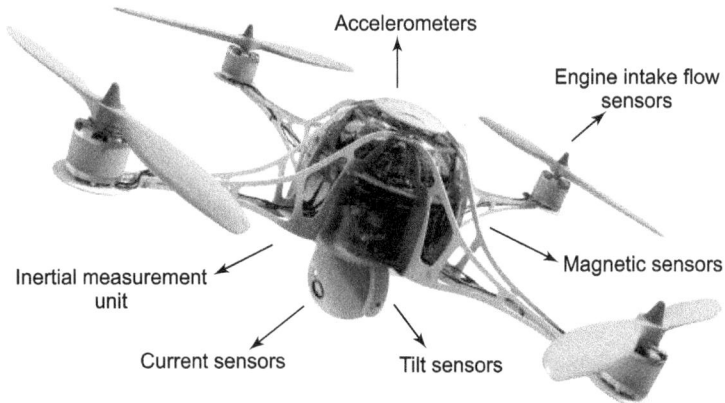

FIGURE 2.15 Various sensors used in modern UAVs.

2.5.10.1 Accelerometer

The accelerometers are used to determine the position and orientation of the UAV in flight. Accelerometer notices the change in the acceleration due to gravity whenever the UAV performs a movement along x-axis (forward and backward) or along the y-axis (lateral). It senses both static gravity acceleration and dynamic acceleration to detect the type of motion. The unit of measurement is g (9.8 m/s^2). The accelerometer senses both linear motion and the direction of the ground by sensing the earth's gravitational pull. So basically, the accelerometer values help determine the roll and pitch movements of the UAV (Gupta et al., 2018).

2.5.10.2 Gyroscope

The gyroscope is a component of the IMU and provides navigational information to the central flight controller. Gyro sensors, also known as angular rate sensors or angular velocity sensors, are devices that measure the rate of rotation of an object about its axis, which is measured in degrees per second or *rpm* as well as changes in the orientation (Gupta *et al.*, 2018). Gyros, and their useful application for stabilizing things, first emerged at the start of the 20th century. In the 1970s, optical gyros emerged. Ring Laser Gyro and Fiber Optic Gyro use the phase shift of light traveling in opposite directions around a fixed path length to detect the angular velocity. These gyros are very accurate but are quite complex and so are relatively large and costly to manufacture. Silicon Sensing was one of the first companies to commercially exploit the potential for solid-state gyroscopes back in the 1980s, with the launch of the Vibrating Structure Gyroscope. In recent years, Vibration gyro sensors have found their way into camera-shake detection systems for compact video and still cameras, motion sensing for video games, and vehicle electronic stability control (anti-skid) systems, among other things.

Vibration gyro sensors sense angular velocity from the Coriolis force applied to a vibrating element. For this reason, the accuracy with which angular velocity is measured differs significantly depending on element material and structural differences. A gyro is based on Newton's second law, just such as an accelerometer. The Coriolis Effect is a force that acts on objects which are in motion relative to a rotating reference frame (Cai *et al.*, 2011). If an external angular velocity is applied on a body of a certain mass moving with a velocity in a particular direction, a force will be generated, called Coriolis force. The Coriolis force is the result of two orthogonal motions, and is orthogonal to both of them (Figure 2.16).

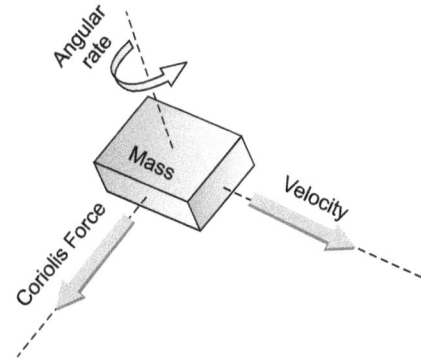

FIGURE 2.16 The Gyro sensor.

Micro-electro-mechanical systems (MEMS) gyroscopes, or strictly speaking MEMS angular rate sensors, are used whenever rate of turn (°/s) sensing is required without a fixed point of reference. All silicon sensing MEMS gyros use a vibrating or resonating ring fabricated using a deep reactive ion etch bulk silicon process (Caughey, 2011). The bulk silicon etch process and its unique ring design enable close tolerance geometrical properties for precise balance and thermal stability. Another advantage of the ring design is its inherent immunity to acceleration-induced rate error. This separates gyros from any other means of measuring rotation, such as a tachometer or potentiometer.

2.5.10.3 Magnetometer

A magnetometer is an instrument that measures magnetism—either magnetization of magnetic material, or the strength and, in some cases, the direction of the magnetic field at a point in space. It acts as a compass to measure the earth's magnetic field, and serves the task of correcting the drift of the gyro and to serve as an ancillary to the GPS system (Gupta *et al.*, 2018).

Basically, the magnetometer sensor measures the earth's magnetic field by using the hall effect (Figure 2.17a) or the magneto-resistive effect (Figure 2.17b). The magneto-resistive-based sensor works using materials, such as FeNi. Such materials, when subjected to magnetic fields, change their resistance, and this resistance is used to obtain values of the magnetic field.

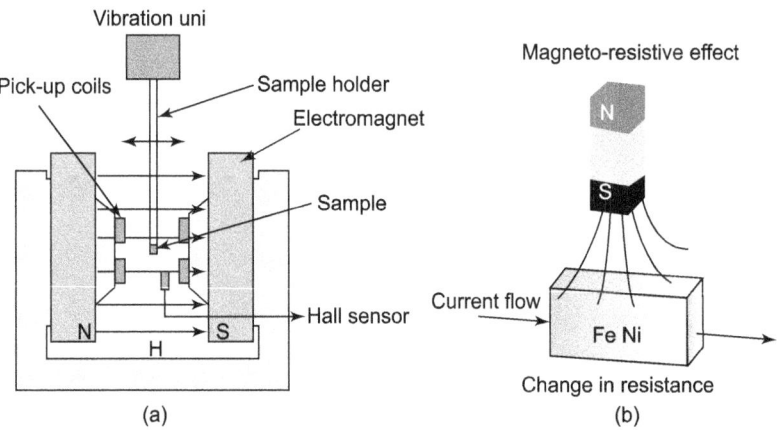

FIGURE 2.17 The Magnetometer (*a*) Hall effect, and (*b*) Resistivity effect.

2.5.10.4 Barometer

The barometer measures the air pressure (Figure 2.18a) as the pressure decreases with an increase in altitude. The barometer is a pressure sensor that senses changes in air pressure and hence can also be used to measure the altitude of the UAV.

2.5.10.5 Inertial Measurement Unit (IMU)

The IMU is an essential component of UAVs. It works by detecting the current rate of acceleration using one or more accelerometers. For accurate measurement of the orientation, velocity, and location of the UAV, IMU is used (Figure 2.18b). The IMU detects changes in rotational attributes, such as pitch, roll, and yaw, using one or more gyroscopes. Some IMUs include a magnetometer to assist with calibration against orientation drift (Vergouw et al., 2016).

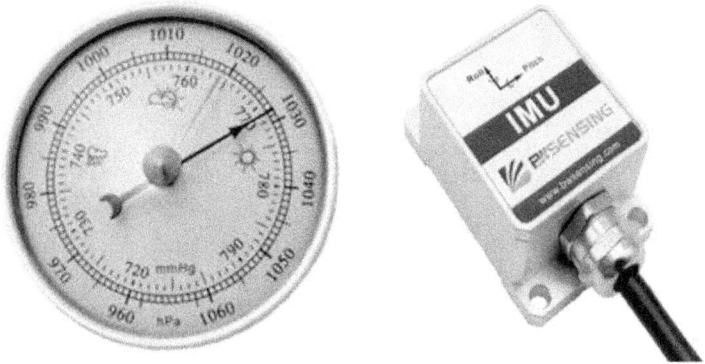

FIGURE 2.18 (*a*) Barometer, and (*b*) Inertial measurement unit.

2.5.10.6 Tilt Sensors

Tilt sensors, combined with gyros and accelerometers, provide input to the flight control system in order to maintain the level of flight. This is extremely important for applications where stability is paramount, e.g., from surveillance to delivery of fragile goods. These types of sensors allow the detection of small variations of movement. It is the gyroscope compensation that allows these tilt sensors to be used in moving devices, such as motor vehicles or UAVs.

2.5.10.7 Time-of-Flight (ToF) Sensors

Depth sensors of the ToF camera sensor emit a short infrared light pulse. The pixels of the camera sensor measures the time in which the infrared light pulse returns back to the UAV/Drone. The ToF cameras are lightweight sensors that provide users with useful information on the distance between the sensor and an object at a sub-millimeter depth resolution.

2.5.10.8 Distance Sensors

Distance sensors are used for the much accurate reading of the altitude and to prevent the UAV from the collision with any external obstacle. Generally, these distance sensors use ultrasonic or light-based system.

2.5.10.9 Chemical Sensors

These types of sensors are particularly useful for the detection of chemicals present in the environment, industry, and emergency response situations. The use of chemical sensors on UAVs is especially critical in preventing the potential exposure of workers to the unknown chemicals present in these situations.

2.5.10.10 Biological Sensors

These sensors, mounted on UAV platforms, can be used in civil and environmental engineering fields to identify specific airborne contaminants or pathogens. They are categorically similar to gas and radiation detection because they require contact between the sensor and the target.

2.5.11 Global Positioning System (GPS)

All sensors as well as cameras and LiDAR, require some degree of accuracy in the UAV's positioning. The absolute accuracy needed for the UAV positioning will heavily depend on the sensor type used and the specific application. GPS receives data from multiple satellites in order to compute the geographical location of the object. The positioning provided by standard GPS units is generally suitable for collecting relatively coarse geospatial data (Jha, 2016).

The RTK satellite navigation is a positioning technique capable of producing survey-grade accurate results down to the centimeter level by measuring the phase of the radio waves sent by GPS satellites (Garg, 2019). The RTK positioning systems have started to be developed for small UAVs in recent years to provide high positioning accuracy. However, achieving such accuracy is only attainable if the position is held for some time (several minutes). The UAV-based RTK GPS has improved over the robust aerial survey used with UAV, as it eliminates the need for having distributed ground control points (GCPs) required for image geo-registration. With the GPS system, specific coordinates for the UAV can also be set to fly. GPS also provides an RTH function that enables the UAV to fly back to the controller in case of loss of connection (Kan *et al.*, 2018).

Other GPS mapping systems with a high degree of accuracy include the Micro Aerial Projects V-Map system, which uses dual-frequency GPS to achieve centimeter-level positional accuracy. RTK and dual-frequency techniques are especially useful for mapping areas, such as deserts that lack identifiable features which could be used to create GCPs (Patrick *et al.*, 2019). However, UAVs equipped with RTK or dual-frequency remains very expensive relative to lower-cost GPS solutions, and they are necessary for high precision mapping projects. Advanced RTK positioning methods, such as those using network-based architectures, have great potential to benefit data collection for all integrated sensors and need further research.

2.5.12 Remote Sensing Sensors

Innovations in sensor technology have fostered the successful implementation of UAVs for several applications. Remote sensing applications are based on the use of active sensors, such as LiDAR and Synthetic Aperture Radar (SAR); and passive sensors, such as red, green, and blue (RGB), multispectral, hyperspectral, and thermal cameras (Garg, 2019), which can be employed in UAVs for data collection.

In 2016, UAVs using ToF sensors came into the market (Corrigan, 2020). The ToF sensors, also known as "Flash LiDAR" can be used on their own or with RGB and regular LiDAR sensors to provide various solutions. Flash LiDAR ToF sensors have a huge advantage over other technologies, as it is able to measure distances to objects within a complete scene in a single shot. The ToF depth ranging sensors can be used for object scanning, indoor navigation, obstacle avoidance, object recognition, tracking objects, measure volumes, reactive altimeters, 3D photography, augmented reality games, and much more.

Despite the popularity of RGB cameras and, to a lesser extent LiDAR, other sensors have also been implemented in UAVs in a variety of fields, as summarized in Table 2.5. The applications listed in this table are not exhaustive. Magnetometers are the only sensor listed in this table capable of probing beneath the ground surface. This is a frontier area for UAVs that has not been investigated extensively yet but has significant relevance to geotechnical infrastructure systems, such as pipelines, levees, earth dams, landfills, and subsurface hazard detection.

TABLE 2.5 The UAV-based Sensors and Corresponding Applications.

Sensor Type	Applications
Biosensor	Agriculture, environmental monitoring
Gas detection	Volcanology, environmental monitoring, climatology
Hyperspectral imaging	Precision agriculture, glaciology, geology, minerals mapping
LiDAR	Civil engineering, glaciology, forestry, precision agriculture, mapping
Magnetometer	Geophysics/geology/geotechnical engineering
Multispectral imaging	Precision agriculture, landcover, water, resource mapping
SAR	Glaciology, geology, soil moisture, mapping
Temperature	Glaciology, water
Thermal imaging	Surveillance and security, infrastructure inspections, water source identification, livestock detection, and heat signature detection

Visual sensors offer still or video data with RGB camera collecting data in standard visual red, green and blue wavelengths (Garg, 2019). Remote sensing images can be studied either in 2D or by creating 3D model. These 3D models can be used to create DEM, DTM, or DSM, which are required in many applications. DEMs generated from high-resolution commercial satellites are at lower accuracy as compared to LiDAR and other sources.

Colomina and Molina (2014) provided an in-depth review of UAV-based photogrammetry with RGB cameras. The review also provides insight into remote sensing in which UAVs have been utilized and includes details of other camera types for UAV-based remote sensing (e.g., multispectral). In recent years, the introduction of new sensors, such as near-infrared, thermal infrared, hyperspectral as well as laser-based sensors has enhanced the efficiency of capturing the additional data by UAV with improved spatial resolution. Table 2.6 describes the fundamental characteristics of some common and/or representative remote sensing sensors.

TABLES 2.6 Some Common and/or Representative Remote Sensing Sensors (Colomina and Molina, 2014).

Common and/or Representative Small Format (SF) and Medium Format (MF) Visible Band Cameras						
Manufacturer and Model	Format Type	Resolution (mp)	Size (mm^2)	Pixel Size (m)	Weight (kg)	Frame Rate (fps)
Phase One iXA 180	MF	CCD 80	53.7 x 40.4	5.2	1.70	0.7
Trimble Q180	MF	CCD 80	53.7 x 40.4	5.2	1.50	
Hasselblad H-4D 60	MF	CCD 40.2	53.7 x 40.4	6.0	1.80	0.7
Sony NEX-7	SF MILC	CMOS 24.3	23.5 x 15.6	3.9	0.35	2.3
Ricoh GXR A16	SF IUC	CMOS 16.2	23.6 x 15.7	4.8	0.35	3

Common and/or Representative Multispectral Cameras for UAVs					
Manufacturer and Model	Resolution (mp)	Size (mm^2)	Pixel Size (m)	Weight (kg)	Spectral Range (nm)
Tertracam MiniMCA-6	CMOS 1.3	6.66 x 5.32	5.2 x 5.2	0.7	450–1050
Quest Innovations Condor-5 UAV-285	CCD 1.4	10.2 x 8.3	7.5 x 8.1	0.8	400–1000

Common and/or Representative Hyperspectral Cameras for UAVs						
Manufacturer and Model	Resolution (mp)	Size (mm^2)	Pixel Size (m)	Weight (kg)	Spectral Range (nm)	Spectral Resolution and Band
Rikola Ltd. Hyperspectral Camera	CMOS	5.6 x 5.6	5.5	0.6	500–900	(40) 10 nm
Headwall Photonics Micro-Hyperspec X-series NIR	Indium Gallium Arsenide (InGaAs)	9.6 x 9.6	30	1.025	900–1700	(62) 12.9 nm

Continued

Common and/or Representative Thermal Cameras for UAVs						
Manufacturer and Model	Resolution (mp)	Size (mm²)	Pixel Size (m)	Weight (gm)	Spectral Range (nm)	Thermal Sensitivity (mK)
FLIR TAU 2 640	Uncooled VOx Microbolo-meter 640 x 512	10.8 x 8.7	17	70	7.5–13.5	<50
Thermoteknix Systems Ltd. Miricle 307K-25	Amor-phous Silicon 640 x 480	16 x 12.8	16	105	8–12	<50
DJI Zenmuse XT	Uncooled VOx Micro-bolo-meter			270	7.5–13.5	

2.5.12.1 RGB and Multispectral Imaging Systems

The use of UAV-mounted multispectral RGB camera (as shown in Figure 2.19) is dominantly used in a large number of applications (Jordan et al., 2018). However, many other types of sensors on UAVs, including hyperspectral imaging, thermal imaging (Nishar et al., 2016), and Synthetic Aperture Radar (SAR) can also play a vital role in multi-scale data collection and enhanced classification (Garg, 2019). Multispectral cameras in UAVs have been used to collect aerial imagery for creating orthoimages and DEM or inspection of large infrastructures, such as bridges or power lines. Some digital cameras, such as the Canon S100, have the ability to track the GPS location of where each image was captured, producing data that can be used to georeference the image with image processing software, although the positional accuracy is not as high as that obtained with the GCPs.

FIGURE 2.19 Multispectral imaging systems.

In 2016, a number of integrated gimbals with optical and digital zoom cameras came into the market. The Walkera have released their latest Voyager 5 camera with an incredible 30x optical zoom camera. The camera includes redundancy systems, such as dual GPS, dual gyroscope, and three battery systems. It also has an optional thermal infrared and low-light night vision camera (Corrigan, 2020). This allows for more industrial uses, such as inspecting microwave towers or wind turbines to get a very detailed look at structures, wires, modules, and components to detect damage.

In October 2016, DJI released the Zenmuse Z30 integrated aerial zoom camera with 30x optical and 6x digital zoom for a total magnification up to 180x. The Zenmuse is compatible with DJI Matrice range of UAVs. The maturity of visible-spectrum cameras has penetrated into several side-technologies, for example, mobile phones. Current smartphones are equipped with high-quality cameras at a reasonably low cost. Multispectral cameras may also include a near-infrared band in addition to RGB bands. Many agriculture-based applications, such as the monitoring of crop yield estimation or crop health monitoring, or crop identification, require near-infrared images with multispectral or hyperspectral cameras (Shamshiri *et al.*, 2018).

2.5.12.2 Hyperspectral Sensors

Independent of the aerial platform, sensors play an important role in data acquisition. Hyperspectral sensors take a large number of images in narrow spectral bands over a continuous spectral range, producing the spectra of all pixels in the scene (Colomina and Molina, 2014). Multispectral imagery generally ranges from 5 to 12 bands that are represented in the form of pixels, whereas hyperspectral imagery consists of a much higher band number, more than hundreds, taken in a narrower bandwidth (5–20 nm, each). Hyperspectral sensors, therefore, extract more detailed information than multispectral sensors because an entire spectrum is acquired at each pixel (Garg, 2019). Figure 2.20 represents the differences between multi and hyperspectral imaging.

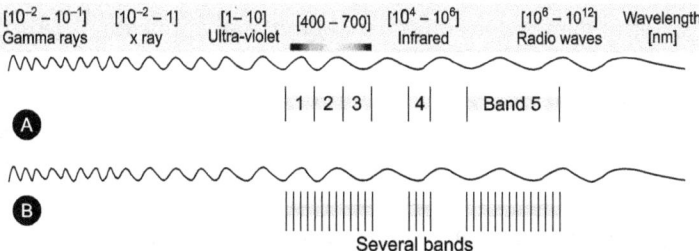

FIGURE 2.20 Spectrum representation for (*a*) Multispectral bands, and (*b*) Hyperspectral bands.

Compared with others, hyperspectral imaging sensors can effectively capture more detail in spectral and spatial ranges. RGB imaging does not provide spectral information beyond the visible spectrum, which is of high importance for characterizing the chemical and physical properties of a specimen. Spectroscopy is a proximity technology mainly used for sensing tiny areas (e.g., leaf spots), aiming the acquisition of spectral samples without spatial definition, which is very much possible through hyperspectral images. Hyperspectral sensors can help identify minerals and vegetation and are ideal for use in crop health, water quality, and surface composition. Contrary to the visible-spectrum camera developments, the miniaturization process of multi- and hyperspectral cameras is challenging in terms of optics and sensor calibration (Colomina and Molina, 2014).

The most significant differences are given in Table 2.7, which considers spatial discrimination and spectral discrimination.

TABLE 2.7 Main Differences Between Hyperspectral and Multispectral Imaging, Spectroscopy and RGB Imagery.

Techniques	Spectral Information	Spatial Information
Hyperspectral imaging	High	High
Multispectral imaging	Medium	High
Spectroscopy	High	Low
RGB imagery	Low	High

2.5.12.3 Thermal Infrared Sensors

A standard visual RGB sensor collects red, green, and blue wavelengths of light. These wavelengths are visible to the human eye. Thermal sensors capture differences in radiant energy to show the relative temperature of objects (in comparison to one another). To do this, they capture the energy of wavelengths within the long-wave infrared (IR) region of the spectrum (8–12 μm), well beyond the visual wavelengths (Garg, 2019). This enhances the potential uses of UAVs technology for multiple users by providing more spectral data required for a wide range of applications (Austin, 2010).

When long-wave infrared radiation emitted from objects strikes the thermal sensor, it heats up the microbolometer and changes the electrical resistance. These changes are converted to electrical signals and stored as raw data or processed into thermal imagery. For cases in which a visual component (video or images) is needed, the sensor has a thermal imaging (or infrared) camera. These cameras form an image using infrared radiation and differentiate apparent temperature variations using different color palettes.

Thus, thermal images show differences in total radiant energy allowing for temperature calculations (Dering *et al.*, 2019). For example, thermal infrared sensors can be important in surveillance or security applications, such as heat signature detection or forest fire detection (Hoffmann *et al.*, 2016). A thermal camera attached to a UAV can detect poor insulation in a building or areas that need to be repaired by revealing areas of higher temperatures, such as electric switch-gear, sub-stations, or pylons.

Thermal sensors require no external light sources, and they can collect the data at night. Flying in optimal weather conditions is important when capturing quality aerial thermal data. Both wind (over 15 mph) and precipitation can change surface temperatures, and high humidity (over 60 %) will result in a haze in the imagery. Low-weight, small-size sensors, such as those developed by FLIR, were first used in the military context for remote reconnaissance and are becoming more common in applications, such as forest fire monitoring (Banu *et al.*, 2016). The potential uses for thermal IR cameras are vast and include search and rescue, crop and forest health, pipeline inspection, leak detection, geothermal mapping, and building efficiency.

2.5.12.4 Synthetic Aperture Radar (SAR)

The SAR is another remote sensing technology that uses electromagnetic waves between 2.7–4.0 mm (W-band) and 0.9–6 m (VHF band) (Garg, 2019). Airborne SAR sensors have been commonly used during the last decades to monitor different phenomena in medium-scale areas of observation, such as object detection and characterization or topographic mapping. The SAR instrument generates radar imagery for detecting and measuring solid earth vertical deformation to improve prediction models for landslides, levee erosion, and earthquake and volcanic activity (Colomina and Molina, 2014). Table 2.8 presents some SAR software that could be used in UAV-based applications.

TABLE 2.8 Common and/or Representative SARs Used in UAVs (Colomina and Molina, 2014).

Manufacturer	Model	Spectral Bands	Weight (kg)	Transmitted Power (W)	Resolution (m)
IMSAR	NanoSAR B	X and Ku	1.58	1	0.3–5
Fraunhofer FHR W	Miranda	W	-	0.1	0.15
NASA JPL	UAV SAR	L	200	2000	2
SELEX Galileo	PicoSAR	X	10	-	1

The adoption of SAR technology to UAV has still not been done widely. The SAR allows 3D representation and the measurement of terrain deformations useful for applications, such as landslides. The SAR has several capabilities in UAV platforms, such as repeat-pass interferometry, polarimetric measurements, and vertical or circular trajectories that can be used for tomographic purposes. The use of this platform introduces new possibilities in airborne SAR observation due to its inherent flight capabilities and characteristics. The SAR-based UAV platforms have limitations when used in some applications. One of the most important is the undesired deviations from the nominal flight trajectory that may cause defocusing, geometric distortions, or phase errors in the SAR images (Lort *et al.*, 2017).

2.5.12.5 Light Detection and Ranging (LiDAR)

The UAV-based laser scanning or LiDAR is the other active remote sensing (non-imaging) technology in the area of 3D topographical data collection (Garg, 2019). The UAV essentially consists of a laser scanner, INS, GPS receiver, and controller and data recording device (Figure 2.21). Laser scanners can be used in two modes; terrestrial laser scanner (TLS) and airborne laser scanner (ALS), both providing point cloud data.

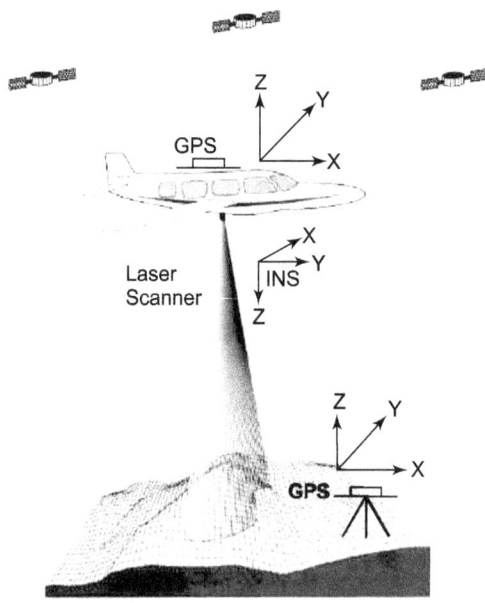

FIGURE 2.21 The UAV LiDAR system.

As the UAV flies over the terrain, ALS transmits laser pulses to the terrain below and detects their return signal from the ground/objects. The distance from the scanner to the terrain can be determined from the time delay between the transmission and the return signal. This distance data is combined with the information from INS, which determines the orientation of the aircraft, and the GPS receiver, which records the (x, y, z) positions of the antenna. Typically, short-range 3D LiDARs designed for low-speed vehicles in urban environments exhibit a range of 50 m or less, while long-range LiDARs meant for high-speed vehicles in highway environments can exhibit a range of 200 m or more. The results from the laser pulses are vector displacements from specific points in the air to points on the ground (Banu *et al.*, 2016). As the laser is capable of generating pulses at the rate of thousands of pulses per second, the output is a dense pattern of measured x, y, z points on the terrain. The LiDAR data of an area is shown in Figure 2.22.

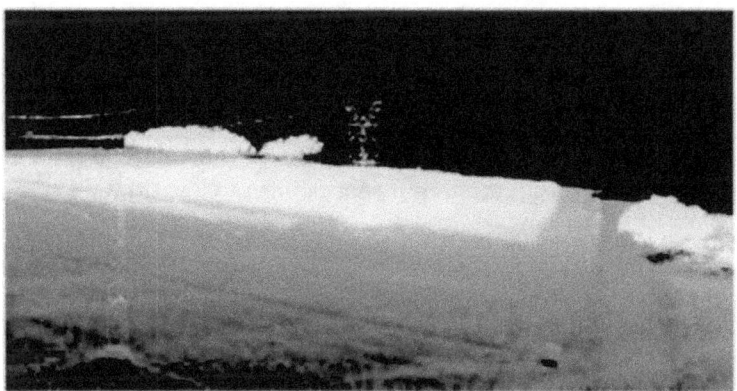

FIGURE 2.22 Output from LiDAR providing elevation information. (*https://delair.aero/portfolio/surveying-and-mapping/*) [*See Plate 1 for color*]

The ALS and TLS differ in terms of data capture mode, typical size area, scanning mechanism, accuracy, and resolution but share many common features, especially those resulting from the laser-ranging technology. Particularly when it comes to 3D point-cloud processing and proper geo-referencing of data, the same algorithms are often applied to both ALS and TLS data. The TLS has proven to be valuable for monitoring dams, tunnels, and areas susceptible to land erosion or landslide, capture complex surfaces of historical buildings and sculptures, and mapping buildings and trees. In contrast, ALS is efficient for monitoring bridges, power lines, forests, etc.

Among non-imaging sensors used in civil engineering applications, the most popular is LiDAR (Nagai *et al.*, 2009). The LiDAR is used to map the surrounding environment to develop a 3D model, as they are capable of 3D mapping without having a GPS. The LiDAR sensors have frequently been used in agriculture for mapping trees, soil, and crops. The LiDAR mapping data can be used to evaluate the impact of agricultural production methods.

The 3D LiDAR enables superior night vision as compared to human drivers and standard cameras. It performs better in poor weather; however, its performance depends on the LiDAR system design (Wijeyasinghe and Ghaffarzadeh, 2020). For example, LiDARs that operate at NIR wavelengths are affected more by foggy conditions than SWIR LiDARs because SWIR light is scattered less by fog than NIR light. The FMCW LiDAR is an emerging technology, and such LiDAR can compete with radar technology by detecting object velocity in real-time. Compared to conventional ToF LiDAR technology, FMCW LiDARs enable better signal coding to minimize the risk of interference from surroundings.

The technical specifications of different airborne LiDAR sensors for UAV are given in Table 2.9. They are specially indicated for rotary-wing UAV due to the higher payload capability and low weight. Within the last decade, due to technology development, the LiDAR sector has been facing a progressive shift from different points of view (Marchi *et al.*, 2018):

1. *Data quality*: improved accuracy in terms of data localization and measures.

2. *Data density*: from low- to high-density (e.g., single-photon counting LiDAR).

3. *Data types*: from a discrete return to a full waveform.

4. *Analysis*: from area-based to single-object (e.g., tree) approaches.

TABLE 2.9 Characteristics of Some Airborne LiDAR Systems (Gonz·lez-Jorge *et al.*, 2017).

System	Weight (kg)	GNSS Accuracy (m)	LiDAR Accuracy (m)	LiDAR Pulse Rate (kHz)	Range (m)	Photo
3DT Scanfly	1.5	0.02	0.03 (includes Velodyne VLP16)	300	100	

System	Weight (kg)	GNSS Accuracy (m)	LiDAR Accuracy (m)	LiDAR Pulse Rate (kHz)	Range (m)	Photo
RIEGL mini VUX-1 UAV	1.6	External GNSS/IMU required	0.015	100	250	
Routescene LiDAR Pod	2.5	0.04	0.03 (includes Velodyne VLP16)	300	100	
Yellow Scan Surveyor	1.6	0.05	0.03 (includes Velodyne HDL32)	300	100	
Yellow Scan Mapper	2.1	0.15	0.1	40	150	

2.5.13 Battery

Many devices need the power to operate. There are different power sources for a UAV; battery, solar, hydro fuel cell, combustion engine, cable-tethered power supply and laser beam power. The most used type of batteries to provide power in smartphones, UAVs, laptops, or tablets is lithium polymer (LiPo). The LiPo batteries have higher capacity, and discharge and are lighter in weight (Park *et al.*, 2017). There are also some disadvantages, such as a shorter lifespan (300–400 cycles), the possibility of ignition (if these are punctured and vents into the air), and the need for extra care for storing, charging, or discharging.

There are two major kinds of batteries: (i) wet cell batteries—which generate power from an electrode and a liquid electrolyte solution, for example, lead-acid batteries, nickel cadmium (Ni-Cd), nickel metal hydride (Ni-Mh) batteries, etc. and (ii) dry cell batteries—which use a paste electrolyte, with only enough moisture to allow current to flow, for example, AA,

AAA, lithium ion (Li-Ion), lithium polymer (LiPo), lithium sulfur (Li-S), etc. (Gupta et al., 2018). A Li-Ion battery is a type of rechargeable battery in which lithium ions move from the negative electrode to the positive electrode during discharge and back when charging. Most of the common source of power today is lithium-based batteries because of their lightweight and higher charge density than the typical lead-acid and Nickel-based batteries. Unfortunately, these batteries are a little more expensive and, in a few cases, have dangerous side effects. Table 2.10 presents the comparison of features of four types of batteries to each other, which show that lithium batteries are better choices.

TABLE 2.10 Comparison of Characteristics of Different Batteries (Hassanalian and Abdelkefi, 2017).

Characteristics	Ni-Cd	Ni-Mh	LiPo	Li-S
Specific Energy (Wh/kg)	40	80	180	350
Energy Density (Wh/l)	100	300	300	350
Specific Power (W/kg)	300	900	2800	600

The battery chosen depends on the motors of UAV. Batteries are the heaviest item on a UAV, so they are usually mounted on the dead center of the UAV to subject the motors to the same load. The capacity of a battery pack is measured in amp-hours (Ah), which indicates the working hours of the battery. The average flight time of a UAV is around 10–20 min on a 2–3 Ah battery. The higher the capacity, the heavier the battery will be. Also, as the battery gets larger, the increase in flight time becomes lesser effective, and eventually, a point is reached when no more flight time is increased with the size (Saha et al., 2011). This is mainly caused by the weight of the battery, and hence the agility of UAV is reduced. A multicharger and a set of two batteries used in Aibot X6 V2 UAV are shown in Figure 2.23, respectively.

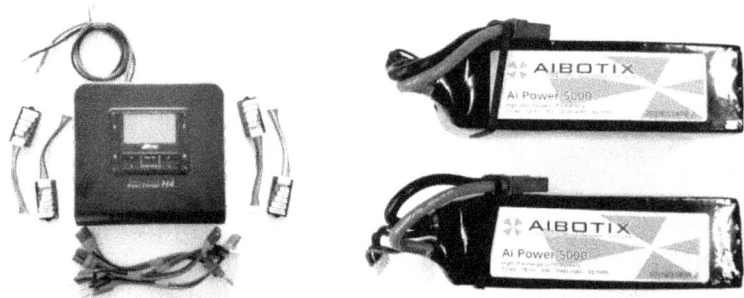

FIGURE 2.23 (a) Multicharger, and (b) Two batteries of UAV.

Another important specification to consider with LiPo batteries is the "C rating." This is an indicator of the maximum continuous discharge rate of a LiPo battery (Park *et al.*, 2017). For example, a 3S 1000mAh 20C LiPo pack can draw a maximum recommendable current of 20A (i.e., 1000 mAh x 20C). The battery might suffer from overheating when discharging at an extremely high current rate. A battery rated at more mAh will give power for a longer time, given the same usage pattern but have a drawback of heavier and larger size.

The performance of UAV communication systems is significantly affected by battery lifetime (flight time) of UAVs. The flight time of a UAV depends on several factors, such as the energy source (e.g., battery, fuel, etc.), type, weight, speed, and trajectory of the UAV (Hui *et al.*, 2006). Table 2.11 presents the battery lifetime of various UVAs. The endurance of the aircraft will heavily depend on aircraft size, weight, and payload weight. For simplicity, the endurance calculation is estimated by dividing the "battery capacity" with the "current."

TABLE 2.11 Battery Lifetime of Various UAVs (Mozaffari *et al.*, 2018).

Size of UAV	Weight of UAV	Example of UAV	Battery Life-time (min)
Micro	<100 g	Kogan Nano Drone	6–8
Very small	100 g– kg	Parrot Disco	45
Small	2 kg–25 kg	DJI Spreading Wings	18
Medium	25 kg–150 kg	Scout B-330 UAV helicopter	180
Large	>150 kg	Predator B	1,800

In general, the total energy consumption of a UAV is composed of two main components: (i) communication-related energy and (ii) propulsion energy (Mozaffari *et al.*, 2018). The communication-related energy is used for various communication functions, such as signal transmission, computations, and signal processing. The propulsion energy pertains to the mechanical energy consumption for movement and hovering of UAVs. Typically, the propulsion energy consumption is significantly more than the communication-related energy consumption.

One of the most active researches worldwide is to gain the maximum flight time of UAVs. For more than a decade, small solar- and battery-powered electric UAVs were the subject of research and development. Increased focus has been placed recently on developing hybrid-powered electric UAVs, especially with the combination of solar energy and battery or fuel cells in a battery-powered system (Rajendran and Smith, 2018).

Fixed-wing UAVs can provide higher flight duration because of their aerodynamic performance. In the case of fixed-wing UAV, it is also possible to integrate energy recovery systems based on solar panels, which results in longer flight duration (Zadeh et al., 2017). Some examples are the Alphabet and Facebook drones, powered with solar energy to deliver high-speed, reliable Internet around the world. On the other hand, rotary-wing systems exhibit more limited flight times, and their operations are circumscribed to typically less than 30 minutes of work.

The HALE UAV provides a cost-effective and persistent capability to collect and disseminate high-quality data across wide areas. In 1995, the maximum loitering time was 24 hours. This grew to 38 hours in 2005 and reached more than 60 hours in 2008, as shown in Table 2.12. Solar-powered-based UAVs have a demonstrated endurance of more than 300 hours (D'Sa et al., 2017).

TABLE 2.12 Growth in Endurance Time Limit (Lucintel, 2011).

UAV	Flight Time
QinetiQ Zephyr Solar Electric (2010)	336 hr 22 min
QinetiQ Zephyr Solar Electric (2008)	82 hr 37 min
Boeing Condor	58 hr 11 min
QinetiQ Zephyr Solar Electric (2007)	54 hr
IAI Heron	52 hr
AC Propulsion Solar Electric	48 hr 11 min
MQ-1 Predator	40 hr 5 min
GNAT-750	40 hr
TAM-5 (2005)	38 hr 52 min
Aerosonde (2005)	38 hr 48 min
TAI Anka (1995)	24 hr

2.5.14 Ground Control Station (GCS)

The GCS, which is also known as the base station (BS), is used to remotely control the UAV. It receives telemetry data, and displays real-time transferred images (Austin, 2010). The GCSs are stationary or transportable hardware/software devices to monitor and command the UAVs (Figure 2.24). Although the word 'ground' is inherent to the concept, a UAV may actually be operated from the ground, sea, or air. The computerized system on the ground is designed to monitor the mission development and eventually operates the UAV and its payload. The GCS enables the interface with the "human intelligence" as any change in the route of the UAV, any eventual error on the aerial platform and/or any outcome of the payload sensors shall be sent to and seen within the GCS (Colomina and Molina, 2014).

FIGURE 2.24 The GCS system (Johnson, 2015).

The ground station is composed of three parts: a ground monitoring station, a remote control transmitter, and a telemetry receiving antenna (Blom, 2011). The GCSs vary in size according to the type and mission of the UAV. In other words, for recreational mini- and microdrones, GCSs are small hand-held transmitters used by hobbyists. For tactical and strategic UAVs, a large self-contained facility with multiple workstations is employed as the GCS. A GCS communicates with the UAV through a wireless link to send commands and receive real-time data, thus creating a virtual cockpit.

The hardware of the ground station is based on the embedded PC architecture and may contain an LCD, a digital video recorder, a membrane keyboard and a spherical mouse, while the software system may consist of windows-based software modules for navigation, video processing, map generation and information processing. The navigation interface, video interface, and task management and status bar interface can be displayed on the LCD and switched in real-time. The graphical interface can realize the real-time display of reconnaissance images and some related position information and store them into the digital video recorder (Austin, 2010). The task management module is used for flight planning before or on flying. In addition, the flight path and related parameters can be stored in real-time and then used for playback. The improvement in embedded computation capabilities and image processing algorithms contributes to applications related to onboard real-time image processing. It would allow using this information in UAV navigation and decreasing the bandwidth needs of the communications datalink with the ground station.

2.6 SUMMARY

The UAVs are a complex amalgamation of mechanics, hardware, and software. A UAV consists of an airframe and a computer system, sensors, GPS, servos, and CPUs. All these elements combined have to pilot the UAV without any human intervention. Current UAV technology offers feasible technical solutions for airframes, flight control, communications, and BSs. In recent years, the increased use of UAVs in several applications is mainly due to the advancement of UAVs' manufacturing technologies and the reducing cost of components, which also make them easily accessible to the public.

In practice, there are many types of UAVs due to their diversified applications. However, there is no single standard for UAV classification. The UAVs can be practically differentiated according to different criteria, such as functionality, weight/payload, size, endurance, wing configuration, control methods, cruising range, flying altitude, maximum speed, energy supplying methods, etc. Broadly, UAVs can be classified into two categories: fixed-wing and rotary-wing, each with its own strengths and weaknesses. For example, fixed-wing UAVs usually have high speed and heavy payload, but they maintain a continuous forward motion to remain aloft; thus are not suitable for stationary applications, such as continuous and close inspection. On the other hand, rotary-wing UAVs, such as quadcopters, though having limited mobility and payload, can move in any direction and stay stationary in the space for continuous monitoring of area/phenomena. Thus, the choice of selection of a UAV critically depends on the applications, apart from other factors.

To improve the ability of UAVs to collect various data/information as well as to maneuver around difficult locations, a variety of different sensors are incorporated into UAVs. These sensors include RGB camera, multispectral, hyperspectral, short/mid-wave range cameras (e.g., thermal), LiDAR, and many more. The specific type of sensor that a user will require for a UAV will depend entirely upon the application in which the UAV will be used and the specific data that will be collected during its flight. For example, thermal images using IR sensors on UAV have been used for inspection of the bridge deck, leakage in the pipeline, construction, mining, electrical, surveillance, firefighting, search and rescue, etc. The development of UAV platforms and various sensors has also motivated many remote sensing applications, such as crop mapping, urban planning, etc., at finer scales and applications requiring 3D model of the terrain. The acuity of photographic sensors to capture high-resolution images will improve, as the endurance and range of UAVs improve.

REVIEW QUESTIONS

Q.1. What do you understand by the aerodynamics of an air vehicle? Explain Lift, Drag, and Thrust used in aerodynamics.

Q.2. Explain Roll, Pitch, and Yaw. What is their role in controlling the stability of an air vehicle?

Q.3. Explain in various brief ways the UAVs can be classified.

Q.4. Describe Nano, Mini, Micro, Close-range, HALE, and MALE UAVs.

Q.5. What do you understand by Fixed-wing and Rotary-wing UAVs? Discuss their salient features.

Q.6. With the help of a diagram, discuss in brief about various components of a UAV.

Q.7. Explain the function of a transmitter and a receiver in UAV operation.

Q.8. Draw a diagram and explain the (i) body frame of multirotor UAVs and (ii) propellers.

Q.9. Discuss the role of a (i) motor, (ii) ESC, (iii) Gimbals, (iv) accelerometer, (v) gyroscope, (vi) magnetometer, (vii) IMU, (viii) battery, and (ix) GCS for operation of UAVs.

Q.10. What is GPS? What is the function of GPS in UAV data collection?

Q.11. Explain in brief the characteristics of various remote sensing sensors used in photogrammetry UAVs. How is LiDAR sensor different from image-based remote sensing sensors?

CHAPTER 3

Autonomous UAVs

3.1 INTRODUCTION

The UAVs are generally broadly divided into two categories; remotely piloted and autonomous vehicles. Increased level of UAV autonomy has resulted in the reduction of the human role for supervision, and therefore autonomous operation is gaining momentum. The autonomous vehicle enables a vehicle to decide on the most appropriate action under the prevailing circumstances and surrounding environment. Although the concept of autonomy is not new, but it is less used in aeronautical applications, mainly due to legal and safety issues. An unmanned aircraft is said to have some level of autonomy when the system software is designed and executes a significant number of real-time decisions to control the vehicle without human interaction (Valvanis and Vachtsevanos, 2015).

Automatic systems are an essential part of modern flight control systems, which can significantly help to increase the efficiency, comfort, and safety of flights. The most obvious advantage is the absence of an onboard human pilot, which enables flying long-enduring missions or operations in hazardous conditions, often described by the term "3D missions" (Dull, Dirty and Dangerous). Dull operations are too monotonous or require excessive endurance for human occupants. Dirty operations are hazardous and could pose a health risk to a human crew. Dangerous operations could result in the loss of life for the onboard pilot (Pascarella *et al.*, 2013).

Due to the increased use of UAVs, it has become necessary to control their movement effectively. Currently, UAVs are not fully capable of re-routing or re-tasked in flight, which is crucial when the mission objectives change, threats evolve, environment changes, or where there is no prior information about the scenario. Due to these limitations, the main concern for UAV's growth is autonomous and intelligent control (Faied *et al.*, 2010). It is imperative that these UAVs are able to conduct their mission with a certain level of autonomy. A fundamental ability of autonomous UAVs is auto-motion/path planning. There are several software available in the commercial market which helps in autonomous path planning of a UAV and collision avoidance system. The use of autonomous UAV, at present, is limited due to existing regulations, technological challenges, and the absence of required hardware and software. In the future, it is expected that their use will be enhanced in a large number of military, civil and commercial applications, as the technology percolates further, and regulations become less stringent with additional safety features, such as "geofencing."

3.2 THE AUTOMATIC AND AUTONOMOUS UAVs

The terms "automatic" and "autonomous" are often used synonymously, which can sometimes be misleading. An automatic system or "autopilot" is designed to fulfill a pre-programmed task, but it cannot place its actions automatically into the context of changing environment and cannot decide the best options among all available options at that time (INOUI, 2009). For example, Paparazzi is an open-source, fully autonomous autopilot system.

Over the past decade, there has been a significant rise in the use of auto-UAVs in military contexts, particularly where surveillance and remote air strikes are needed. The majority of these systems require a remote pilot to guide the aircraft, although much autonomy applies. Most UAVs use radio frequency to communicate between the remote and the UAV, which may be capable of autonomous or semi-autonomous operations. Remote control signals from the operator side can be used for (i) ground control, which is a human operating a radio transmitter, smartphone, computer, or other similar control systems, (ii) remote network system, like satellite data links, cellular and long-term evolution network, and (iii) another aircraft serving as a relay or mobile control station (Faied *et al.*, 2010). Krishnan (2009) identifies four types of autonomy related to UAVs:

(i) *Tele-operation*—which involves continual use of remote control.

(ii) *Pre-programmed autonomy*—which involves a range of pre-programmed computer-directed behaviors.

(iii) *Supervised autonomy*—The UAV operators make corrections when any problem occurs with the UAV.

(iv) *Complete autonomy*—where a robotic/UAV system can identify and handle the problem that arises without any human intervention.

An autopilot is a complete system that enables the UAV/drone to fly autonomously to defined waypoints. Today's UAV uses an automatic ECS in the form of autopilot. The ECS employs a feature called feedback or closed-loop operation (Kan *et al.*, 2018). The flight path, attitude, altitude, and airspeed of UAV are measured and electrically fed back and compared to (subtracted from) the desired state. The difference, or error signal, is amplified and used to position the appropriate control surface, which, in turn, creates a force to cause the UAV to return to the desired state, driving the error signal to zero. The autopilots listed in Table 3.1 are just a few examples from a wide spectrum of UAV autopilots. These autopilots may have additional sensors to demonstrate different capabilities.

TABLE 3.1 Some Autopilots and their Salient Features (Colomina and Molina, 2014).

Product	Company	Sensors	Weight (g)*	Size (cm)	Platform
osFlexPilot	Airware	INS/GPS/AS/MM	200	11 x 5.5 x 2.8	Fixed-wing, VTOL
osFlexQuad	Airware	INS/GPS/AS/MM	32	6.73 x 6.73 x 3	Multirotor
MP2128	MicroPilot	INS/GPS/BA/AS	24	10 x 4 x 1.5	Fixed-wing, VTOL
MP2028	MicroPilot	INS/GPS/BA/AS	28	10 x 4 x 1.5	Fixed-wing, VTOL
Piccolo Nano	CloudCap Technologies	GPS/AS/BA	65	4.6 x 7.6 x 2	Small platforms
Piccolo II	CloudCap Technologies	INS/GPS/BA/AS	226	14.2 x 4.6 x 6.2	Fixed-wing, VTOL
VECTOR	UAV Navigation	INS/GPS/MM/BA/AS	180		Mini and Tactical
Ardu Pilot Mega 2.5**	3DRobotics	INS/MM/BA	17	6.7 x 4 x 1	Fixed-wing, VTOL

*Does not include the weight of GPS antennas,
**Does not include GPS receiver, power module, and telemetry module
AS, airspeed sensor; BA, baro-altimeter; MM, magnetometer.

An autonomous system has the capability to select among multiple possible options in order to achieve its goals. An autonomous operation is one during which a remotely piloted aircraft is operating without pilot intervention in-flight management (INOUI, 2009). Basically, a UAV is automatically piloted by an embedded computer, called the flight control system (FCS) (Jones *et al.*, 1993). This system reads information from a wide variety of sensors (e.g., accelerometers, gyros, GPS, pressure sensors) and drives the UAV mission along with a pre-determined flight plan (Kan *et al.*, 2018).

The technical systems that claim to be autonomous must be able to make decisions and react to events without direct interventions by humans. Therefore, some fundamental elements are common to all autonomous vehicles. These elements include the ability of sensing & perceiving the environment, the ability of analyzing, communicating, planning and decision making using onboard computers, as well as actions that requires vehicle control algorithms (Faied *et al.*, 2010).

Autonomous systems are therefore considered to be intelligent. Intelligence can be defined as the ability to discover knowledge and use it to do something. The decision to choose the intelligent action is based on the current knowledge base, the current internal and external situation, and the internally defined criteria and rules. There may be many possible ways to achieve the goals in a dynamic terrain/environment with those defined criteria and rules. The real challenge for autonomous systems is not finding just any solution to a problem, but a good or ideally the best one (Stenger *et al.*, 2012).

Typically, the level of control by a human operator and the capability to implement the decision are directly related to the system's automatic level. With an increasing level of autonomy, freedom of the system to take decision rises, and the level of human intervention/interaction decreases. A higher degree of autonomy allows a lower workload for human operators and enables a human operator to control several UAVs at the same time (Huang *et al.*, 2007). Recent technological advancements in the field of UAVs have designed autonomous systems that require less and less human intervention and carry a higher level of onboard intelligence. Ideally, autonomous vehicles should not rely too much on human inputs but rely only on the information gathered by their onboard sensors.

The autonomous UAV operation is still not recommended for industrial or commercial use in many regions, where regulations do not permit beyond

VLOS or exceed certain distance and altitude limits (Al-Mousa *et al.*, 2019). But their use in conjunction with terrestrial sensors can allow for a range of automatic data gathering systems where infrastructure networks are not available to allow direct connectivity to the internet. This is likely to enhance the monitoring applications in dangerous and remote locations, such as in mining, civil infrastructure, etc.

Veres and Molnar (2010) outlined five essential features that would make a system autonomous: (i) efficient energy source, (ii) structural hardware, (iii) computing hardware, (iv) sensors and actuators, and (v) autonomous software. The remotely operated vehicles share the first four features. Autonomous systems can optimize their route in a goal-oriented manner under unforeseen situations. A fully autonomous vehicle can consider its position as well as its environment to act appropriately to combat unexpected events. For a successful mission, the autonomous systems should continuously monitor its resources and record the tasks to maximize mission performance during the operation.

3.3   ARCHITECTURE OF AN AUTONOMOUS SYSTEM

In autonomous entities, the term *agent* is frequently used. An *agent* is anything that can perceive and influence its environment by using sensors and actuators (Russell and Norvig, 2010). For a UAV, sensors can be air sensors, position sensors, or cameras that would collect information about the environment in which the *agent* acts. An actuator is any means of affecting the environment of the *agent* (Stenger *et al.*, 2012). The systems of an *agent* can be viewed at different levels of abstraction, as shown in Figure 3.1. The lowest level is responsible for the physical implementation of the system (i.e., computer, robot, UAV, etc.). It includes the hardware and low-level software functions needed for basic computational tasks. The *agent architecture* level forms the interface to the *physical level* and comprises the software structure of the *agent* as well as the routines and algorithms to acquire, represent and process the knowledge. The *knowledge level* at the top incorporates all the facts of the environment and the behavioral rules the *agent* needs to know in order to act according to its design. The knowledge base consists of both implemented and acquired knowledge used in general cognitive capabilities, such as language processing and specific domains (Laird, 2012).

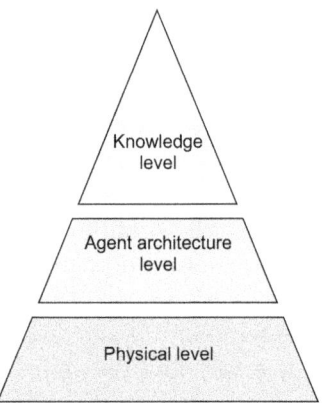

FIGURE 3.1 Abstraction hierarchy of an agent (Stenger *et al.*, 2012).

In order to ensure that autonomous systems are effective and function across multiple vehicle types, a framework needs to be created which is flexible enough in design and functionality. It must provide an open and easily modifiable architecture across a range of vehicles and sensors' capabilities. The autonomous system algorithms must run within this framework to ensure maximum flexibility while understanding the capabilities of a UAV (Eaton *et al.*, 2016).

Figure 3.2 provides an architecture for an autonomous UAV system. It has five primary functions: mission management, vehicle management, sensors management, communications management, and safety management, which are required for any system, regardless of mission and vehicle types. The advantage of this architecture system is that it has a modular functionality that can be adjusted for specific vehicles but can be common also across many vehicle types. The autonomous algorithms will reside within the mission management functionality and will be dependent upon common interfaces and architecture.

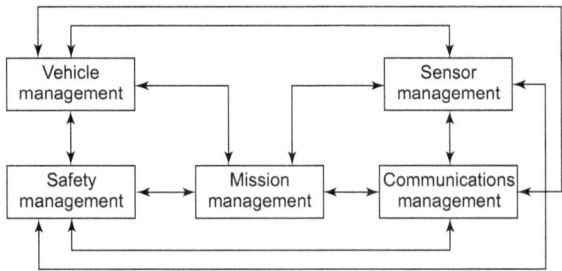

FIGURE 3.2 Architecture of an autonomous system.

Figure 3.3 provides a detailed definition of the system architecture with key critical functionalities within the primary management systems. The functions and capabilities within each management area could be changed depending on the vehicle type used (Eaton *et al.*, 2016). However, the interface to the mission management system needs to remain consistent. The key to the architecture is that each primary functional area has a controller that manages the overall function, but capabilities can be added or removed in a modular structure without affecting the larger system, depending upon the mission and system requirements. Additionally, depending on the mission task and systems onboard, the controller could enable or disable any resident capability to improve the performance of mission objectives without changes in the software.

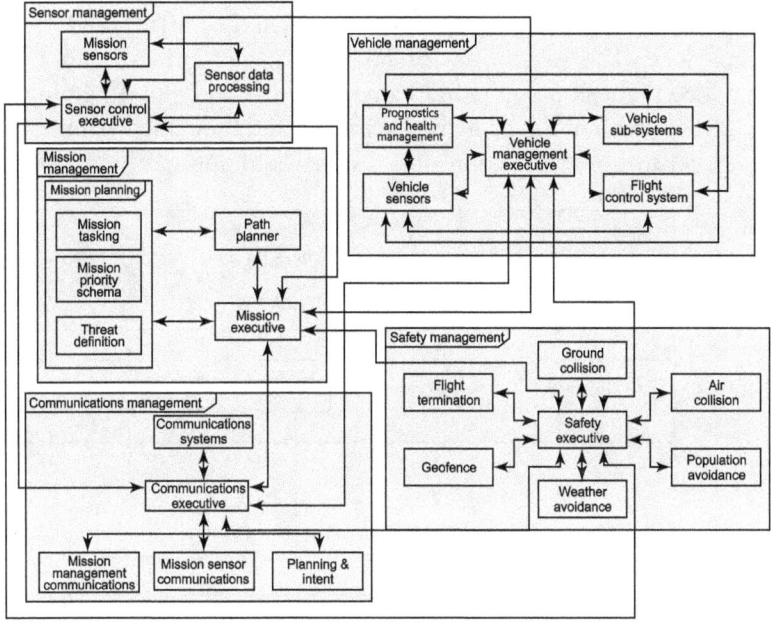

FIGURE 3.3 Detailed system architecture (Eaton *et al.*, 2016).

3.4 THE CONCEPT OF AUTONOMY

During the last decade, significant efforts have been made on UAV autonomy. Autonomy in this discipline can be defined as the capacity of sensing, interpreting, and acting upon unexpected or unforeseen changes in environment

(Huang *et al.*, 2007). Such UAVs require autonomous path planning capabilities that respond to changes in the environmental conditions, e.g., the existing and predicted wind (Stenger *et al.*, 2012). Tedious and repetitive tasks, such as surveying an area continuously for a long time, could be effectively performed autonomously. Small, inexpensive long-endurance UAVs can offer significant advantages for such tasks.

In general, the concept of autonomy may involve all mission phases, from automated flight planning (and re-planning, if necessary) to real-time guidance, navigation, and control. A two-tiered structure of an autonomous UAV model has been developed by Vidyadharan *et al.* (2017) to serve to execute different tasks critical to Structural Health Monitoring. The proposed model is shown in Figure 3.4, which describes a two-tier hierarchical structure for a multicopter UAS. It has five primary sub-systems in its first-tier: propulsion, power, Ground Navigation and Control (GNC), structures, and payload. The second tier of sub-systems comprises motors, battery, IMU, GPS, flight computer, 3D camera, arms, body, and landing gear (Laired, 2012). Each of the primary disciplines and their sub-systems, namely; GNC, structures, power, propulsion, and payload, interact.

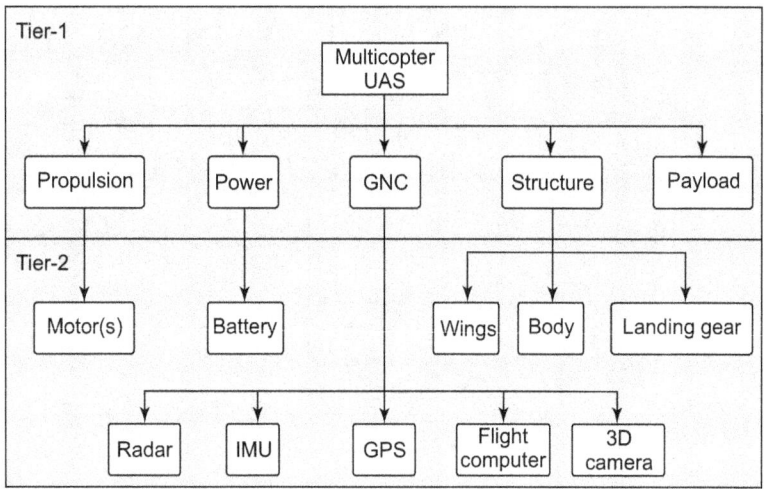

FIGURE 3.4 Hierarchical decomposition of a multicopter UAS (Vidyadharan *et al.*, 2017).

Long-term remote missions greatly benefit from autonomous system, which requires a means to respond to environmental dynamics. Given measurement of key environmental variables (world states), autonomy implies an efficient response to these measurements without further external input.

A response is "efficient" if it serves the mission goal. The planning and trajectory generation algorithms may constitute high-level components of a complete the autonomy architecture, as shown in Figure 3.5.

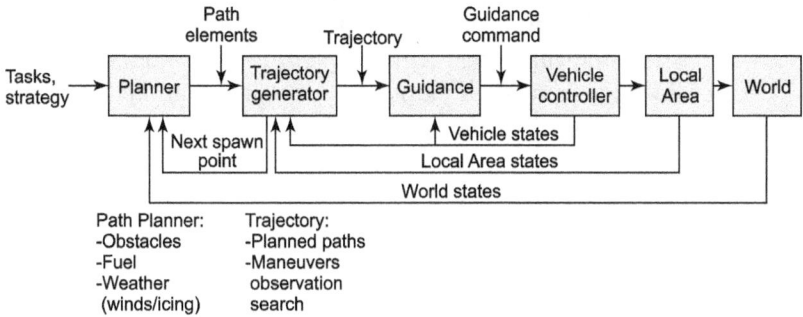

FIGURE 3.5 An autonomous system (Rubio et al., 2004).

Presently, many UAVs are able to carry out on their pre-planned mission; however, human supervision is still needed, especially in highly dynamic environments (Valvanis and Vachtsevanos, 2015). For this reason, the analysis of the execution of plans in terms of their correctness is imperative. Recently, an effort has been made to use small low-altitude UAVs for applications, such as small packet delivery, urban surveillance, and infrastructure inspection. In these scenarios, reliable and accurate flight autonomy is one of the key factors to explore the potential of UAVs and maximize their advantages (Boskovic et al., 2002).

Higher autonomy is expected to improve UAVs' operation in the following areas:

(i) *Energy consumption*: optimize the power sources and vehicle's power consumption to enhance its capabilities in long-endurance missions.

(ii) *Risk-free optimal navigation*: navigating and positioning correctly with minimum position calibration error.

(iii) *Robust decision making*: capable of making efficient decisions when facing different challenging situations.

(iv) *Continue the mission*: continue its mission alone instead of making a premature return to the base station, in case of limited or lost communication.

3.5 VARIOUS MEASURES OF UAV AUTONOMY

The UAVs have been classified as per the grade of autonomous level (AL). The levels range from simple remotely piloted vehicles (level one) to fully autonomous, human-like vehicles (level ten) at a scale of 10 (INOUI, 2009). In order to determine the degree of autonomy, several operational areas are taken into account, such as perception and situational awareness, analysis and coordination, and decision making. Generally, autonomy for UAVs can be defined at three primary levels; fully autonomous, semi-autonomous, and fully human-operated systems.

3.5.1 Fully Autonomous Operations

The UAV is adjusted and stabilized by feedback control, and can autonomously route its trajectory to assigned waypoints. At this level, the UAV can locate its position on the map and follow the defined path without operator's interaction (Huang *et al.*, 2007). Given a sequence of waypoints and prior environmental information, the path planner provides trajectories to safely guide the UAV to move through the waypoints until it reaches its destination. After visiting each waypoint, the time spent to reach that waypoint is compared with the expected time. Suppose the UAV is behind its pre-determined schedule. In that case, re-planning, task re-scheduling, or mission objective re-prioritization is performed to maximize the chance of success in delivery of mission considering the remaining tasks, mission objectives, and available resources. Such UAVs take advantage of several routines that provide capabilities, such as real-time path planning, real-time situation awareness, real-time contingency management, and task scheduling (Kendoul, 2013).

3.5.2 Semi-autonomous Operations

Such UAVs have human operator over the site with semi-autonomous and autonomous path planning. They usually fail to perform the contingency management, path re-planning or task re-scheduling. The UAV's navigation system is capable of indicating intermediate waypoints, which allows the human operator to define main targets without giving details. The system also can recognize the situation, planning a trajectory between points, autonomous departure, return, and collision avoidance. However, the human operator decides to re-arrange the tasks to compensate the lost time (if any) and re-plan the mission scenario (INOUI, 2009).

3.5.3 Fully Human Operated

Such a UAV system is fully controlled and monitored by a human operator, and all the decisions are taken by the operator considering inputs, some previous experiences, and knowledge (Kendoul, 2013). Let us understand its operation to cover four points A, B, C, and D. The human operator starts to move the UAV from point A, plans to visit point B, and then C, and finally reaches point D before running out of battery. If an unforeseen situation arises in the midway of moving from point A to B (such as sudden strong wind or water turbulence), the severity of situation will be tackled based on operator's intelligence. The operator will now decide how to direct the movement of UAV and safely pilot it toward point B, and then decides to skip visiting point C for on-time arrival at point D, so as to compensate the wasted time in the movement from point A to B. Thus, these systems rely completely on human's decision-making to deal with any unexpected circumstance in a continuously changing environment.

3.5.4 Sheridan Scale for Autonomy

Sheridan and Verplank (1978) introduced a scale metric, known as Sheridan's scale, with ten levels of autonomy in which the lowest level represents an entity that is entirely controlled by a human (teleoperated), and the highest level represents an entity that is completely autonomous and does not require human input. The various levels of autonomy presented by Sheridan and Verplank are as follows:

1. Fully controlled by operator.
2. Computer offers action alternatives.
3. Computer narrows down the choices.
4. Computer suggests a single action.
5. Computer executes that action upon the operator's approval.
6. Operator has limited time to veto the computer's decision before its automatic execution.
7. Computer executes commands automatically, and then the operator would be informed.
8. Computer informs the operator after automatic execution if the operator requests it.

9. Computer informs the operator after automatic execution only if it decides to.
10. Fully controlled by the computer (ignoring the operator).

3.5.5 UAV Rating and Evaluation

Autonomous mobile robotic technology has rapidly evolved over the last decade and has begun to impact multiple sectors across various industries. Although there are multiple criteria and metrics that have been created for the evaluation of UAV autonomy, Vidyadharan *et al.* (2017) have proposed a method of the Autonomy and Technological Readiness Assessment (ATRA), which can be retrofitted into the larger model for decision analysis of UAV operations. The two parameters used to evaluate a system's autonomy and robustness include autonomy level (AL) and technology readiness level (TRL). These two parameters are combined to form the ATRA, which can be used to provide a holistic view of the software component of UAV. The ATRA framework can also be used to measure the maturity and robustness of the UAV.

The AL is a set of metrics that are used to systematically evaluate and measure the autonomous capabilities and limitations of a UAV (Huang *et al.*, 2007). These metrics range from 1 to 10, where 1 represents the accession of autonomous behavior from remote-controlled UAV with rudimentary first order flight control up, and 10 indicates fully autonomous capabilities with no involvement of human-pilot decision-making at all, as given in Table 3.2. The AL of a UAV is mainly based on its GNC functions, which are primary attributes to enabling autonomy in the hardware component of a UAV. The GNC capabilities of the UAV are directly proportional to the autonomous capabilities of the UAV.

TABLE 3.2 List of Autonomy Levels with Descriptions (Huang *et al.*, 2017).

Level	Autonomy Levels Description
1	Remote control
2	Automatic flight control
3	System fault adaptive
4	GPS-assisted navigation
5	Path planning & execution
6	Real-time path planning

Level	Autonomy Levels Description
7	Dynamic mission planning
8	Real-time collaborative missing planning
9	Swarm group decision making
10	Full autonomous

The next parameter considered is the TRL, a set of metrics used by many companies and government agencies to systematically evaluate and assess the technology's reliability and maturity in autonomous UAVs. This metric also ranges from 1 to 10, and is related to the advancement of systems' development, integration, validation, and deployment. The range indicates from the formulation of basic principles of the system up to its deployment and full operational status. The TRL of a UAV is mainly based on its systems' integration, test environment, mission scenario, and performance, which indicate the reliability of the combination of autonomy and hardware that can be implemented in the UAV (Kendoul, 2013). The levels of TRLs are shown in Table 3.3.

TABLE 3.3 List of Technology Readiness Levels with Description (Kendoul, 2013).

Technology Readiness Levels	
Level	Description
1	Basic principles
2	Application formation
3	Technology concepts and research
4	Technology development and proof of concept
5	Low fidelity component testing (Labs)
6	System integration and flight tests
7	Prototype demonstration and operation
8	Prototype operation in realistic mission scenario
9	UAV mission deployment
10	Fully operational status

Vidyadharan *et al.* (2017) developed an ATRA graph (shown in Figure 3.6) to demonstrate the utility of AL and TRL evaluations on a hexacopter UAV, created by assembling off-the-shelf components. The highlighted block of

AL 5 and TRL 8 with a thick line indicates the extent of safe operational autonomous capabilities of the UAV. The "+" and "−" signs in each of the block denote whether that AL and TRL have been implemented on the system. In this graph, the TRL remains below 6 for all ALs above 6, due to the fact that it does not fulfill the requirements needed to be implemented on the UAV. For a system to be able to be implemented, the TRL of the UAV must be at least 7. Therefore, a UAV with TRL below 7 may only have a "−" sign; this threshold is denoted using a thick line under the TRL 7 row. Such computation would help in rating the AL of UAVs.

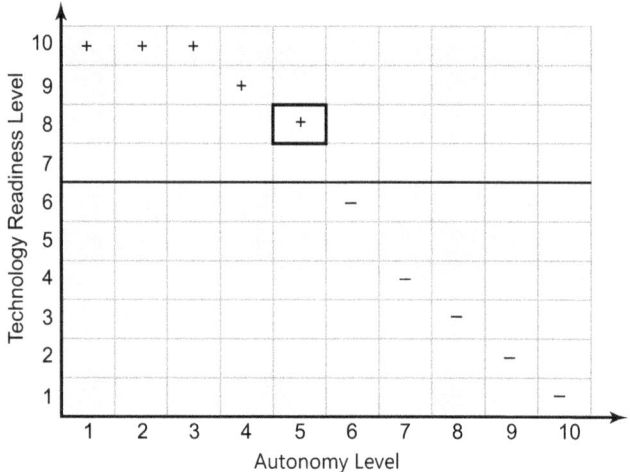

FIGURE 3.6 The ATRA graph for hexacopter UAV (Vidyadharan *et al.*, 2017).

3.6 SIMULATION OF THE ENVIRONMENT

Simulation is an important aspect of any modern development process. Therefore, designing a comprehensive simulation framework is considered as an important part for the development of the autonomous UAV (Stenger *et al.*, 2012). Simulation is a very important and cost-effective tool, especially if an actual real-life implementation is costly or associated with hazardous conditions. Moreover, simulators have other advantages, like paving the way to the adoption of new technologies, lowering training costs, and, most importantly, enabling researchers to experiment with fewer resources and time (Al-Mousa *et al.*, 2019). In order to validate the behavior of an autonomous system, the simulation environment provides the decision finding system with all relevant data, and simulates the execution of the generated action plans.

In autonomous UAV simulation systems, the interaction between a UAV and its possibly complex surrounding environment must be considered. The availability of computer simulation resources benefited the development and integration of autonomous UAVs in real-life applications, as they reduce cost, risks, development, and testing time, as well as increase the safety levels. However, it becomes difficult to compare different algorithms using multiple UAV simulation environments that were simulated before using different simulation platforms. Therefore, there is a need to understand using a simulator before using it with any type of UAVs. The simulator should implement different steps, such as path planning, obstacle avoidance, communication protocols, and much more. Table 3.4 presents a summary of the main relevant flight simulators with their highlights and limitations.

TABLE 3.4 Flight Simulators with their Main Features and Limitations (Al-Mousa *et al.*, 2019).

Simulator	Features	Limitations
Drone Watch and Rescue (DWR)	Low-cost. Easily distributable simulator (built using web technologies). 2-Level architecture (server–client). Simulating missions carried out by multiple UAVs, while allowing for data extraction.	High system requirements are needed.
Real-time Multi- UAV Simulator (RMUS)	Supports multiple UAVs. Data collection and control. Supports the direct communication links between UAVs. Functions as both a testing and validation mechanism.	Separate framework for communication simulation; CommLibX. Constrained environment.

(Continued)

Simulator	Features	Limitations
AirSim	Proposed to collect data to be used by machine learning algorithms. Cross-platform support (Linux and Windows). Open-source. Enables simulation of a host of physical phenomena. Enables the definition of different types of vehicles.	Underlying platform "Unreal Engine" is not well supported. Does not simulate collision response nor advanced ground interaction models. Does not simulate various oddities in camera sensors except those directly available in "Unreal Engine." Does not have advanced noise and lens models, which are needed to simulate the degradation of GPS signals. Does not simulate advanced wind effects and thermal simulations for fixed-wing UAVs.
Gazebo	Includes a physics engine, a host of sensor models and the ability to create 3-D virtual worlds. Can be used to build different types of robots, such as manipulator's arms.	Limited legacy physics engine in terms of photorealism.
RotorS	Enables developing and simulating control and state estimation algorithms for micro unmanned aerial vehicles (MAVs). It is based on Gazebo plugins and the Gazebo physics engine.	Intended for the design of MAVs.
Ether	Multi-agent simulation. Evaluation of mission planning models for teams of UAVs.	The overhead for users to build their own Emods of their tested algorithms.

Simulator	Features	Limitations
Virtual Air Traffic Simulation (VATSIM) Network	Free online platform. Provides the user with virtual skies, flying all kinds of aircraft all over the world. Provides resources, training, and software.	More for pilots training and flying skills improvements not UAVs test.
Lockheed Martin Prepar3D	Allows users to create training scenarios across aviation, maritime and ground domains. Used by pilots, academic, commercial organizations and military. Customizable atmosphere. Easy configuration of hardware across laptops, desktops, and multi-monitor environments. Multiple view and multiplayer.	Commercial software.

The simulated autonomous UAV can be equipped with different levels of autonomy, following simple rules to highly complex actions. All simulation activities are visualized and recorded, which give the developer an opportunity to monitor, validate and further optimize the functionalities of the autonomous system, if needed. Using the simulation framework, the autonomous UAV will be exposed first to a nominal mission, where few unexpected events for an application might occur, and then to a stressed mission, where the originally planned mission will be altered due to frequently occurring unforeseen events. Especially in the stresses mission, the autonomous UAV can demonstrate its capabilities to make decisions even with incomplete knowledge (Stenger et al., 2012).

The nominal mission requires one or more UAVs to continuously monitor the region. Here, the GCS acquires planning tasks, like issuing a detailed mission description, and manages the overall mission for the UAV deployment. As such, it has access to all UAVs which are principally assigned for the mission. The stress mission is more important as a nominal mission is unlikely to occur in reality. Consequently, the robustness of the autonomous UAV is tested when confronted which typical stressors (situations), such as system failures or other aerial traffic. A selection of stressors that can be applied to test the reliability and suitability of the UAV include moderate system failure, close traffic, adverse weather, critical level of resources (e.g., fuel), and unavailable landing place (Rodriguez 2019).

3.7 PATH PLANNING

For UAV path planning, a big challenge lies in selecting a proper combination of hardware and software (Eaton *et al.*, 2016). Choosing the right software in relation to quality and efficiency is of central importance. Path planning has two sub-components: mission planning and flight planning, as shown in Figure 3.7 (Pascarella *et al.*, 2013). The mission planning aims at an optimal scheduling (assignment and ordering) of the mission targets to cover and work on a strategic horizon. On the other hand, flight planning software enables to thoroughly scheme the mission according to the specific environment (e.g., mountains, forest, water, etc.). Instead of using 2D, a detailed 3D flight plan, including speed, altitude, etc., will normally greatly impact the results. The data processing software would create the results of point clouds, as DEM or DTM, at certain accuracy (Valvanis and Vachtsevanos, 2015). However, many companies today offer so-called "end-to-end solution" with complementary components and software. The focus now also lies on the data analysis part of the process, which will support the workflow to ensure a high level of automation and data analysis, and thus saves resources.

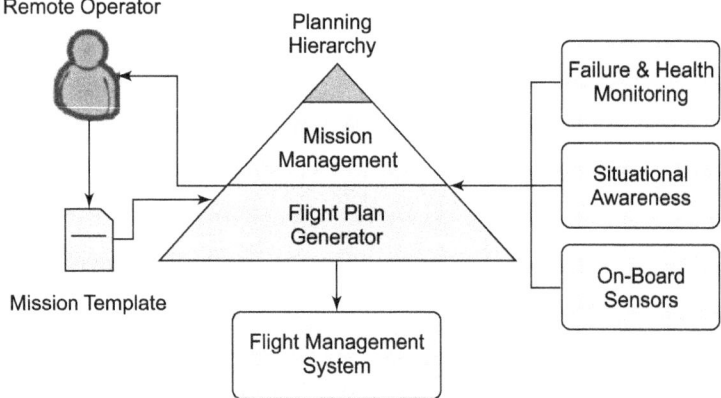

FIGURE 3.7 Path planning components (Pascarella *et al.*, 2013).

The autonomous UAV planning sub-system processes a path plan that allows the UAV to reach the target area requested by the remote operator. It is an optimization problem since the planning sub-system is expected to produce plans which maximize the mission effectiveness, and at the same time, considers a set of constraints (Pascarella et al., 2013). Constraints can be further classified as:

- *Mission constraints*: These are related to the mission objectives (start waypoint, end waypoint, target areas, etc.).
- *System constraints:* These are related to the system configuration and state (data link coverage, vehicle performances, system failures, etc.).
- *Path constraints:* These are related to the environment features (no-fly zones, meteorological hazards, air traffic, etc.). Path constraints can also be divided into barriers and threats. The former are space regions that cannot be crossed for no reason and are also known as hard constraints. The latter marks hazardous space regions that threaten the reliability (mission success) or even safety (the vehicle survival). Threats are considered to be soft constraints.

The mission planning must process an optimal mission plan, that is, an optimal route, which consists of (i) mission waypoints that are directly related to the mission targets coverage and (ii) cross waypoints that are intermediate waypoints and are inserted by the planner to avoid "no-fly zone" or high-risk threat areas. It provides a feasibility assessment of the proposed mission. The mission planning must include some management functions, such as the implicit constraints assessment and the logic for re-planning. As shown in Figure 3.7, the mission planning also receives the following information:

- *The mission template:* that is supplied by the remote operator and that formally specifies the user's request.
- *The situational awareness data*: that feeds the evaluation of implicit constraints related to the environment state.
- *The health monitoring data:* that feeds the evaluation of the implicit constraints related to the system state.
- *The vehicle sensor data*: is needed to track the mission.

3.7.1 Mission Planning

The primary parameters involved in the operations of a UAV include (i) the mission scenario, (ii) the mission objective, and (iii) the characteristics of aerial vehicle, as shown in Figure 3.8 (Vidyadharan et al., 2017). The mission scenario is defined by the meteorological conditions and the structure type, while the mission objective aims at the data acquisition method used and mission tasks. The characteristics of UAVs are divided by its size, class, and AL and TRL. The structure type and the data acquisition method consist of attributes and metrics that help determine the complexity of the mission (Eaton et *al.*, 2016).

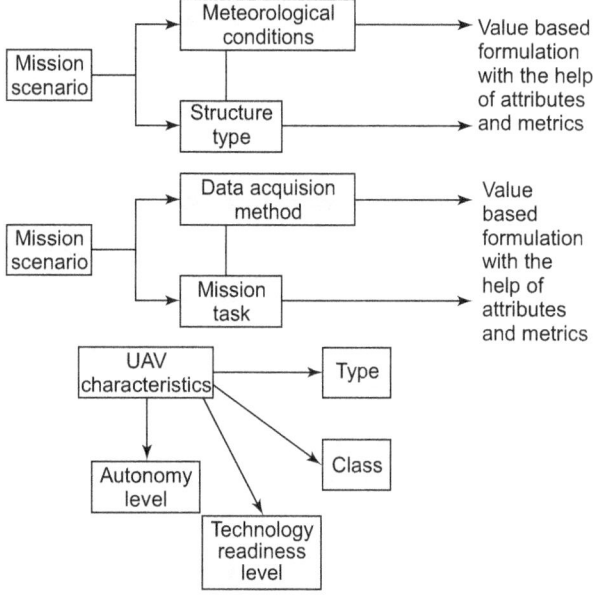

FIGURE 3.8 Parameters for mission planning (Vidyadharan et al., 2017).

Mission complexity is a key factor with respect to a potential mission scenario. It is classified into three categories: *mission*, *team*, and *communication management* (Stegner *et al.*, 2012). *Mission management* describes the ability for a given mission to have a high amount of diverse capabilities and goals. This gives the *agent* the opportunity to act upon a broad number of possibilities. Hence, the mission puts no or few constraints on the UAV's capabilities. *On the other hand, team management describes the extent to which members or agents of the team are able to cooperate and* maintain themselves. Their capability to communicate and perhaps start a formation within the mission is included in *communication management*. Based on these definitions, the mission fulfills the requirements, resulting in a complex mission to test the *agent's* capabilities.

3.7.2 Flight Planning

Typically, a UAV mission starts with flight planning. Although each autonomous UAV's mission is unique, almost similar steps and processes are normally followed (Eaton et al., 2016). These steps would, however, depend on a specific flight planning algorithm. Flight planning uses a reference map or satellite image to define the extent of the flight area. It also requires the desired flying altitude, overlap, focal length and orientation of the camera, and desired flight path (Hugenholtz et al., 2016). The flight planning algorithm then computes the overlapping stereo-images required to be taken to cover the area of interest. During the flight planning process, the algorithm can adjust various parameters until the operator is satisfied with the flight plan (Valvanis and Vachtsevanos, 2015). For example, the camera shutter speed settings must satisfy different lighting conditions. If the exposure time is too short, the images might be too dark to discriminate among all key features of interest, but the images will be either blurred or bright if the exposure is too long.

Autonomous UAV uses basic heuristics to travel to a known target while avoiding various static obstacles. A sensor system is used to provide feedback about the obstacles in the UAV's local environment. This information is then processed at each sampling time and used as one of the factors to determine the movement of the autonomous UAV for obstacle avoidance and path planning to the target. A path planning strategy must consider the dynamic and kinematic constraints of the vehicle and information about obstacles to generate a path between a starting point and a target location (Shakhatreh et al., 2017).

In several scenarios, mission objectives rely on the capability of an autonomous UAV to maneuver in a dynamic and changing environment or in areas with no prior knowledge of the surroundings (Boskovic et al., 2002). For example, a UAV may encounter sudden forest fires, which cause the destruction of large land, heavy damage, and loss of lives (Vergouw et al., 2016). The extent and intensity of forest fires can change rapidly depending on environmental conditions, and become difficult to combat, so real-time data are required to be obtained for faster decision making, as situations can become worst due to fluctuating wind, terrain, conditions, etc. (Gupta et al., 2016). In all these scenarios, the mission could be improved by autonomous path planning vehicles. Most effective path planning algorithms rely on complete prior knowledge of the environment; therefore, it limits autonomous UAVs' capabilities. Furthermore, these algorithms cannot attain fairly accurate and real-time solutions for such complex and dynamic environments;

therefore, a more sophisticated method to path planning is needed to not only allow for maneuvering in an unknown environment but also is necessary to move in a dynamic and changing environment in real-time.

Given the importance of path planning in the overall success of an autonomous UAV mission, many approaches have been used to solve the task of path planning problems. For most problems involving operation in uncertain environmental changes, dynamic path planning is essentially required. The well-known direct method of optimal control theory, called Inverse Dynamics in the Virtual Domain method, is used to develop and test an onboard real-time trajectory generator for autonomous UAV deployment (Yazdani et al., 2017). A real-time Differential Evolution-based motion planner, proposed by Zadeh et al. (2017), can also be used to provide a time/battery efficient operation for an autonomous UAV in a dynamic time-varying environment. Another elaborated Evolution-based online path planner/re-planner is introduced by Zadeh et al. (2018) for autonomous UAV path planning in an uncertain terrain to ensure safe deployment.

The generated flight plan is uploaded to the UAV autopilot. The autopilot uses the instructions contained in the flight plan to find climb rates and positional adjustments that enable the UAV to follow the planned path as closely as possible (Pastor, 2007). The autopilot reads the adjustments from the GNSS and the IMU several times per second throughout the flight. After the flight completion, the autopilot downloads a log file, which contains information about the recorded UAV 3D placements throughout the flight and information about the camera when it was triggered. The information in a log file is used to provide an initial estimate for image center positions and camera orientations, which are then used as inputs to recover the exact positions of surface points.

3.8 CONTROLLING THE MISSION PLANNING

The mission planning and control station is the "nerve center" of the entire UAV system. It controls the launch, flight, and recovery of the air vehicle; receives and processes data from the internal sensors of the flight systems and the external sensors of the payload; and controls the operation of the payload. To accomplish its system functions, the mission planning and control station incorporates the following sub-systems (Fahlstrom and Gleason, 2012):

1. Air vehicle status readouts and controls.
2. Payload data displays and controls.
3. Map displays for mission planning and for monitoring the location and flight path of the air vehicle. The ground terminal of a data link transmits commands to the air vehicle & payload, and receives status information & payload data from the air vehicle.
4. Computer provides an interface between the operator(s) and the air vehicle, and controls the data link and data flow between the air vehicle and the mission planning and control station. It may also perform the navigation function for the system, and some of the "outer loop" (less time-sensitive) calculations associated with the autopilot and payload control functions.
5. Communications links to other organizations for command & control and for dissemination of information gathered by the UAV.

The mission planning and control station displays two types of information to the operators. Display basic status information, such as position, altitude, heading, airspeed, and fuel remaining for air vehicle control. The second type of information to be displayed consists of the data gathered by the onboard sensors of the payloads. These displays can have many varied features, depending on the nature of the sensors and how the information is to be used. The planning and operation functions of a mission planning and control station can be described as follows (Fahlstrom and Gleason, 2012):

- *Planning:* It includes process tasking messages, study mission area maps and designate flight routes (waypoints, speeds, altitudes), and provides the operator with the plan.
- *Operation*: It includes load mission plan information. It launches UAV, monitors UAV position, controls UAV, controls and monitors mission payload, recommends changes to a flight plan, provides information to operator, saves sensor information when required, recovers UAV, and reproduces hard copy or digital tapes or disks of sensor data.

The UAV is an under-actuated system having six degrees of freedom (position and orientation) and multiple control inputs (e.g., rotor speed). It also has multivariable, non-linear, and strong coupling characteristics, all of which make its flight control design very robust. There are options available for the automation of operator's function, ranging from no automation to full automation. The UAV operator is responsible for flying the air vehicle from one point to another and continues to fly the air vehicle successfully, dealing with any change in environment, such as gusts, wind shear, or turbulence, if needed (Valvanis and Vachtsevanos, 2015).

Full automation of the piloting function requires a detailed flight plan. But many times, it is necessary to change the flight plan while the mission is in progress. For example, there might be various possible reasons for a change in the UAV flight plan, such as loss of power, loss of command uplink of the data link, loss of GPS navigation (if used), payload malfunction, weather change, and structural damage in UAV, where malfunctioning of the sensor payload has a higher priority in the pre-planned mission. Some of these events, such as loss of power, loss of data link, changes in-flight characteristics, payload malfunctions, and loss of GPS can be recognized in a straight-forward manner by the computers/software on the UAV, while others may or may not be easily determined in a fully automatic UAV (Pastor, 2007). This would lead to a significant amount of control in the automation of mission plans to accommodate unexpected situations. Therefore, it is important to search the location at which the UAV attempts a crash landing or deliberately creates a crash to minimize the damage on the ground. It can be achieved by diving the UAV steeply into a body of water or some open area without people. Such cases would force a change in the flight plan that might be very hard to plan in advance.

In case the power is lost, the control system may alert a human operator and allow the human to determine and direct the response. In such cases, the autopilot can immediately establish a minimum descent rate in-flight mode, and the computers installed at the GCS can identify the areas on the ground where the UAV can reach without power and have "safe" crashing sites within those areas. The sensors can be used to look at possible crash sites, and the human can then evaluate the information available and apply judgment to decide where to crash the UAV. Once that decision is made, piloting to the crash landing or deliberate dive into an open area can be achieved with the automation available in the UAV, depending on how that particular system is designed and how its operators are trained.

3.9 CONTROLLING THE PAYLOADS

The UAV can be operated with great automation if the mission plan is not required to be changed or modified due to payload failures. Payload sensors on the UAV monitor the information sensed by them and downlink and/or record the information as a function of time and location. If there is no requirement for real-time or near real-time response, the mission plan will consist of flying UAV to some specified flight plan while operating the sensors, and, at most, monitoring the operation of the sensors (Valvanis and Vachtsevanos, 2015).

Sensors looking for chemical or biological agents or for radiation may be quite capable of detecting any things that they are looking for in a fully automatic UAV. Once that detection has been made, a computer could look up rules to perform the task. For example, the rules might guide the computer to interrupt its pre-planned flight in order to collect and map the cluster distribution or repeat the observations. The software could adapt the mapping process related to the intensities measured by these sensors in order to determine the geometry of the chemical, biological, or radiological contamination (Zadeh *et al.*, 2018).

Imaging payloads present a special challenge for the automation of the operator's function. In particular, imaging sensors may not be able to "notice" anything without a human in the loop. Suppose the only function of these sensors is to downlink and/or record images of pre-planned areas, with no real-time interpretation. In that case, the function of the operator is simply to point the sensor in the correct direction and turn it on and off. Those functions are easy to automate. Similarly, there are some missions in which an imaging sensor (e.g., Radar system) might automatically detect objects of interest with reasonable accuracy. Radar system, augmented by a scanning laser rangefinder, may have the capability to reliably detect moving objects. For "automatic target detection," the images need to be downlinked to a human operator in real-time, and if the sensor has variable magnification and viewing angle, the operator can control the view angle pointing and magnification to take a closer look to better detect the objects of interest (Pastor, 2007).

3.10 FLIGHT SAFETY OPERATION

Flight safety is a critical aspect for UAV usage, as aviation is more focused on low accident rates and high safety rates. The agent-based simulation environment has proven to be a valuable tool for evaluating the safety function of the autonomous agents during the flight. Onboard autonomous sub-systems could provide faster reactions to external changes due to their direct access to sensor data. However, there are certain limitations to effectively use autonomous

UAVs (Zadeh *et al.*, 2018). Available bandwidth is a limited resource in most circumstances, and communications with the ground station may not be available in emergent situations. Data link is another big constraint for using autonomous UAVs. When flying UAV under satellite coverage, navigation performance achieved by standalone or differential GNSS/inertial technology typically may fulfill the flight requirements. But in challenging scenarios, the autonomous flight is hindered by navigation issues. Navigation in such environments may usually be tackled by using additional measurements, i.e., visual-aided navigation, radio beacon, positioning based on phone signals, and actuator outputs (Azar *et al.*, 2018; Taha and Shoufan, 2019).

The significant growth and widespread application of UAVs present a major safety challenge for regulators around the world. This includes both safety on the ground (to prevent UAVs from falling to Earth and injuring people and/or damaging property) and safety in the air (to prevent midair collisions). Current regulations limit the low-altitude operations (below 400 ft or 120 m) to the VLOS of a human pilot who is always controlling the UAV. Nonetheless, many enterprises are currently exploring the potential of autonomous UAVs, both BVLOS and without a pilot's direct control (Takacs *et al.*, 2018). Figure 3.9 shows the example of operation modes of autonomous UAVs. Examples of such applications include parcel delivery, medical supply delivery, remote and large-scale infrastructure monitoring, and surveillance. Aviation authorities around the globe have initiated programs to define the rules of UAV operation, in relation to address the safety issues regarding commercial use of UAVs.

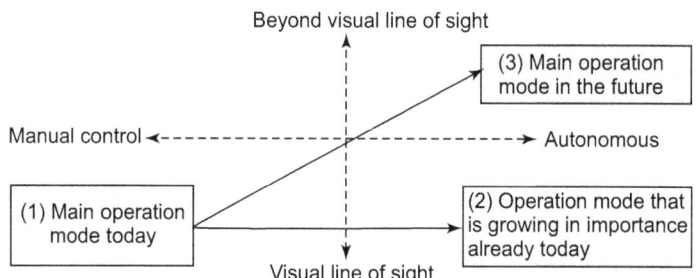

FIGURE 3.9 Operation modes of UAVs (Takacs *et al.*, 2018).

3.10.1 Geofencing

Geofences are virtual boundaries set up within physical locations where UAVs can be detected if they reach certain delimited areas. The UAV will need a reliable navigation system (i.e., GPS) and autopilot software to create and interact with a fence (Kandoul, 2013). Geofencing cannot be done with manually controlled UAVs and DIY autopilots. Geofencing can be set up using GPS, Wi-Fi, bluetooth, cellular data, geomagnetic or Radiofrequency identification (RFID). All of them can be "paired" with geofencing, resulting in precise and accurate data as compared to using the standalone GPS technology. This technology keeps a UAV away from accidentally venturing into a sensitive "no-fly zone," for example, an airport, nuclear power plants, or a high-profile event (Jones et al., 1993).

In a critical operational scenario, a boundary is drawn by the user to confine the movement of the UAV. A standard geofence is generally circular, with the biggest at-risk area forming the nucleus, but it can be polygon or rectangular. In circular type fencing, once the user enters the radius, a circle is drawn at user's point of interest. In polygon type geofencing, the user will select different points and a polygon is drawn between those points and check whether the UAV is within the boundary. In geofencing, if a UAV crosses the defined boundary, an alarm sounds with a message box is displayed so that the user can operate it safely (Santoso et al., 2018).

The main benefit of geofencing is the fact that it doesn't require building/constructing anything physical in the real world, which is important for UAVs to fly over fences (Veres and Molnar, 2010). Instead, it is a digital system of coordinates of defined boundary which rests on the top of a map. One of the best things about geofencing is the fact that the technology can be defined virtually anywhere. This technology right now uses Google and Apple databases (known as native positioning), which means that even mapping work is not required.

One of the major applications for geofencing is security; when anyone enters or leaves a particular area, an alert or text message is sent to the users. In military applications, this can be used when the enemy vehicles enter into a country's boundary; it gives an alarm sound to alert the officers to take necessary actions. More modern geofencing uses include "Smart Home" functionality, such as heating at your home will turn itself off when your phone pings at a certain distance away from home, and back on when you are returning home (Sinsley et al., 2008).

Autonomous UAVs have built-in geofencing software, so they can't fly over airports or other restricted areas (Santoso et al., 2018). DJI, the market leader, has been using geofencing in compatible UAVs since 2013. The DJI has developed a mobile app called "DJI GO app" for fencing. DJI's version of geofencing, called "Geospatial Environment Online," offers users up-to-the-minute information where flight is banned, or limitations have been applied. This includes large public gatherings, natural disaster areas, and other events which warrant restrictions. The Parrot company, the manufacturer of UAV, also has geofencing in its ANAFI software, but pilots can turn it off. The Yuneec UAV has also implemented a geofence system that has "in compliance with FAA No-Fly Zone Regulations" focusing on preventing flights near commercial airports. The geofencing features of Yuneec prevent UAV operators from flying over heights of 400 ft and beyond distances of more than 300ft outwards from the pilot's position. The aircraft's GPS also draws a "Smart Circle" 26 ft in diameter around the pilot during take-off and landing, as shown in Figure 3.10.

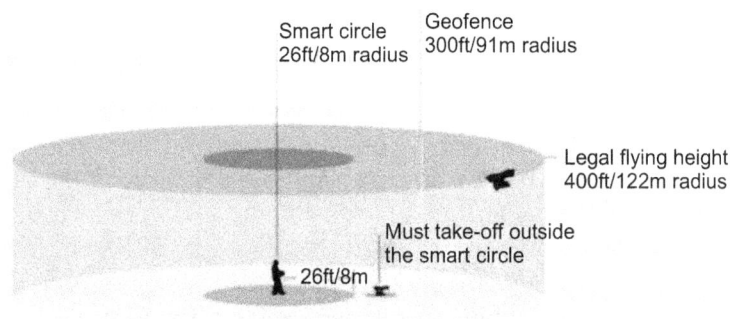

FIGURE 3.10 Geofencing concept by UAV.

3.10.2 Collision Avoidance System

One of the challenges faced by UAVs is in the area of collision avoidance. The UAVs can collide with obstacles which are either moving or stationary objects, in an indoor or outdoor environment (Shakhatreh et al., 2018). During UAV flights, it is important to avoid accidents with these obstacles using collision avoidance techniques.

The use of UAVs in an indoor environment is a challenging task, and it needs higher requirements than an outdoor environment. It is very difficult to use GPS to avoid collisions in an indoor environment, which is usually a GPS-denied environment. Moreover, RF signals cannot be used in this environment, as these signals could be reflected and degraded by the indoor obstacles and walls. Many collision avoidance approaches have been proposed to avoid collisions by UAVs, such as geometric approaches (Park *et al.*, 2008), path planning approaches (Alexopoulos, 2013), potential field approaches (Mujumdar and Padhi, 2009), and vision-based approaches (Saunders, 2009). Vision-based methods (using cameras and optical sensors), and onboard sensors-based methods (using IR, ultrasonic, laser scanner, etc.), and the combination of both can be utilized for collision avoidance in an indoor environment (Luo *et al.*, 2013).

The geometric approaches require UAVs to sense and extract information about the environment and obstacles, such as position, speed and size of the obstacles. Vision-based collision avoidance methods using cameras suffer from heavy computational operations for image processing. One of the best possible solutions is to combine with a vision-based approach that uses a passive device to detect the obstacles. The UAVs could be integrated with vision sensors, laser range finder sensors, IR, and/or ultrasound sensors to avoid collisions in all directions (Chee and Zhong 2013). Most of the intelligent UAV systems have in-built collision avoidance technology.

3.10.3 UAV Detection Technology

UAVs pose a significant challenge in terms of security and privacy within society. Appropriate technology can be deployed to support the safe operations of UAVs. Various sensors and methods are receiving attention to detect and disable UAVs used for malicious purposes. Table 3.5 presents the data related to the effectiveness of sensors to various environmental conditions for the detection of UAVs. Radar can be used to detect and locate a UAV in space. However, radar detection can be unreliable, as adverse weather conditions affect the reflected wavelength, distorting the wave. Radar also fails to identify the UAV type, and suffers from high false-positive rates, since it cannot distinguish between birds and UAVs (Nassi *et al.*, 2019).

TABLE 3.5 Ability of Various Sensors for UAV Detection (Nassi *et al.*, 2019).

Factors	RF		Optical			Acoustical
	Active Radar	RF Scanners	Visible	IR	LiDAR	
Light	Yes	Yes	Yes		Yes	Yes
Darkness	Yes	Yes	Yes	Yes	Yes	Yes
Noise	Yes	Yes	Yes	Yes	Yes	
Birds		Yes				Yes

(Continued)

Factors	RF		Optical			Acoustical
	Active Radar	RF Scanners	Visible	IR	LiDAR	
UAV Identification		Yes	Yes	Limited		Yes
Autonomous UAV Detection			Yes	Yes	Yes	Yes
Multiple UAV Detection	Yes	Only if UAVs use different channels	Yes	Yes	Yes	Only different types of UAVs
Cost		Yes	Yes			Yes
Long range Detection	Yes	Yes	Require lens		Yes	
Immunity to NLOS		Yes				Yes
Locating	Yes	Multiple	Yes	Yes	Yes	Multiple

The RF scanners and spectrum analyzers are primarily used to detect UAV radio signatures by detecting bands that are known to be used by UAVs and other radio signatures. They can be used for classifying a suspicious transmission as an FPV channel to locate a UAV (Eriksson 2018). RF scanners can be very effective at detecting the presence of a UAV and identifying its type by comparing them to known used bands; however, RF scanners suffer from an inability to accurately locate a UAV in space unless they are triangulated. Based on optical cameras, visible frequencies are used to detect a UAV and its trajectory from a single video stream by detecting motion cues. They also suffer from high false-negative rates due to the increasing number of UAV models, the use of non-commercial UAVs, and ambient darkness. In order to address the compromised UAV detection rate in dark conditions, thermal cameras can be used.

Acoustic detection methods are not dependent on the line-of-sight or the size of the target UAV. Many studies have suggested using a microphone array to detect UAVs by analyzing the noise of the rotors. The UAVs can be detected based on comparing a UAV's captured acoustic signature with other signatures stored in a database of previously collected sound signatures. While acoustic methods can be used to identify a UAV and locate its presence (using multiple distributed microphones), relying on acoustic signature methods for UAV detection suffers from false-negative detections due to the increasing number of UAV models and false-positive detections due to ambient noise (Qu *et al.*, 2016). It is seen from Table 3.5, that there is no single sensor capable of fulfilling all the requirements. In order to overcome the limitations that arise from using a single sensor, sensor fusion methods combining acoustic and optical methods can be used.

Effective, safe UAV operation supported by technologies would enable (i) UAV detection, classification, and tracking, (ii) UAV interdiction, and (iii) evidence collection in the case of violation (Taha and Shoufan, 2019). In addition, UAVs can have onboard preventive technologies to support safe operation, such as sense & avoid, geofencing, parachuting systems, and mechanisms against different attacks, such as jamming or hijacking of the control signal. Figure 3.11 classifies the technologies into four main categories: deployed to support safe UAV operations. A major task toward the use of these technologies is to detect, classify, and identify the UAVs in the sky. The general advantages and disadvantages of different detection technologies are summarized in Table 3.6.

TABLE 3.6 Advantages and Disadvantages of Different UAV Detection Technologies (Guvenc et al., 2018).

Detection Technology	Advantages	Disadvantages
Radar	Low-cost Frequency-Modulated Continuous Wave radars are resistant to fog, cloud and dust as opposed to visual detection; and less prone to noise than acoustic detection. Radar does not require a line-of-sight (LOS). Higher frequency radars, such as mmWave radars, offer higher resolution in range and enable capturing Micro-Doppler Signature.	UAVs have small Radar Cross Sections, which make the detection more demanding. The mmWave has higher path loss, which limits UAV's detection range.

(Continued)

Detection Technology	Advantages	Disadvantages
Acoustic	It does not require a LOS, so it works in low-visibility environments. Low-cost depending on the employed microphone arrays.	Sensitive to ambient noise, especially in loud areas. Wind condition affects detection performance. Requires a database of acoustic signatures for different UAVs for training and testing.
Visual	Low-cost depending on the utilized cameras and optical sensors or reusing existing surveillance cameras. Human assessment of detection results using screens is easier than other modalities.	Level of visibility is affected by dust, fog, cloud and daytime. High cost thermal, laser-based, and wide field-of-view cameras may be required. LOS is necessary.
RF signal-based detection	Low-cost RF-sensors. No LOS is required. Long detection range.	Not suitable for detecting UAVs flying autonomously without any communication channels. It requires training to learn RF signal signatures.

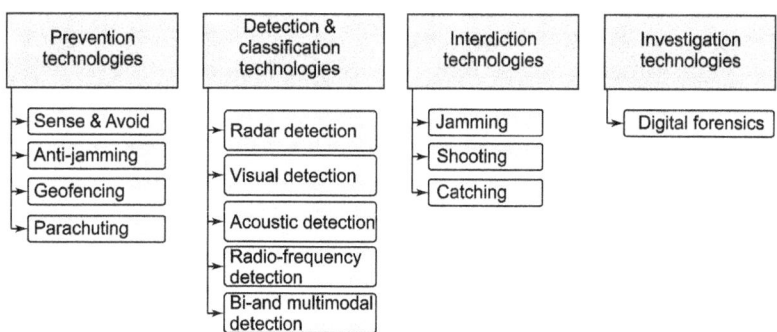

FIGURE 3.11 Different technologies applied to support safe UAV operations (Taha and Shoufan, 2019).

As clear from the Table, no single technology is perfect for the detection and classification of UAVs. Therefore, several authors suggested the use of bi-modal and multimodal systems with promising results (Zunino, 2016, Liu et al., 2017). Regardless of the technology used, all proposed solutions are achieved for a statically located detection system. In modern cities, the RF and the radar signal have limitations because many obstacles can deteriorate the detection capability (e.g., high-rise buildings), blocking the view and acoustic detection is also more difficult due to high noise levels. Distributed and collaborative detection systems using wide-area solutions or city-wide surveillance sensors can provide an effective solution for UAV detection and its classification.

3.10.4 Air Traffic Management

The overall architecture of drones for air traffic management and safe operations is presented in Figure 3.12. The traffic management system (in the center) handles drone's flight approvals and is also responsible for coordinating with manned aviation and its air traffic control systems. It will ensure compliance with low-altitude airspace regulations. The UAV traffic management or drone traffic management system will make further decision based on data provided by the supplemental data services, such as weather services, and provide information to the drone operators (Takacs *et al.*, 2018). The operator will connect to the traffic management system to seek flight approval and is responsible to periodically report the status of drone flights. It can be facilitated by providing telemetry reports to aid the drone traffic management system in keeping an up-to-date status of the airspace.

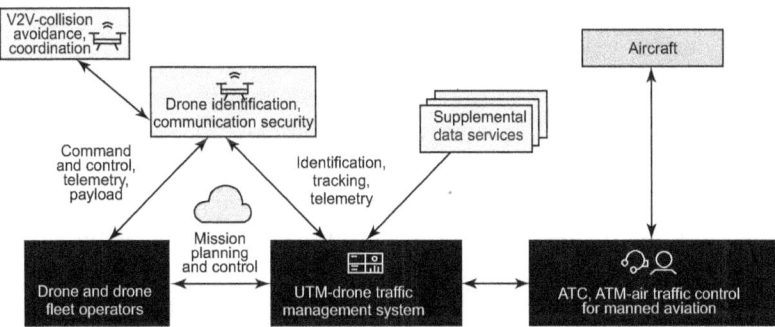

FIGURE 3.12 System elements supporting safe operations of drones (Takacs *et al.*, 2018).

Authorities would require some mechanisms to uniquely identify each drone and its operators. The identification needs to be secure and tamper-resistant. Two main categories for drone identification are considered. First one is *local solutions* which rely on a broadcast signal sent from the drone and can be read in the area where the drone is operating. The second one is *network publishing solutions* which rely on using network connectivity to publish data on a remote server, potentially to the drone traffic management system. In addition, drones positions over time need to be monitored to ensure that drone traffic management systems have up-to-date information and are able to locate any drone in the airspace. Communication security is already inherent within the mobile networks (GSMA, 2018), and can be provided at many levels, from encryption on the radio link to higher-layer security mechanisms.

Drone tracking is supported by the mobile positioning service and can be queried from the mobile network and integrated into the drone traffic management systems. Mobile networks can assist with drone identification, authorization, and geofencing (ARC, 2017). Mobile networks are equipped with a variety of tools to identify and authorize users and devices that can access the networks. Communication is also needed for the management to support authentication and authorization, whereas command and control communication is needed to operate the drone. For collision avoidance, drones may require means to communicate with other nearby drones in a vehicle-to-vehicle (V2V) manner.

3.11 PATH PLANNING ALGORITHMS

Drones/UAVs, in some ways, are like flying computers. Software in relation to quality and efficiency is needed for autopilot drone/UAV path planning. Path planning defines the methodology that an autonomous robot must complete moving from an initial location to a final location, deploying its own resources like sensors, actuators, and strategies while avoiding obstacles during the trip. Software for path planning is of utmost importance, as it can differentiate between flight planning and data management software (Zheng *et al.*, 2014). Flight planning software enables to thoroughly scheme the mission according to the specific environment (e.g., mountains, forest, water, urban, etc.), while data processing and management software may create point clouds, orthophotos, DEM, DTM, 3D maps, etc. Figure 3.13 shows an example of an auto-path generated by the software.

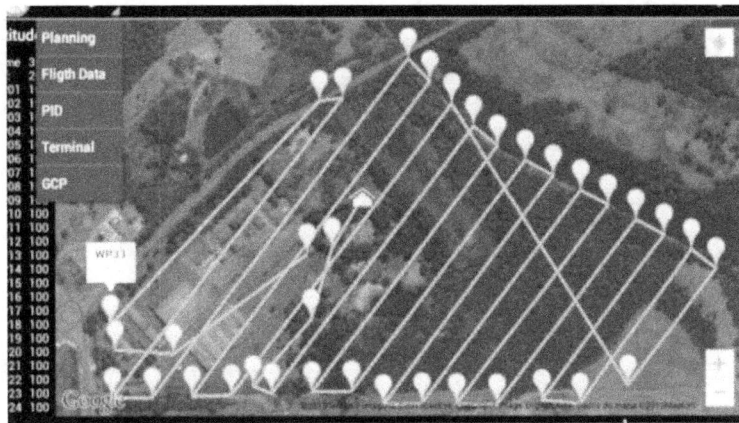

FIGURE 3.13 The flight plan automatically calculated by software (Source: http://www.eyeondrones.com/a-breakthrough-in-autonomous-drones/).

The usage of UAVs that can operate independently in dynamic and complex working environments has become increasingly widespread. Autonomous robots and vehicles are being used to perform missions in hazardous environments, such as operations in nuclear power plants and surveillance of enemy forces on the battle field. One of the main challenges for intelligent UAV development for such application is path planning in adverse environments. The problem of planning a path in these applications is to find a collision-free path in an environment with static or dynamic obstacles. It is important that path planning UAV software copes up with emergencies in a complex environment, avoiding the obstacles.

Path planning software can handle varying numbers of objectives, complex tasks, different constraints, and uncertainty to make the scenarios realistic (Jaradat et al., 2009). Because path planning algorithms have been studied for many years, excellent strategies are developed for path planning in obstacle environments. The UAV may have two types of algorithms to prevent obstacles: traditional algorithm and intelligent algorithm. The traditional route planning algorithm determines an ideal or time-based path between the required sites under certain restrictions. The path planning methods that have been studied and utilized most include; Visibility graphs, Voronoi diagrams, Cell decomposition, Artificial Potential Fields (APF), and Fuzzy logic. The last two methods fall under the intelligent algorithm category. These methods have their own advantages and disadvantages, as enumerated in Table 3.7.

TABLE 3.7 Advantages and Disadvantages of Common Path Planning Techniques.

	Visibility Graphs	Voronoi Method	Approx. Cell Decomposition	Artificial Potential Fields	Fuzzy Logic
Optimal	Very useful				
Complex obstacles			Very useful	Useful	Very useful
Higher dimensions			Very useful	Very useful	Very useful
Computable online				Very useful	Very useful
Uncertainty		Useful	Useful	Very useful	Very useful
Finds impossibilities	Very useful	Very useful			
Ease to implement				Very useful	Very useful

The biggest benefit to using Visibility graphs is that they are optimal. This is due to the fact that all paths are formed along the edges of obstacles. However, it is also one of the disadvantages of this method. If there is any uncertainty about the size or shape of the obstacle, the path can easily become infeasible. On the other hand, Voronoi diagrams are able to handle a little uncertainty well because they create paths that inherently steer clear of the obstacles. However, they are not optimal and can often create paths that intuitively do not make much sense. An approximate cell decomposition method has disadvantages as it requires complete prior information and all obstacles to be polygons. While this is more practical to implement than the previous two methods, it can be difficult for very complex environments. Furthermore, if the problem is not solvable, this method will not recognize the path (Sabo and Cohen, 2012). For example, if obstacles block all possible paths, this algorithm will iteratively search for a solution (creating smaller cells and searching for a path) indefinitely.

Much focus has also been on intelligent strategies, like APF, as these techniques have shown excellent results for comparable situations. Also, this method is quick to compute, and therefore, practical to do online. In addition to APF as an intelligent control method, Fuzzy logic has shown excellent path tracking and collision avoidance for both mobile robots and UAVs (Hui *et al.*, 2006). All of these benefits, and the fact that optimal solutions are computationally intractable for complex scenarios, confirm intelligent control methods (e.g., APF and Fuzzy logic) as the practical solution to these problems (Jaradat *et al.*, 2009).

A highly demanded task for UAVs is 3D autonomous navigation (either in static or dynamic environments) that can optimize the route and minimize the computational cost. Some representative techniques used in path planning methods and based on continuous and discrete environment sampling include Rapidly-exploring Random Tree, Probabilistic Road Map, Voronoi diagrams, and APF (Samaniego *et al.*, 2019). The Rapidly-exploring Random Tree and Probabilistic Road Map techniques make random explorations (continuous sampling) of the defined environment. The Rapidly-exploring Random Tree is an expensive algorithm in terms of computational cost when searching for feasible solutions in cluttered environments. Once the Probabilistic Road Map is made, a methodological base built on nodes must be invoked to define the lowest cost path. The main disadvantage of the Voronoi diagram by applying Djiktra's algorithm is that it is an offline method. Finally, the APF algorithm presents little computational complexity, although it tends to fall into local minimums.

Evolutionary computation (EC) techniques have been successfully applied to compute near-optimal paths for UAVs. These methods for path planning are based on a search technique that resembles the evolutionary process. The EC adapts to changes without the need to recompute a completely new solution. The EC search technique is not as efficient in finding an optimal solution, but its important advantage is that a viable solution is available at any time and that it approaches an optimum as more processing time is available. Therefore, it is computationally efficient and allows for easy integration with maneuvers triggered by the changing environment. With the use of EC path planning, the execution of a local maneuver is simply interpreted as an environmental change.

The EC has shown promising results for path planning where complex environments play an important role. The path planning concept employed by Rubio *et al.* (2004) using EC is presented in Figure 3.14. The algorithm performs a random search for a feasible solution by mutating, evaluating, and selecting members with higher fitness within all possible vehicle paths. This cycle is repeated until a criterion is met.

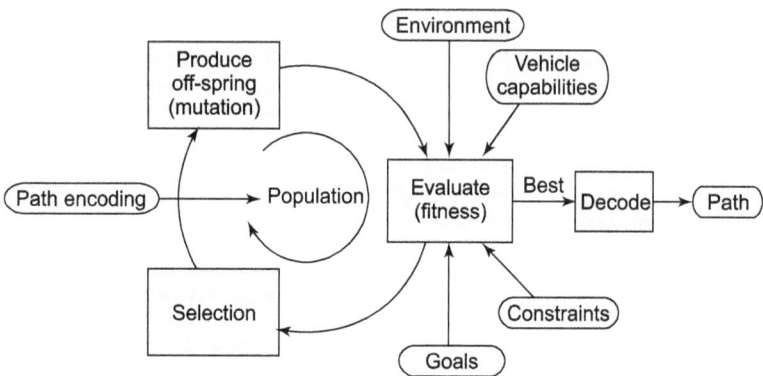

FIGURE 3.14 Evolutionary computation based path planning algorithm (Rubio *et al.*, 2004).

In addition to the above, there are many software (open source and commercial) used for flight planning and control of autonomous vehicles. Some of these are briefly described below.

3.11.1 Open-Source Software

The following open-source software may be used for UAVs data capture and control some (or all) of flight autonomously using pre-programmed coordinates or real-time data (Zheng *et al.*, 2014, Baker 2018, Reudenbach, 2018).

3.11.1.1 ArduPilot

ArduPilot is licensed under GPLv3, and its source code is available on GitHub. ArduPilot founded the Dronecode Project but was split from it in 2016 over disagreements around open-source licensing. ArduPilot claims that it is the most advanced, full-featured, and reliable open-source autopilot software available. It has been installed in a very large number of drones and other UAVs, including airplanes, helicopters, boats, and submarines. Its features include advanced data-logging, analysis, and simulation tools. It is supported by a broad ecosystem of third-party sensors, companion computers, and communication systems.

3.11.1.2 Dronecode

Most UAVs use Linux, and a few use MS Windows. The Linux Foundation launched a project in 2014 called the Dronecode Project. The Dronecode Project is an open-source, collaborative project which brings together existing and future open-source UAV projects under a non-profit structure governed by the Linux Foundation. The project works to build a common open-source platform for UAV development. It has been updated many times, but today it serves as the governance structure for the components of the overall platform. These components include the PX4 autopilot FCS, the MAVLink, and MAVDSK robotics communication toolkit, and the QGroundControl users interface for flight control, mission planning, and configuration; all of which have individual GitHub repositories.

3.11.1.3 DronePan

DronePan is an iOS app (with an Android app) that automates panoramic photography with DJI drones. The DronePan temporarily takes control of the drone/UAV heading and camera angle. It begins shooting 15 to 25 photos automatically with the proper overlap required for an aerial panorama. After completing the panorama, it can fly to other locations to shoot more panoramas. DronePan is licensed under the GPLv3 license, with its source code available on GitHub.

3.11.1.4 Flone

Flone basically turns a smartphone into a drone. It combines a digitally fabricated airframe with software that allows an Android smartphone on the ground to control the one strapped onto the airframe via bluetooth. It is licensed under GPLv3, and its source code resides on GitHub. As the project is based in Spain, most of the documentation and materials are in Spanish.

3.11.1.5 LibrePilot

The LibrePilot software suite is designed to control multicopters and other radio-controlled drones. The project roots lie in the Open Pilot UAV software project, and its goals are to support research & development of software & hardware for vehicle control and stabilization, and UAVs and robotics applications. It has an open and collaborative environment, and the main project is hosted on BitBucket. It is licensed under GPLv3.

3.11.1.6 Mission Planner

It is an open-source software to direct the UAV, schedule flight missions, monitor the state of the aircraft in operation, and generate telemetry records.

3.11.1.7 OpenDroneMap

OpenDroneMap software is designed to run in Linux but can be run with Docker to avoid the need to exact configuration environment the project was built for. The software is an open-source toolkit for processing aerial imagery. The OpenDroneMap can be used by a drone/UAV to capture the overhead image of an area of interest. It is used to convert the simple images into 3D geographic data that can be used in combination with other geographic datasets. It is used for processing the raw drone imagery to other useful products, such as point clouds, DEM, DSM, textured DSM, orthorectified imagery, etc. A beta release is available on GitHub under a GPLv3 license, along with a sample data set.

3.11.1.8 Paparazzi UAV

Paparazzi UAV is a GPLv2 licensed project that combines both the software and hardware needed to build and fly an open-source vehicle under an open license. Its primary focus is autonomous flight, and it's portable, allowing operators to easily take their devices into the field and program their flights across a series of waypoints. Source code and releases of the software components can be found on GitHub, and tutorials for adapting it to off-the-shelf or custom-built hardware can be found on the project's wiki.

3.11.1.9 Pix4Dcapture

It is a mobile app that plans and controls drone flights using mobile phones. Pix4Dcapture supports drones/UAV from DJI, Parrot, and Yuneec; three of the biggest drone manufacturers in the market. The Pix4Dcapture app is available with different missions providing the flexibility required for the projects to: (i) easily define the size of a mission to map areas of all sizes, (ii) customize mapping parameters, like the image overlap, camera angle, and flight altitude according to needs, and (iii) start the mission and monitor it live using the mapview and cameraview (*https://www.pix4d.com/product/pix4dcapture*).

3.11.1.10 PX4

The PX4 is an open-source flight control software for drones and other unmanned vehicles. It provides a flexible set of tools for drone developers to share technologies and to create tailored solutions for drone/UAV applications. The PX4 provides a standard to deliver hardware support and software stack, allowing building and maintaining hardware & software in a scalable way (*https://px4.io/*).

Historically, the PX4 grew from the PIXHAWK project at ETH Zurich, where the PIXHAWK MAV was specifically designed as a research platform for computer vision-based flight control. The software is used by some of the world's most innovative companies across various drone industry applications. The open-source community around the PIXHAWK open autopilot hardware and the PX4 flight stack is the largest industry-backed development community in the drone space today. The PX4 is part of Dronecode, a non-profit organization, administered by the Linux Foundation to foster the use of open-source software on flying vehicles. Dronecode also hosts QGroundControl, MAVLink, and the MAVSDK.

3.11.1.11 RealFlight

It is a drone/UAV flight simulator for learning and practice of flight maneuvers with multiple aircraft models, useful to improve the operators' skills for drone/UAV flights.

3.11.1.12 Unmanned Aerial Vehicle R-based mission planning (uavRmp)

The uavRmp package is designed for autonomous mission planning. It is a simple and open-source planning tool for monitoring flights of low-budget drones based on R. It provides an easy workflow for planning autonomous surveys, including battery-dependent task splitting, saving departures, and each monitoring chunk's approaches. It belongs to the uavR package family that provides more functionality for the pre- and post-processing as well as analysis of the derived data (*https://github.com/gisma/uavRmp*).

3.11.2 Commercial Software

Some of the commercial software used for path planning are briefly explained below.

3.11.2.1 DroneDeploy

There are numerous advantages of using DroneDeploy, which include automating mapping and photo flights for DJI drones using DroneDeploy's mobile app. The DroneDeploy Enterprise features are used to plan automatic flights, maintain compliant flight logs, cloud-based map processing, and storage, share maps and connect with other tools. In addition, it tracks hundreds of pilots and collaborators at a glance *via* the Admin Panel, delivers data in the format, such as JPG, TIFF, OBJ, LAS, SHP and DXF, and builds custom applications. For other details, see Chapter 5.

3.11.2.2 Universal Ground Control Software (UgCS)

The UgCS is leading software for drone surveys and supports a variety of drones/UAV from DJI to custom-built Ardupilot or PX4 and a wide range of mapping, industrial and public safety applications. The UgCS is a 3D mission planning software for plan and fly mission without an internet connection. It creates routes from the KML file, built-in "no-fly zones" for major airports. The UgCS features a mission planner with a Google Earth, like a 3D interface for UAV mission planning, enabling the navigation of the environment more easily. A 3D mission planning environment gives more control allowing viewing the created flight plan from all angles, taking into account any obstacles, such as terrain or buildings (*https://www.ugcs.com/*).

It uses either pre-installed or imports more precise DEM data to increase the accuracy and safety of flight missions with the terrain. The UgCS has an in-build geotagging tool, allowing synchronizing the timestamps of the images with the autopilot's telemetry data recorded during the flight. After matching the geotags with coordinates and correct altitudes, the images can be used for further processing.

3.11.2.3 WingtraPilot

The WingtraPilot is the intuitive drone/UAV software to manage the data acquisition process, prepare flight plans, and monitor and revise missions during flight. After confirming the mission, the WingtraOne takes off, captures aerial images and lands without any human interaction. It has an automatic Return To Home (RTH) function (*https://wingtra.com/mapping-drone-wingtraone/drone-flight-planning-software/*).

3.11.3 Simultaneous Localization and Mapping (SLAM) Algorithm

The SLAM algorithm is a hotspot in robot and computer vision research and is considered as one of the key technologies for automatic navigation in unknown environments (BIS Research, 2018). It is used in computer vision technologies that receive visual data from the physical world with the help of numerous sensors installed in the devices. The SLAM technology converts this data in a different form, making it easier for the machines to understand and interpret the data through visual points (Liu *et al.*, 2012). It generates the map of a UAV's surroundings and locates the vehicle in that map at the same time.

The goal of SLAM is to provide a continual estimate of where a moving autonomous UAV is located along with its posture within the environment, both vertically and horizontally, while also constructing a 3D map as the UAV navigates its environment. The SLAM, which is also used extensively in augmented/virtual reality applications, can use a variety of sensing and location techniques. It also uses the depth sensors, such as LiDAR, GPS, to gather a series of views (such as 3D snapshots of its environment), with approximate position and distance. Based on this estimate, the aerial system can navigate over the terrain without GPS, indoors, and outdoors (Qu *et al.*, 2018). Small UAVs render them suitable for use in urban environments and indoor applications. The uncertainty of GPS signal in such environments leads to navigation problems for UAVs. In the absence of GPS signal, vision-based airborne SLAM techniques have emerged to allow navigation.

There are many similarities between the SLAM and the Structure from Motion (SfM) techniques. Both can estimate the localizations and orientations of a monocular camera and sparse features. Non-linear optimization is widely used in SLAM and SfM algorithms (Kaess *et al.*, 2012). The SLAM is a pure work of deep learning and machine learning that could be used in mining, warfare, autonomous cars, construction sites, after an earthquake, or other major disaster to map the area where it is impossible to go easily. The UAV leverages an AI to plan its path to the goal. The UAV, with the application of AI, refines its approach as it moves through the environment and gathers more information about the obstacles in its way. The SLAM could be used in warehouses, construction sites, etc., and for cleaning tasks done by robotic vacuum cleaners or UAVs. The popularity of SLAM is connected with the need for indoor applications of mobile robotics.

The biggest problem of SLAM is that some algorithms are easily converging to a local minimum which usually produces a wrong estimate. Liu *et al.* (2012) proposed improved SLAM algorithms based on convex relaxation to avoid convergence to local minima for different applications. According to a market research report by BIS Research (2018), the SLAM technology market was US$ 50 million in 2017 and is estimated to reach US$ 8.23 billion by 2027.

3.12 INTELLIGENT FLIGHT CONTROL SYSTEMS

FCSs play a vital role in determining the maneuvering capability of a UAV. Intelligent flight control is defined as autonomous, adaptive control algorithm that can find non-trivial solutions to control problems using trivial strategies. There is a worldwide increase in unmanned aviation, but the integration of unmanned aircraft in the current airspace is difficult due to safety regulations. While conventional commercial aviation autopilots are not adaptive (the human pilot has to act if something unexpected happens), autonomous systems should be able to adapt to changing circumstances themselves. That is where intelligent FCSs are needed.

The intelligent FCS is a next-generation FCS designed to provide increased safety of air vehicles as well as to optimize the air vehicle's performance under normal conditions (Kendoul, 2013). While, there is no universal consensus on how to define or measure an intelligent system, there are several characteristic traits that an intelligent controller might have, including adaptability, learning capability, non-linearity, autonomous symbol interpretation, and goal-oriented and knowledge-based behaviors (Menouar *et al.*, 2017). These intelligent UAVs have to use sophisticated learning and decision-making with the help of machine learning. Reinforcement learning and approximate dynamic programming methods are also based on human learning and the concept of building a value function.

The intelligent flight control design involves classical analyses of aircraft stability, control, flying qualities, and human psychology and physiology. It is responsible for mission-level control of each participating autonomous UAV (Sinsley *et al.*, 2008). Figure 3.15 shows an overview of intelligent controller architecture and its environment. The intelligent controller has two main modules, perception and response. It receives real world data from the sensors and payload inputs. The intelligent controller operates with a control loop where the perception module processes the input data, and the response module acts on the results generated by the perception module. Actions generated by the response module are sent as messages to the UAV system control components. This cycle continues until the program ends.

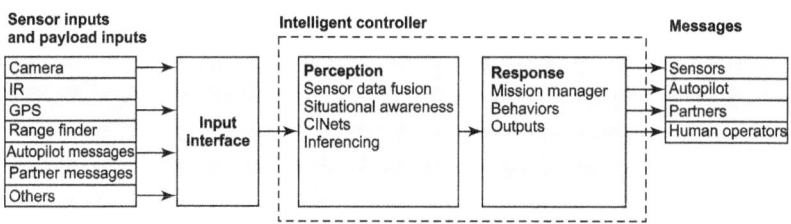

FIGURE 3.15 Overview of an intelligent controller (Sinsley et al., 2008).

All latest UAVs have intelligent flight controllers and modes, such as Follow Me, Active Tracking, Waypoints, RTH, and many others. The Phantom-4 Pro from DJI has the most autonomous intelligent flight modes of any UAV. For example, it has the following intelligent flight modes; Active Track (profile, spotlight, circle), Draw Waypoints, TapFly, Terrain Follow Mode, Tripod Mode, Gesture Mode, S-Mode (Sport), P-Mode (Position), A-Mode (Attitude), Beginner Mode, Course Lock, Home Lock, and Obstacle

Avoidance. Each mode has specific work and utility (Corrigan, 2020). The UAV is calibrated with the location of GPS satellites. If more than 6 GPS satellites are tracked, it allows the UAV to fly in Ready To Fly (RTF) mode. In order to increase flight safety and prevent accidents in restricted areas, the latest UAVs from DJI and other manufacturers include a "no-fly zone" feature. Manufacturers can change the "no-fly zone" UAV technology using UAV firmware updates (Uerkwitz et al., 2016).

The GPS allows the UAV and remote control system to know exactly its flight location. A Home point is also set, and this is the location the UAV will return to, if the UAV and the remote control system stop connecting. This is also known as "fail-safe." The UAV will signal on the remote controller display (Feist, 2020): (i) that enough GNSS satellites have been detected and the UAV is RTF, (ii) the current position and location of the UAV in relation to the operator, and (iii) the Home point for RTH safety feature. Most of the latest UAVs have three types of technology for RTH, as follows (Corrigan, 2020);

(i) Operator's initiated RTH by pressing the button on the remote controller or in an app.

(ii) Low battery level where the UAV will fly back automatically back to Home point.

(iii) Loss of transmission between the UAV and remote controller with the UAV flying back automatically to its Home point.

Many UAVs are now equipped with obstacle detection and collision avoidance system. The vision systems of UAVs use obstacle detection sensors to scan the surroundings, while algorithms produce the images into 3D maps allowing the flight controller to sense and avoid the objects (Jha, 2016). These systems fuse one or more of the following sensors to sense and avoid obstacles; Vision sensor, Ultrasonic, Infrared, LiDAR, Time of Flight (ToF), and Monocular vision. The DJI Mavic-2 Pro and Mavic-2 Zoom UAVs have obstacle sensing on all 6 sides. The DJI Mavic-2 uses both Vision and Infrared sensors fused into a vision system, known as Omni-directional obstacle sensing. It slows down when an obstacle is sensed. It will stop and hover, then fly backward and ascend upwards until no obstacle is sensed. This is known as the *Advanced Pilot Assistance System* on the DJI Mavic-2 and Mavic Air UAVs (Corrigan, 2020). Matthew et al. (2018) presented a good description and summary of intelligent control systems.

3.13 INNOVATIVE TECHNOLOGICAL UAVs/DRONES

Some highly advanced UAVs commercially available are, such as the DJI Mavic-2, Mavic Air, Phantom-4 Pro, Inspire-2, Walkera Voyager-5, DJI Inspire-1, Pro Quadcopter, ALTA Drone, Faucon's Drone, Passport (Hover camera), DJI Mavic Pro, Acecore, and DJI Phantom-4 Pro. The Flirtey company is a UAV-based business that has been working on developing drone delivery services to carry small packages to clinics and hospitals in underdeveloped countries. The specifications of some advanced UAVs are given in Table 3.8.

The DJI drones have a huge demand in professional drone market. The latest DJI drones are getting smaller, smarter and more affordable. The DJI, 3D Robotics (3DR), SenseFly (acquired by Parrot), Yuneec, and Trimble are the top five most penetrated UAV makers among commercial UAV operators in the US (Uerkwitz *et al.*, 2016). Some of the advanced drones are as follows: (Corrigan, 2020, *https://airdronecraze.com/new-drones/*):

ANAFI—It is manufactured by the French company Parrot, designed to meet four essential needs: image quality, flight performance, foldability, and ease of use. It has a 4K HD camera with 2.8x lossless zoom and a 21MP sensor. It is equipped with a 3-axis gimbal stabilizer that has 180 degrees of vertical tilt freedom to look straight down, and straight up. It has a 25 min flight time and a range of up to 2.5 miles. The drone can be charged with the smart battery with a power bank or with the USB Type-C charging system from a smartphone or computer.

DJI Mavic Mini—It is a preferred micro UAV in the market, as it weighs only 249 g (8.78 ounces) and can film in 2.7k (2720 × 1530 at /25/30 pixels). It offers up to 30 min of flight time. It does not have any obstacle avoidance but uses GPS receivers for location and downward visual sensors to detect the ground below. It will automatically "return home" if the drone loses connection to the controller or detects a low battery level.

DJI Mavic-2—It is a foldable drone and is available as a Pro or Zoom model; both have 4k cameras and include collision avoidance on all sides.

DJI Mavic-2 Enterprise (M2E)—It is available with a zoom or thermal camera and includes special accessories, such as a beacon, spotlights, and loudspeaker. It is specifically designed for search and rescue or similar work.

DJI Mavic Air—This is a lightweight, and small-sized professional foldable UAV with a 4k camera, providing a very stable flight. It has collision avoidance technology and face recognition technology. It can also fly using hand gestures.

TABLE 3.8 Specifications of Some Popular Commercial UAVs/Drones (Nassi et al., 2019).

Manufacturer	Name	FPV		Functionality				Characteristics							
		Wi-Fi	Analog	Follow Me	RTH	Automatic Navigation	Smart Capture	Flight Range (km)	Speed (km/h)	Flight Time (m)	Weight (g)	Altitude (km)	Video Resolution	Date	Price (US $)
DJI	Matrice 600Pro		Yes	Yes	Yes	Yes		5	65	32–38	9500	2.5–4.5	4K	2016	4999
	Mavic 2 Pro	Yes	Yes	Yes	Yes	Yes	Yes	8	72	31	907	6	4K	2018	1499
	Spark	Yes		Yes	Yes	Yes	Yes	2	50	16	300	4	1080p	2017	399
	Inspire 2		Yes	Yes	Yes	Yes		7	94	23–27	3440	2.5–5	4K	2016	2999
	Phantom 4 Pro		Yes	Yes	Yes	Yes	Yes	7	72	30	1388		4K	2016	1499
	Mavic Air	Yes	Yes	Yes	Yes	Yes	Yes	2	68.4	21	430	5	4K	2018	799
	Tello	Yes						0.1	28.8	13	80	0.1	720p	2018	99
Parrot	Bebop 2	Yes		Yes	Yes	Yes		2	64	30	525	0.1	1080p	2017	399
	Anafi	Yes		Yes	Yes	Yes		4	53	25	312	4	4K	2018	599
Yuneec	Mantis Q		Yes	Yes	Yes	Yes		1.5	70	33	479		4K	2018	499
	Typhoon 4K		Yes	Yes	Yes	Yes		0.6	30	25	200	0.1	4K	2015	499
	Typhoon H Plus		Yes	Yes	Yes	Yes		1.2	48	30	1995	0.5	4K	2018	1899
Skydio	R1	Yes	Yes	Yes				0.1	40	16	997	0.1	4K	2018	1999
Rabing	Rabing Mini	Yes						0.1	20	10	90	0.1	720p	2017	70
HASA KEE	H1	Yes						0.05	20	7	50	0.05	0.3mp	2018	83
Holy Stone	HS190		Yes					0.05	15	7	25	0.05	0.3mp	2017	35
Tozo	Q2020		Yes					0.045	12	8	50	0.045	No	2017	60

DJI Phantom-4 Pro—It is a multipurpose UAV, and can be used for 4k aerial filming, photography and photogrammetry. It has automatic collision control, which will alert when the drone has come too close to an obstacle. There is also a Sport Mode which will make the drone fly at a super-fast speed of over 45 mph. This drone also has ActiveTrack technology, enabling the drone pilot to select another moving object, for example, a car, cyclist, or another drone, and the Phantom 4 will actually follow it without assistance from a tracker chip or beam.

DJI Inspire-2—It is a multipurpose UAV for professional 5k aerial filming, photography, photogrammetry, multispectral and thermal imaging.

DJI Matrice 600—This multirotor UAV is a true aerial cinematography platform with options to mount 7 different cameras.

DJI Matrice 200—This quadcopter has 32-min flight time, with onboard FPV camera. Two more cameras, for example, thermal and zoom camera, can be mounted simultaneously. Another camera can be mounted on top of UAV, which makes surveying bridges real easy. It provides centimeter-level positioning accuracy and is resistant to magnetic interference. It is sensitive to 6 directions of collision avoidance using ToF laser, ultrasonic, and vision sensors.

DIDION—It has an inflatable structure, and the perforation-resistant fabric makes it rugged and easy to fold for transport. It is fully waterproof and can land & take-off on water due to the compressed gas of the inflatable structure. It can be fitted with a wide variety of sensors for search & rescue missions, industrial inspection, and military missions.

3DR Solo—It is developed by 3DRobotics for complete camera control. This quadcopter is the world's first "smart drone." It can actually isolate the camera movements from the actual movement of the drone itself. The 3DR Solo drone will also allow streaming live video directly to a mobile device through the use of a remote controller.

EVO—It is manufactured by Autel Robotics, a foldable quadcopter packed full of intelligent flight features and equipped with a 4K camera. It has intelligent flight modes and obstacle avoidance due to two front & two downward cameras (computer vision), and rear (IR sensor) for backward obstacle avoidance during selfie shots. The 3D software of EVO creates a 3D environment and reacts to obstacles that are on the way, and the quadcopter will actually go around them.

Yaneec Mantis Q—It is a foldable compact drone that can be used with some simple voice commands, including take-off, take a picture, record video, wake up, return home, and take a selfie. In improved facial recognition and gesture mode, it enables you to take your picture within a range of 13 feet. Down-facing dual sonar sensors and IR detection make it safe enough to fly indoors and outdoors.

Yuneec Typhoon-H Pro—It uses "realsense" collision avoidance technology. It has CGO3+ 4K-resolution, which is very useful for professional aerial photography and filming. It can be used for longer flight duration. The camera can take vivid 12 MP snapshots in any 360-degree direction. It will stream live video directly to the mobile device through the use of a remote controller.

Walkera Voyager-5—This UAV includes camera options with 30x optical zoom, TIR, and a low-light night vision camera.

Walkera Vitus Starlight—It is a small-sized UAV featuring collision avoidance sensors and a low-light night vision camera.

Most UAVs/drones currently are used in intelligence operations, such as surveillance or reconnaissance missions. Besides strike missions, drones have also been used for electronic attack, and to take down enemy air defense stations, networks nodes, and communication relays. Drones are also useful for scoping out regions where civilians may be in need of rescue. Some of the latest drones include RQ11 Raven, AeroVironment Wasp III, AeroVironment RQ-20 Puma, RQ-16 T-Hawk, Predators, MQ-1C Grey Eagle, MQ-9 Reaper, RQ-7 Shadow, and RQ-4 Global Hawk. There have been swift advances in the latest drone technology in the military. This means that military drones manufactured for the battlefield can become much smaller, go higher, and faster. They will be increasingly used due to decreasing costs and less space required for storage. They can also fly a longer length of distance, and thus helpful when the target is several km away. These types of drones will have a size ranging from one lb. to over 40,000 thousand lbs. and can cost from US $1000s to $1 million or more (*https:// thewiredshopper.com/top-5-latest-technology-drone/*).

3.14 COST OF UAVS/DRONES

There are multi-rotors, fixed-wing, and hybrids UAVs; the last one is a mix of the first two configurations. All of them allow users to carry various sensors to inspect structures or map/survey certain areas. Typically, managing operation is a manual process that goes from setting up the UAV for each mission to changing and managing the batteries all the way to processing the raw data. Semi-automated solutions for data processing require powerful hardware–software or equivalent cloud solutions.

The perfect platform fulfills all specific mission requirements according to flight time, payload capacity, range, communication, deployment time, weather resistance, and ease of use. All of this adds essential value to the utility of an individual sensor. Typical digital cameras are available on the market starting from US$ 500. On the other hand, laser scanners are available at a much higher price. Although the most basic laser scanners are available for approximately US$ 8,000, UAV-mountable and mission-relevant laser scanner systems (including IMU and GPS unit) cost more than US$ 25,000. Figure 3.16 shows the range of cots of various UAVs (in US$). These costs keep on varying depending upon the advancement of hardware and software technology.

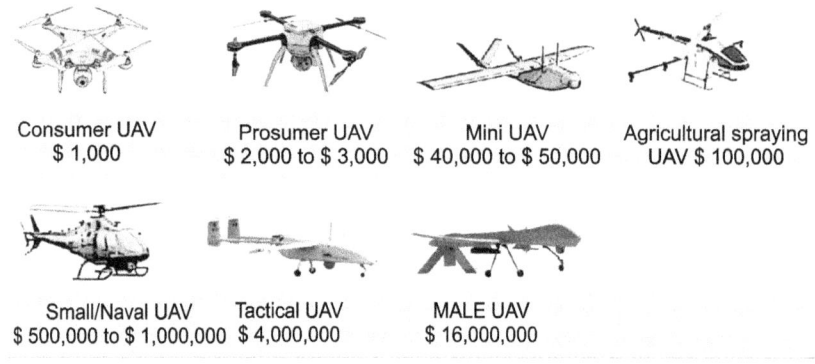

FIGURE 3.16 Range of cost for various UAVs (Goldman Sachs, 2016).

3.15 SUMMARY

The autonomous vehicle can fly in an autonomous way and operates in a wide range of missions and emergencies that can be controlled from a ground base station. These UAVs are able to make decisions and react to events without direct intervention by humans. There are some fundamental aspects that are common to all autonomous vehicles. These aspects include sensing and perceiving the environment, analyzing the sensed information, communicating, planning and decision making, and acting using control algorithms and actuators. The flight planning of autonomous UAVs addresses the air vehicle, mission planning, and control, several types of mission payloads, data links, including their interaction with mission performance and launch & recovery concepts. Although some UAVs are able to perform in increasingly complex environments, most UAVs are not fully autonomous; instead, they are mostly operated remotely by humans. To make UAVs fully autonomous, many technological and algorithmic developments are still needed. For instance, UAVs will need to improve their sensing of obstacles and subsequent avoidance.

There are several path planning and motion planning algorithms that could be used for autonomous UAVs, but these algorithms have various needs for specific applications. Path planning can be simplified by navigating a UAV to a distance point safely to avoid collisions along the way. The ability to plan collision-free paths in complex environments is an important element of UAV autonomy. Geofencing and collision avoidance technologies will further boost the applications of autonomous UAVs. The mission and payload controls are the main bottlenecks that may prevent autonomous UAV development in civilian applications. An important task for autonomous UAVs involves the detection of vehicles and other objects on the ground. Fuzzy logic has been used for the autonomous control of UAVs. The research community is interested in conceiving and applying new algorithms to control or navigate these vehicles autonomously.

A current area of research on the operation of autonomous aerial vehicles is their safe landing and recovery. As most UAV navigation methods rely on GPS signals, many UAVs cannot land properly in the absence of such signals. With the use of vision and image recognition technology, the position and posture of the UAV in 3D can be estimated. As the need for future autonomous flight grows, and systems mature, a high-level framework of design to integrate autonomous systems into existing and new vehicles is needed. Work continues to be performed in developing control capabilities and sophisticated algorithms required to enable autonomous flight. Autonomous UAV is still in experimental stages at the moment. Also, a shortage of skilled onsite crew member is a bigger problem.

REVIEW QUESTIONS

Q.1. Explain the term "automatic" and "autonomous" UAVs with reference to their basic differences. What are the benefits and applications of autonomous UAVs?

Q.2. With the help of a diagram, discuss in brief the architecture of an autonomous UAV.

Q.3. Discuss various methods to quantify the autonomous level of a UAV.

Q.4. Explain why simulation of mission planning of UAV is important?

Q.5. What do you understand by path planning of a UAV? What is involved in path planning?

Q.6. Explain flight planning of a UAV mission. How do you control the entire mission?

Q.7. Discuss in brief the basic flight safety operations to be considered before the flight of a UAV.

Q.8. What is the importance of collision avoidance technology used in autonomous UAVs?

Q.9. Describe the role of flying object detection technology in autonomous UAVs.

Q.10. Explain the utility of path planning algorithms for autonomous UAVs. List some open-source software available for path planning.

Q.11. Describe the role of the SLAM algorithm in autonomous operations of a UAV.

Q.12. Discuss the role and function of an intelligent flight control system.

Q.13. What do you understand by Smart drones/UAVs? Describe in brief four of them.

CHAPTER 4

COMMUNICATION INFRASTRUCTURE OF UAVS

4.1 INTRODUCTION

Communication is the exchange of commands and data between UAV's onboard systems and ground systems. The most critic subsystem in UAVs is communication (Zeng *et al.*, 2019). Although UAVs are manufactured by many companies, information/data link security problem still continues. Besides data transmission over long distances, it is important to transmit accurate information with a high transfer data rate at the right point and without any outside interference.

The GCS unit controls the UAV and payloads. Data link uses an RF transmission to transmit and receive information to and from the UAV. These transmissions can include location, remaining flight time, distance and location to target, distance to the pilot, location of the pilot, payload information, airspeed, altitude, and many other parameters (Al-Mousa *et al.*, 2019). This data link can also transmit live video/images/data from the UAV back to the GCS so the operator on the ground can observe what the UAV camera is seeing. The data link portion of the UAV platform also happens to be the most vulnerable in detection and counter-measures.

In order to have an effective UAVs-based system for several applications, it is essential to have a capability to communicate efficiently among each other (UAV-to-UAV) and existing on-ground infrastructure networks and the Internet (Jawhar *et al.*, 2017). The UAVs operating individually and flying in uncontrolled space would not require a high data rate since only limited communication with the main control station would be needed. The UAVs operating in emergency situations and deployed in heavier air traffic areas would require constant and frequent communication with the control stations in order to coordinate flights and avoid possible collisions with the other aircraft. The communication messages in such an environment are usually small, requiring a limited amount of bandwidth (Gupta et al., 2016). However, due to the real-time nature of the communication, the reliability and delay requirements, in this case, would be stringent. In other applications, the needed communication bandwidth can range from low rates in simple sensor readings to considerably higher rates (several Mbps) in the case of high-quality images or videos. In the latter case, some UAV platforms use LOS microwave links with higher data rates.

The swarm, a set of aerial robots, can perform a collective task for specific applications from interactions among UAVs and between UAVs and their components. Effective and reliable communication is the foundation for swarm-based applications (Campion et al., 2019). There are other communication technologies; however, internet of things (IoT) is the most important. The IoT-based sensors system in coordination with UAV are expected to boost the applications in many areas where real-time data/information is necessary for decision making or deriving action plans (Sharma and Garg, 2019).

4.2 UAV COMMUNICATION SYSTEM

In communication link, two components are of prime importance; UAV and GCS. In general, a UAV communication link can send both control commands from the GCS to the UAV and receives data about the flight on the downlink (Zeng et al., 2019). A bidirectional link can be established in order to provide communication between the UAV and GCS, as shown in Figure 4.1. A communication link between these two components provides long-range operations and a continuous, stable, reliable, and faster data rate.

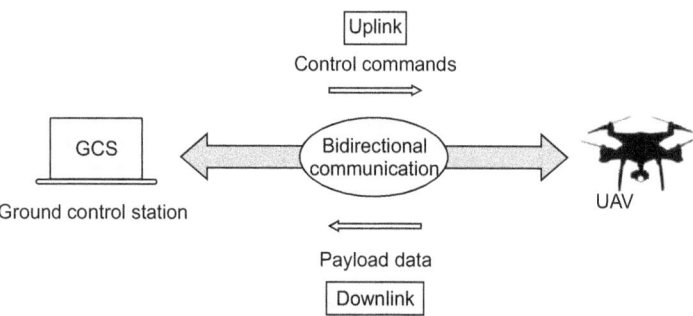

FIGURE 4.1 The UAV communication link.

The components of the UAV communication system are shown in Figure 4.2. The main component is the micro-controller or flight controller, which is the core for all the functions of a UAV. It manages failsafe, autopilot, waypoints, autonomous, and many other functions. This micro-controller interprets the input received from the receiver, GPS, and IMU, including an accelerometer and a gyroscope, as well as onboard sensors (Azevedo et al., 2019). The GCS provides the relevant data about the UAV, such as speed, attitude, altitude, location, yaw, pitch, roll, warnings, and other information.

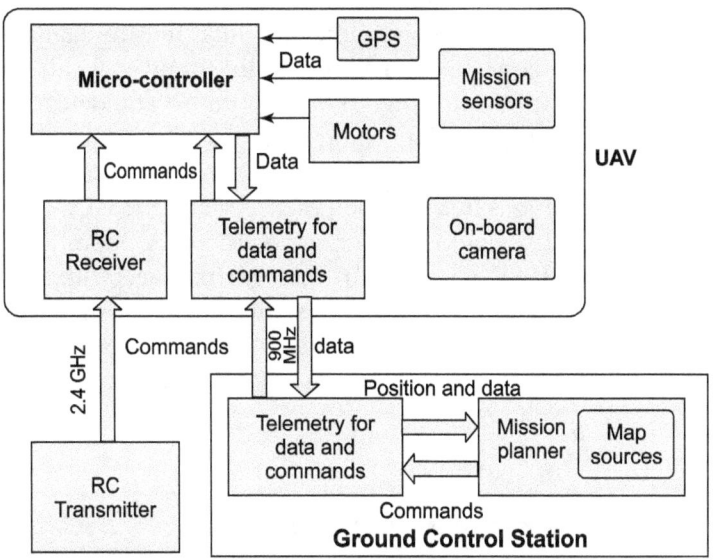

FIGURE 4.2 Components of UAV communication system (Atoev et al., 2018).

A UAV has two links: data link and communication link. The data link that operates in the frequency range from 150 MHz to 1.5 GHz ensures the transmission of data between the UAV and GCS (Atoev et al., 2018). The communication link uses a 2.4 GHz frequency between the transmitter and receiver to control the UAV. The transmitter and receiver both must work on the same frequency. The UAVs have exclusive use of their own frequency allocation due to the longer range and potentially worse consequences of radio interference. Initially, UAVs scan the range of frequencies within the 2.4 GHz band and use only the narrowband frequency that is not in use by another UAV. As a result, many UAVs can utilize a 2.4 GHz frequency band simultaneously. This feature is very useful when a number of UAVs are used as flying base stations in wireless cellular networks to serve an arbitrarily located set of users. Moreover, typical UAVs use multiple radio interfaces to maintain a continuous connection with essential links to GCSs, other UAVs, and satellite relays (Chandrasekharan, 2016).

There are various frequencies used in the data link system. The frequencies are used based on UAV types as well as the functionality of the UAV. Each radio frequency has its common use based on its capability of range and penetration. The lower the frequency, the greater the penetration into obstacles. In other words, the longer the wavelength, the lower will be the frequency, and the longer will be the penetration range.

In a UAV, several communication links may be available, for instance, RF-links, Satcom links, or wireless links; however, all links may not be available at the same time (Petkovic and Narandzic, 2019). Moreover, the cost of using each link could vary. Therefore, a mix of communication mechanisms (radio modems, satcom, microwave links, etc.) is required to have a continuous link between the UAV and the GCS. The adopted communication link would also depend on the UAV operation range (Johnson, 2015). The UAV missions are categorized according to their distance from the GCS into LOS missions where control signals can be sent and received *via* direct radio waves, and BVLOS missions where the UAV is controlled *via* satellite communications or a relaying aircraft which can be a UAV itself.

4.3   TYPES OF COMMUNICATION

There are four main types of UAV communication services: (i) UAV-to-UAV for data and control links; (ii) UAV-to-GCS for control and commands link; (iii) UAV–to-Ground wireless nodes for UAV-aided data dissemination and collection; and (iv) UAV-to-Satellite system (Zeng *et al.*, 2016). Figure 4.3 shows the wireless communication architecture for UAV networking. The UAV wireless communication architecture consists of two basic types of communication links; the control link, which is referred to as data link and Control and Non-Payload Communications (CNPC) link (Chandrasekharan, 2016).

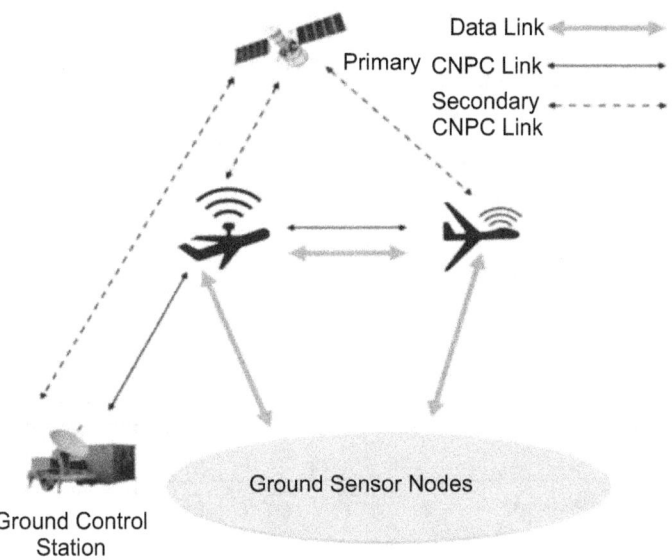

FIGURE 4.3 Types of UAV communication.

In general, CNPC is used for UAV safety and collision avoidance approach of sense-and-avoid information among UAVs. CNCP provides more reliable and accurate information for UAVs operations (Sawalmeh and Othman, 2018). In the CNCP link, control data are exchanged in the communication services of UAV-to-UAV, UAV-to-GCS, and GCS-to-UAVs. For safety purposes, a low data rate is required for the exchange of control data in the control link. There are two types of CNCP available: the primary CNPC link, the preferred control link, and the secondary CNPC link. The former link can be used *via* satellite as a backup link to enhance reliability and robustness. The primary CNPC link is established directly during take-off and landing. On the other hand, the secondary CNPC link can be established *via* satellite when the UAV is in operation (Zeng et al., 2016). The UAV data link connections, types, and basics characteristics are summarized in Table 4.1.

TABLE 4.1 Data-links of UAV System (Dimc and Magister, 2006).

		Data Link					
		Function	Connection	Direction	Application Type	Rate Requirement	Operating Frequency
Communication stream to & from	The UAV	Flight Control	UCS ➞ Autopilot	Up	TC	Low < 30 Kb/s	HF, VHF/UHF
		Task-Payload Control	UCS; ➞ Payload	Up	TC	Low < 30 Kb/s	HF, VHF/UHF
		Flight Status	Autopilot ➞ UCS	Down	TM	High < 1Mb/s	VHF/UHF
		Task-Payload Data	Payload ➞ UCS	Down	TM, TV	Broadband > 1Mb/s	L, S, C, X, Ku
		UAV Flight Progress Reporting	UAV9 ➞ Traffic	Two-way	TM	High < 1Mb/s	VHF/UHF, L, S
			UAV ↔ ATC	Down	TM	High < 1Mb/s	VHF/UHF, L, S
	The UAV Operator	Traffic Coordination	UAV Operator ↔ ATC	Two-way	WM, Voice	Low < 30 Kb/s	HF, VHF/UHF
		Mission Task Coordination	UAV Operator ↔ Command Post	Two-way	WM, Voice	Low < 30 Kb/s	HF, VHF/UHF, L,S, C, X, Ku

Note: **TC** Tele-Command; **TM** Telemetry; **TV** Television (Visual & IR Spectrum); **WM** Written Message
HF 1-30 MHz; **VHF/UHF** 30-1000 MHz; **L-, S-Band** 1-2 GHz; **C-Band** 5 GHz; **X-Band** 10 GHz; **Ku-Band** 15 GHz.

The two of most important subjects of a UAV communication system are data link protection and data security. The provision of sufficient communication exchange between the UAV and its control stations is influenced by environmental or self-implied conditions and hostile jamming (Dimc and Magister, 2006). The UAV uplink and downlink communication streams must be protected against intentional counteractive actions. Typical threats to UAV's data–links are: (a) eavesdropping, (b) information corruption, and (c) masquerade (Kaushal and Kaddoum, 2017). Counter-measures against UAV data link threats are the strict procedures to access the UAV data link and authentication procedures between all communicating entities while data are coded using comprehensive cryptic algorithms. Direct attacks on

UAVs, such as signal spoofing and identity hacking, can cause serious damage to the system. Attacks on GCSs are more fatal than others because they issue commands to actual devices and collect all the data from the UAVs they control (Mansfield et al., 2013). Such attacks normally involve malwares, viruses, and/or keyloggers.

As commercial and military UAVs are increasing in the airspace, the current bandwidth management policies limit the effective use of UAV systems. There is an increasing need for the development of novel techniques to increase the bandwidth throughout for command & control (C2) communication systems and data link signals. Frequency-hopping RF, non-RF, or bandwidth-efficient modulation methods are considered as novel approaches for future UAV-related projects. Network architectures, including protocols and interfaces, must be flexible enough to allow interoperability. To efficiently incorporate the use of UAVs into national network-centric warfare capabilities, the network architectures and related technologies must be redesigned (Mozaffari et al., 2018).

Use of UAV-based aerial networks is a promising solution to enable fast, flexible, and reliable wireless communications. Since UAVs do not require highly constrained and expensive infrastructure (e.g., cables), they can easily fly and dynamically change their positions to provide on-demand communications to critical locations or people in emergent situations. For instance, UAVs can be deployed as mobile aerial base stations in order to deliver broadband connectivity to areas where terrestrial wireless infrastructure is damaged due to disaster activity. The flying UAVs can continuously provide full coverage to a given area within the shortest possible time. Therefore, the use of UAV mounted base stations can also be considered as an appropriate solution for providing fast and ubiquitous connectivity in public safety scenarios (Mozaffari et al., 2018).

In the telecommunications field, one of the most promising applications of UAVs is to use them as support equipment, aimed at extending the capacity or coverage of wireless systems through the deployment of an aerial communication network. Their low power consumption and the high number of connectivity elements that they support (communication chips, antennas, etc.) are among the most valuable features to perform this task (Guillen-Perez et al., 2016). Thus, these devices are able to deploy a wireless network acting as network nodes within the system architecture and allowing the end-users/things to gain connectivity through them. This type of UAV application is known as UAV-assisted cellular communications. Such networks are still in their early stages for supporting communication and sensing applications.

4.4 WIRELESS SENSOR NETWORK (WSN) SYSTEM

An essential enabling technology of UAV is wireless communication. On the one hand, UAVs need to exchange critical information with various groups, such as remote pilots, nearby aerial vehicles, and air traffic controllers, to ensure safe, reliable, and efficient flight operation through CNPC (Martinez-Caro and Cano, 2019). On the other hand, depending on the mission, UAVs may have to transmit and/or receive mission-related data on time, such as aerial images, high-speed video, and data packets for relaying, to/from various ground entities, such as UAV operators, end-users, or ground gateways. This is known as payload communication (Gu et al., 2015). Compared to CNPC, UAV payload communication usually has a much higher data rate requirement. For instance, to support the transmission of "full high-definition" video from the UAV to the ground user, the transmission rate is about several Mbps, while for 4K video, it is higher than 30 Mbps. The rate requirement for UAV serving as aerial communication platforms can even be higher (Zeng et al., 2019).

To meet both the CNPC and payload communication requirements in several UAV applications, proper wireless technologies are needed to achieve seamless connectivity and high reliability throughout for both air-to-air and air-to-ground wireless communications in 3D environments. The main advantages of wireless communication over terrestrial, satellite, or HAP include on-demand deployment, fast response, low cost, more flexible in reconfiguration and movement, and short-distance LOS communication (Jawhar et al., 2014). There are four main communication technologies: (i) Direct link, (ii) Satellite, (iii) Ad hoc network, and (iv) Cellular network. These are listed and compared in Table 4.2 with reference to their advantages as well as disadvantages.

TABLE 4.2 Comparison of Wireless Technologies for UAV Communication (Zeng et al., 2019).

Technology	Description	Advantages	Disadvantages
Direct link	Direct point-to-point communication with ground node	Simple, low cost	Limited range, low data rate, vulnerable to interference, non-scalable
Satellite	Communication and Internet access *via* satellite	Global coverage	Costly, heavy/bulky/energy consuming communication equipment, high latency, large signal attenuation

Technology	Description	Advantages	Disadvantages
Ad hoc network	Dynamically self-organizing and infrastructure-free network	Robust and adaptable, support for high mobility	Costly, low spectrum efficiency, intermittent connectivity, complex routing protocol
Cellular network	Enabling UAV communications by using cellular infrastructure and technologies	Almost ubiquitous accessibility, cost-effective, superior performance, and scalability	Unavailable in remote areas, potential interference with terrestrial communications

When UAVs are used as flying, aerial base stations, they can support the connectivity of existing terrestrial wireless networks, such as cellular and broadband networks. Compared to conventional, terrestrial base stations, the advantage of using UAVs as flying base stations is their ability to adjust their altitude, avoid obstacles, and enhance the likelihood of establishing LOS communication links to ground users. The UAV base stations can effectively complement the existing cellular systems by providing additional capacity to hotspot areas and delivering network coverage in rural areas that are otherwise difficult to reach. In regions or countries were building a complete cellular infrastructure is expensive, deploying UAVs becomes highly beneficial as it removes the need for expensive towers and infrastructure deployment (Motlagh et al., 2016).

The use of UAVs is rapidly growing in wireless systems. Conventional UAV-centric research has typically focused on issues of navigation, control, and autonomy. The recent advances in technology make it possible to widely deploy UAVs/drones, small aircraft, balloons, and airships for wireless communication purposes. UAVs can be used as aerial base stations to enhance the coverage, capacity, reliability, and energy efficiency of wireless networks (Liu et al., 2018) and operate as flying mobile terminals within a cellular network. Such cellular-connected UAVs enable several applications ranging from real-time video streaming to goods delivery, but it requires a number of technical challenges to use them effectively. For instance, while using a UAV base station (BS), the key design considerations include performance characterization, optimal 3D deployment of UAVs, wireless and computational resource allocation, flight time & trajectory optimization, and network planning. Due to their inherent attributes, such as mobility, flexibility, and adaptive altitude, UAV BSs can effectively complement existing cellular systems by providing additional capacity to hotspot areas.

Figure 4.4 (Salah-Ddine *et al.*, 2012) shows an integrated WSN architecture (i.e., a WSN integrated with external networks) capturing architecture-level optimizations. Sensor nodes are distributed in a sensor field to observe a phenomenon of interest (i.e., environment, vehicle, object, etc.). Sensor nodes in the sensor field form an ad hoc wireless network and transmit the sensed information (data or statistics) gathered via attached sensors about the observed phenomenon to a BS or sink node. The sink node relays the collected data to the remote users *via* an arbitrary computer communication network, such as a gateway and associated communication network. Since different applications require different communication network infrastructures to efficiently transfer sensed data, the WSN can optimize the communication architecture by determining the appropriate topology (number and distribution of sensors within the WSN) and communication infrastructure (e.g., gateway nodes) to meet the application's requirements.

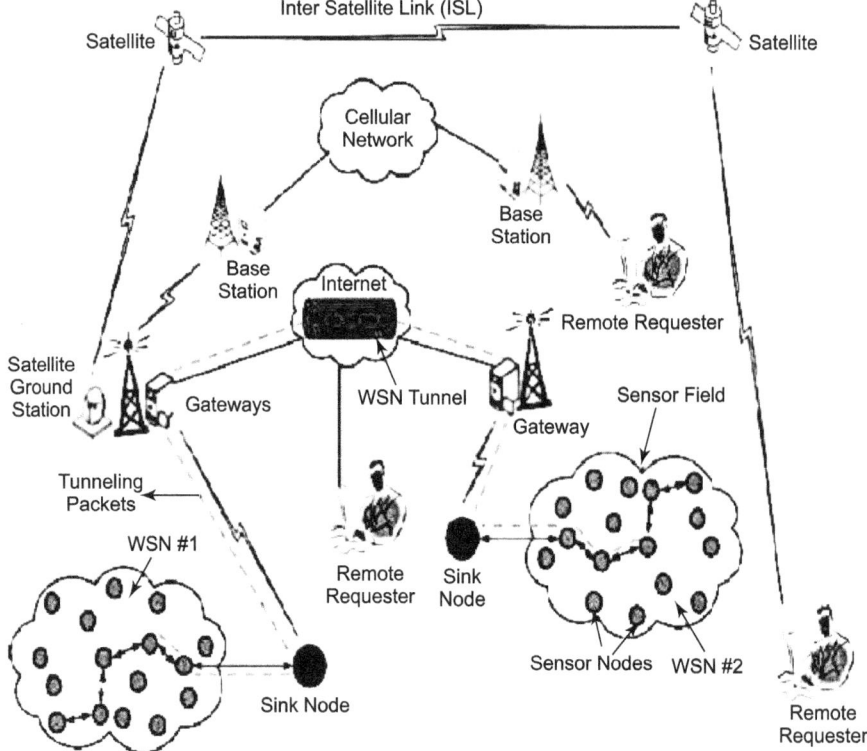

FIGURE 4.4 An architecture of UAV-WSN (Salah-Ddine *et al.*, 2012).

Networks distributed over large spatial areas can be expensive and time consuming to maintain. Recently, there have been significant efforts made to improve incorporating UAVs with WSN to both provide power and also to collect data. The spatial and temporal flexibility of wireless sensors coupled with their computational capabilities are ideal attributes for utilizing networks of wireless sensors with UAVs. Data collection and sensor node integration by UAV makes using widely distributed sensor networks more useful in many applications (Mozaffari et al., 2018).

In particular, key challenges and requirements here would be to design UAV-assisted IoT networks. In recent years, the IoT has made remarkable progress, and is regarded as the most promising technology for many applications (Sharma and Garg, 2019). However, there are no BSs and Wi-Fi stations in many areas, which prevents the applications of the IoT. That means the data acquired through the WSN cannot be transmitted using wireless communications. An alternative solution is to employ UAV to communicate with the WSN in large areas to get real-time data for processing and analysis (Sylvester, 2018).

One of the important applications of UAVs is in IoT (Zeng et al., 2016) whose devices often have small transmit power and may not be able to communicate over a long range. The UAVs can also serve as wireless relays for improving connectivity and coverage of ground wireless devices. In addition, they can also be used for surveillance scenarios or smart city applications, which are the key applications for the IoT. Wireless sensors are a critical component of state-of-the-art structural health and infrastructure monitoring systems. These methods can be extended to UAV-based methods for crack detection and quantification in structures (Mozaffari et al., 2019).

The UAVs can be used to provide wireless coverage during emergency cases where each UAV serves as an aerial wireless BS when the cellular network goes down (Zhang and Duan, 2019). They can also be used to supplement the ground BS in order to provide better coverage and higher data rates for users (Gu et al., 2015). UAVs can be used as gateway nodes in remote geographic or disaster-stricken areas to provide connectivity to backbone networks, communication infrastructure, or the Internet. Figure 4.5a presents two examples of UAV-aided ubiquitous coverage within the serving area; (i) rapid service recovery after infrastructure failure and (ii) BS offloading at hotspot when the cellular network service is not available (Shakhatreh et al., 2018). While Figure 4.5b shows that UAVs can be utilized as relay nodes to provide wireless connectivity between two or more distant users without a reliable direct communication link.

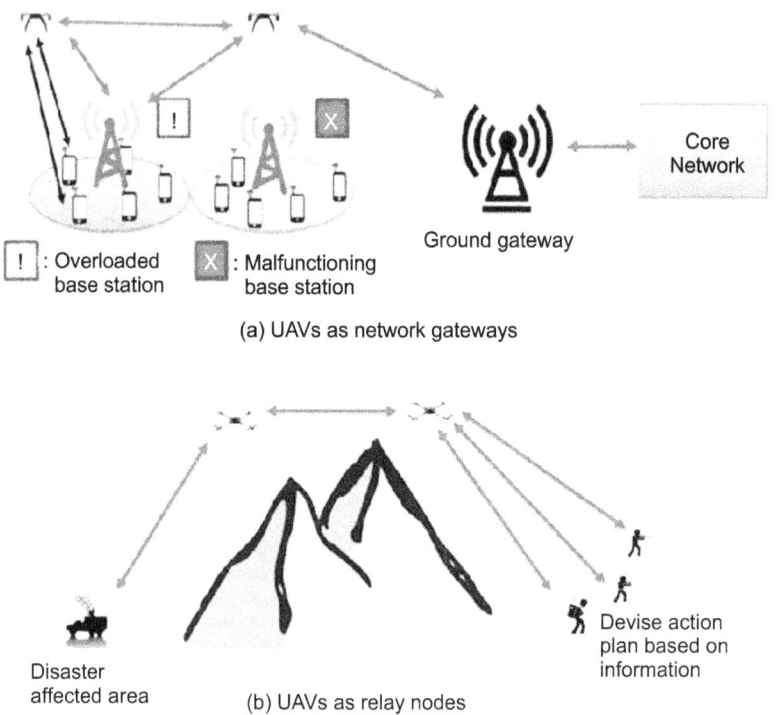

FIGURE 4.5 Example of UAV-aided wireless communications.

4.5 FREE SPACE OPTICAL (FSO) APPROACH

The FSO technology can be used to provide wireless connectivity to remote places by utilizing UAVs, where physical access to 3G or 4G network is either minimal or not present. The FSO represents the optical wireless signal transmission from the IR band spectrum in outdoor environments (Kaushal and Kaddoum, 2017). It is a communications technology in which information is transmitted through the atmosphere on either modulated laser beams or modulated LED beams. Optical wireless communication constitutes a key technology for 5G wireless networks. It alleviates the RF spectrum crunch problem by enabling communications in the visible, infrared, and ultraviolet optical frequency bands for a variety of indoor and outdoor applications. The RF-based UAVs as aerial transceivers can cause interference to the existing terrestrial wireless networks (Safi *et al.*, 2019).

Figure 4.6 illustrates the concept to avoid radio interference and obtain a high data rate of transmission. The Figure proposes to establish FSO-based fronthaul/backhaul links, employing UAVs equipped with optical transceivers, as a promising approach for 5G and beyond wireless networks. Outdoor infrared optical wireless communication systems are widely referred to as FSO communications that rely on line-of-sight transmissions of narrow laser beams. The FSO technology is license-free, easy-to-deploy, cost-effective, and capable of delivering very high data rates. The FSO-UAV-based solutions are proposed to provide efficient backhauling/fronthauling to small cell base stations (SBSs). In this context, UAVs transport the data traffic between the SBSs and the core network through highly directive FSO links. In Figure 4.6, the proposed FSO-UAV framework complements the terrestrial wireless and wired solutions while delivering very high data rates (Petkovic and Narandzic, 2019).

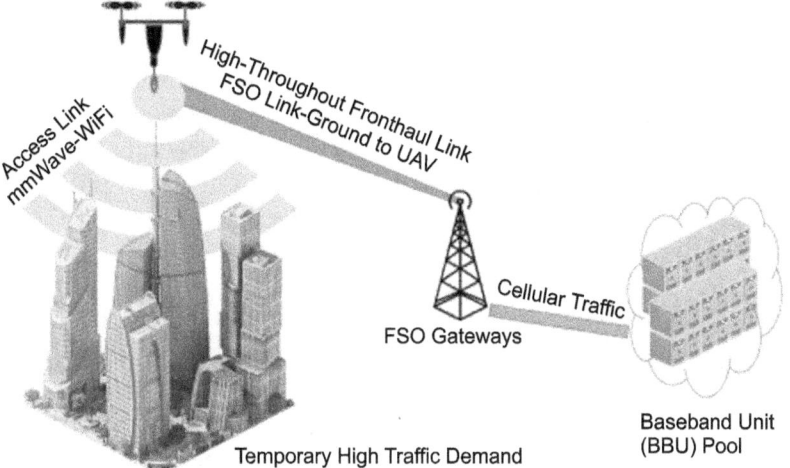

FIGURE 4.6 Ground-to-UAV FSO fronthaul link for 5G and beyond wireless networks (Safi *et al.*, 2019).

There are various applications of FSO; one of these is the high bandwidth communications in the UAV scenarios. The UAVs may require many sensors for different applications, like surveillance for a military mission (e.g., battlefield, observation behind the enemy lines) or a civil mission (e.g., monitoring of a traffic jam, disaster relief, or to broadcast critical data, for example, at some sport events) (Leitgeb *et al.*, 2007). Hence, a lot of data is produced, which has to be delivered to either another UAV or a ground station. Because of the high data rate and the avoidance of monitoring and jamming of the optical signal, the FSO link is more advantageous than an RF-link. The FSO technology over a UAV can be utilized in armed forces, where military wireless communications demand secure transmission of information in the battlefield. Remote sensing UAVs can utilize this technology to disseminate a large amount of images and videos to the fighting forces, mostly in real-time. Facebook plans to provide wireless connectivity *via* FSO links to remote areas by utilizing solar-powered high-altitude UAVs (Kaushal and Kaddoum, 2017).

For areas where deployment of UAVs is impractical or uneconomical, geostationary earth orbit and low earth orbit satellites can be utilized to provide wireless connectivity to the ground users using the FSO links. Establishing a FSO link to a flying vehicle is challenging because of the additional degrees of freedom that have to be considered. Before transmitting data via FSO link, the link itself has to be set up (Leitgeb *et al.*, 2007). The main challenge facing UAV-FSO technology is the high blockage probability of the vertical FSO link due to weather conditions. A disadvantage of the FSO technique is the alignment of two FSO-systems and the effects caused by the atmosphere. Therefore, the link budget calculation (including all the noise sources) is important to apply the right power for the transmitter laser. Furthermore, a specific wavelength is necessary to keep the link budget low (Petkovic and Narandzic, 2019). Millimeter-wave (mmWave) technology can be an alternative to provide high data rate wireless communications for UAV networks (Xiao *et al.*, 2016).

4.6   THE FPV APPROACH

Modern UAVs provide First Person View (FPV) technology to send a live video stream from the UAV's video camera to the pilot (operator) via a GCS on the ground (dedicated controller, smartphone, VR glasses, smartwatch). The technology consists of a video camera mounted on the UAV, which broadcasts the live video to the operator on the ground (Corrigan, 2020). The FPV technology uses a radio signal to transmit and receive the live video. The UAV has a multi-band wireless FPV transmitter built-in along with an antenna. Depending on the UAV, the receiver of the live video signals can be a remote control unit, a computer, tablet, or smartphone device. This live video feed is related to the strength of the signal between the ground controls on the UAV. The ground operator flies the UAV as if he/she was onboard the aircraft. The FPV allows the UAV to fly much higher from looking at the air vehicle from the ground. It also allows more precise flying, especially around obstacles. The FPV allows UAV to fly very easily indoors, or through forests, and around high-rise dense buildings (Yu *et al.*, 2016).

The FPV channel also allows an operator to control a UAV using a GCS. A typical FPV channel consists of an uplink and a downlink, as shown in Figure 4.7. A video downlink is used for video streaming using the data captured by the UAV's camera and sent to the pilot's GCS screen. The video streaming process usually consists of digitizing the captured picture to binary representation by a Complementary Metal Oxide Semiconductor sensor, followed by video compression, encryption, and modulation in real-time (Nassi *et al.*, 2019). The uplink process usually consists of digitizing the joystick's movements (or smartphone) to binary commands, which is followed by encryption and modulation. By its nature, the amount of data that is sent over a video downlink is much greater than that sent by a remote control downlink.

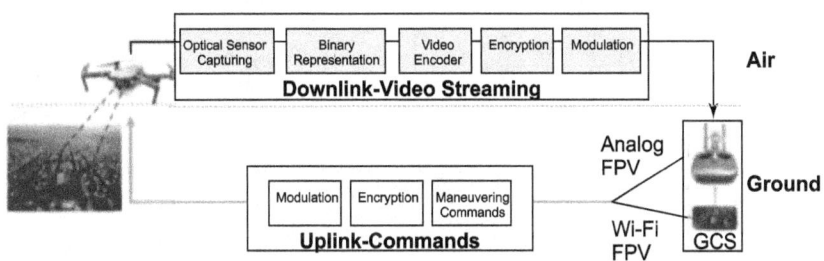

FIGURE 4.7 The FPV channel- downlink and uplink (Nassi *et al.*, 2019).

The latest DJI Mavic-2 has an FPV live video range of 5 miles (8 km) with a 1080 pixels quality video transmission. Other drones, such as the DJI Mavic and Phantom-4 Pro, can transmit live video up to 4.3 miles (7 km). The Phantom-4 Pro and Inspire-2 use the latest DJI Lightbridge-2 transmission system (Vergouw *et al.*, 2016). Drones, such as the DJI Mavic, use integrated controllers and intelligent algorithms to set a new standard for wireless high definition image transmission by lowering latency and increasing maximum range and reliability. Live video and maximizing the range of the transmission is fascinating drone technology.

4.7 CYBERSECURITY APPROACH

The main concerns of utilizing UAVs for data gathering and wireless delivery are cyber liability and hacking (Vattapparamban et al., 2016). The secure operation of a UAV also ensures the protection of the UAV system against cyber-physical threats resulting from intentional or unintentional actions. Various components of the UAV systems provide a large attack surface for malicious intruders, which brings huge cybersecurity challenges to the UAV systems. Figure 4.8 depicts examples of cyber-attacks against different components of UAV systems (Javaid et al., 2012). Most of the identified cyber-attacks on UAVs can potentially lead to taking control of or crashing them. Cyber-attacks have identified that target both the flight controller and GCS and the communication data link.

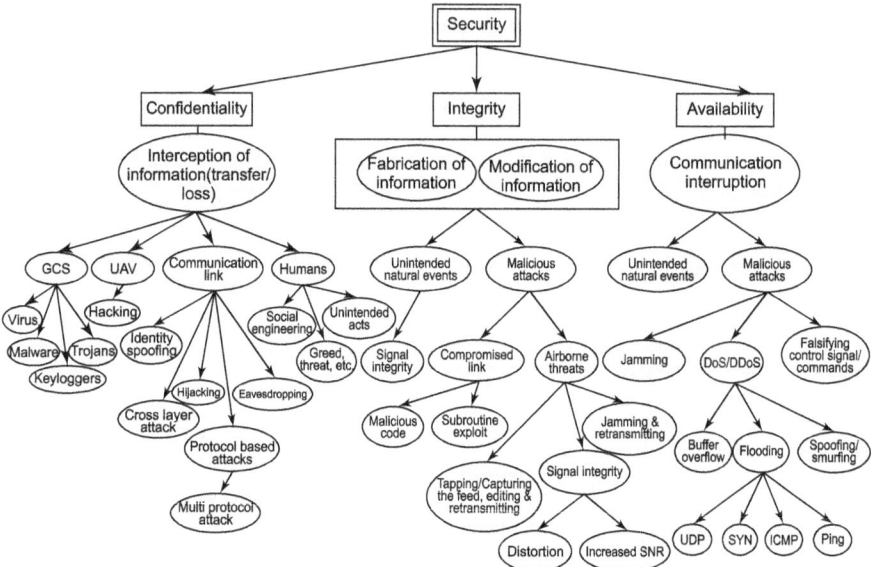

FIGURE 4.8 The UAV system components affected by cybersecurity threats (Javaid et al., 2012).

The primary source of compromising the security of communication links between various system components is through network attacks, such as hacking, eavesdropping, identity spoofing, cross-layer attacks, and multi-protocol attacks. Clearly, all these attacks might not be applicable to each of the links available in the system, as shown in this Figure. Threat analysis is considered an important aspect of ensuring the security of a system because it can actually lead to the discovery of a vulnerability in the system.

The operation of the flight controller is solely dependent on the information received from the ground control station via the data link and acquired by its sensors from the surrounding environment. Accordingly, attacks on both the flight controller and ground control station that do not involve the data link are possible only if the attacker can access and manipulate the internal system communication or can fabricate the sensed physical properties in the surrounding environment (Mansfield et al., 2013). Due to the almost complete reliance of UAV operations on multiple inputs from the external environment, most of the attacks commence by external malicious modification of such inputs. Another type of attack aims to violate the confidentiality and integrity of the communication between the UAV and the ground control station on the data link. Usually, sophisticated attacks aiming to take control of the UAV to combine one or more of the presented attacks on both the flight controller and the data link (Ghosh et al., 2012).

In order to face the cybersecurity challenges of UAV systems, it is important to understand the attack vectors of general UAV systems, and then, based on the attack vectors and the potential capabilities of attackers, security threats may be assessed. Based on the attack vectors and capabilities of attackers and the work, the possible attacks against UAV systems are classified into different categories. Various attacks could be applied to the communications links between different entities of UAV systems, such as eavesdropping and session hijacking (Altawy and Youssef, 2016).

4.8 SWARM APPROACH

A swarm or fleet of UAVs is a set of aerial robots capable of performing a collective task for a specific goal created from interactions between unmanned aircraft and their environment. This innovative approach emerged in the field of intelligence of artificial swarms (Campion et al., 2019). It was inspired by swarms of bees, ants, flocks of birds, and other species in nature. The main idea behind swarm is that UAVs make their own decisions through the information shared. All this is possible due to the number of UAVs that could fly at the same time and be controlled as if they were a single unit, either manually, that is, by remote control operations, or autonomously by using processors deployed on the UAVs. Thus, they would be capable of occupying a larger territory which is one of the greatest advantages of this technology (Shakhatreh et al., 2016).

In most cases, each UAV in a swarm is simultaneously controlled by a GCS, and traditional UAV swarms use a computer as a GCS running ground control software. Higher levels of autonomy would allow UAVs to make decisions using onboard processing power. Communication between UAVs allows the swarm to adjust behavior in response to real-time information.

Real-time information collection makes UAVs swarm well-suited for searching over broad areas for mobile or other hard-to-find units (Kallenborn and Bleek, 2019).

In UAV swarms, there are (i) single-layered swarms with every UAV being its own leader and (ii) multi-layered swarms with dedicated leader UAV at every layer. Multi-layered UAVs report to their leader UAV at a higher layer, while a ground-based server station is the highest layer in this hierarchy. In each swarm, every UAV can have dedicated data collection and processing tasks with sufficient computing capability to execute these tasks in real-time. Its central processing takes place on the more performant server/BS or even in the cloud. The utility of swarming is that many UAVs perform tasks and thus become the greatest advantage by adding more functionality as compared to the limitations of a single UAV having a limited payload and limited flight time. Advantages to swarm include time savings, reduction in man-hours, reduction in labor, and a reduction in other costs (Chandrasekharan, 2016).

Figure 4.9 shows a hierarchical swarm of single-layered quadcopters in a leader-subordinate flying/operation model (Tahir *et al.*, 2019). It controls the movement and formation of the whole swarm through the leader UAV, using an intuitive remote control interface. Its flying principle is associated with a remote user/operator and a wireless communication system between the operator and the swarm. Generally, the hierarchy can be much more complex, such as consisting of multiple clusters, each having its own leader(s) and multiple layers of leaders forming "super clusters" of different sizes. Each UAV in the swarm can directly communicate with its peers at the same level of the hierarchy and with its immediate leader UAV(s). Leader UAV at the highest level of hierarchy communicates with the ground-based server, sharing the data collected and pre-processed by the swarm and distributing downwards mission objectives provided by the server.

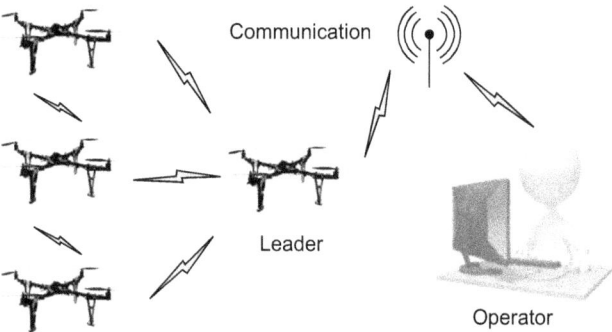

FIGURE 4.9 Concept of swarm UAVs.

Current demonstrations of UAV swarm utilize one of two general forms of swarm communication architecture (Campion *et al.*, 2019). The two forms are infrastructure-based swarm architecture and flying ad hoc network-based architecture.

4.8.1 Infrastructure-based Swarm Architecture

Infrastructure-based swarm architecture is the most common architecture for UAV swarms (Chandrasekharan, 2016), as shown in Figure 4.10. It consists of a GCS that receives telemetry information from all UAVs in the swarm and sends commands back to each UAV individually. These UAV swarms are considered to be semi-autonomous as they still require a direction from central control to complete the assigned task. The GCS software already contains the basic infrastructure-based swarm capabilities (Shakhatreh *et al.*, 2016).

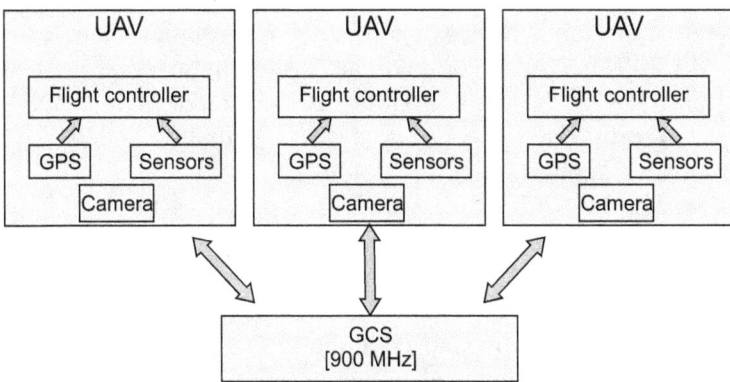

FIGURE 4.10 Infrastructure-based swarm architecture (Campion *et al.*, 2019).

Infrastructure-based swarm architecture is dependent upon the GCS for coordination of all UAVs. In the event of an attack or failure to any operation of the GCS, operability of the entire swarm is compromised. Additionally, infrastructure-based methods require all UAVs to be within the propagation range of the GCS. This dependency causes a lack of system redundancy. Due to the payload capacities of UAVs, the hardware necessary to establish reliable communication with infrastructure may limit the utility of infrastructure-based swarms. Additionally, there is a lack of distributed decisions, as the GCS coordinates the decision making of all UAVs based on computations and algorithms developed in the GCS (Chandrasekharan, 2016). However, the advantages of infrastructure-based swarming are that

(i) optimization and computations can be conducted in real-time by a GCS via a higher performance computer than could reasonably be carried by a UAV and (ii) networking between UAVs needs not to be established.

4.8.2 Flying Ad hoc Networks (FANETs)-based Swarm Architecture

Nowadays, there is a trend to use a network of small UAVs rather than large UAVs. The main reason is that small UAVs are relatively cheaper. Moreover, a team or swarm of UAVs is more flexible and efficient than only one aircraft. Those small UAVs have to exchange information about group formation, mission execution, and control parameters. The Flying Ad hoc Networks (FANETs) are composed of aerial devices that can communicate between UAV-to-UAV and other ground-based UAV-to-ground devices (Pires et al., 2016).

Mobile Ad hoc Networks (MANETs) are composed of mobile nodes, which present random movements. The MANETs are networks composed of mobile devices, such as laptops, cellular phones, sensors, etc. The MANET can rely on or not in an existent network infrastructure. However, most MANETs do not use infrastructure mode. The MANETs can be specialized in terms of connectivity, node mobility models, routing processes, services, applications, etc.

The FANETs can be seen as a subset of the well-known concept of MANETs (Zafar and Khan, 2016). The MANETs composed of terrestrial vehicles as mobile nodes are known as Vehicle Ad hoc Networks (VANETs). Similar to VANETs, whose components are cars, buses, ambulances, etc., incorporating embedded communication devices, in FANETs, UAVs are wirelessly interconnected, either directly or using intermediate nodes. Thus, only a small set of nodes need to be connected to the base station/satellite. Even though the FANETs share common features with the MANETs and VANETs, there are several unique characteristics that make them different, such as mobility, topology changes, radio propagation, and energy constraints. Figure 4.11 shows all the three; MANET, VANET, and FANET structures.

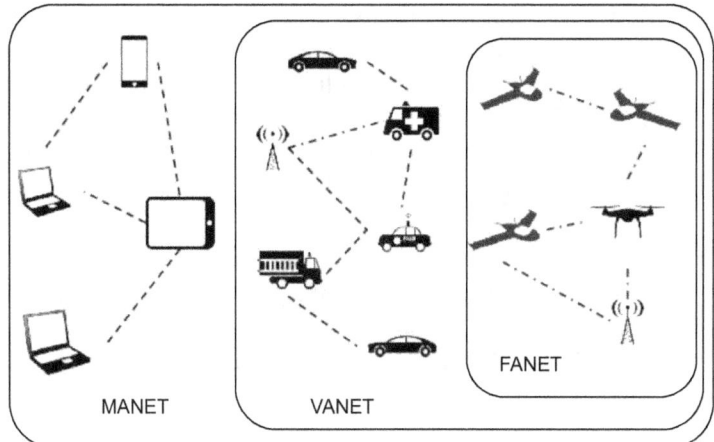

FIGURE 4.11 The MANET, VANET and FANET structures (Pires *et al.*, 2016).

The FANETs can operate either independently and by transmitting the flow received from land-based devices to a remote server or support other types of networks (Figure 4.12), such as via satellite or cellular, if they are overloaded or unavailable. Such technology can play an essential role in the next generation of cellular networks, offering future support to 5G networks (Mozaffari *et al.*, 2016) with very high speed and minimum latency. Thus, FANETs present themselves as a low-cost, scalable solution for maintenance and expansion of the Internet infrastructure worldwide, and it is essential to research, simulate, and validate their utility in different applications, taking into account the limitations of both UAVs and network itself (Azevedo *et al.*, 2019).

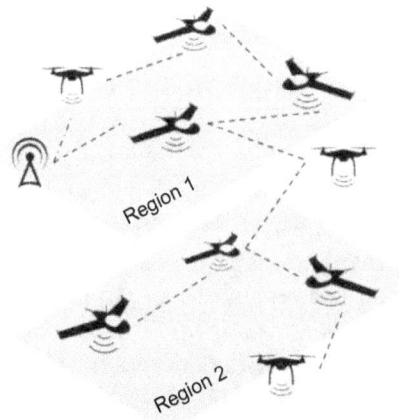

FIGURE 4.12 The FANET structure (Bekmezci et al., 2013).

With their mobility, lack of central control, and self-organizing nature, FANETs can expand the connectivity and communication range in geographical areas having limited cellular infrastructure (Zafar and Khan, 2016). The FANETs play an important role in various applications, such as traffic monitoring, border surveillance, disaster management, agricultural management, wildfire management, and relay networks (Bekmezci et al., 2013). In particular, a relaying network of UAVs maintains reliable communication links between remote transmitters and receivers that cannot directly communicate due to obstacles or their long separation distance. Compared to a single UAV, a FANET with multiple small UAVs has several advantages, as listed in Table 4.3 (Bekmezci et al., 2016).

TABLE 4.3 The Advantages of a FANET with Multiple Small UAVs.

S. No.	Parameter	Advantages
1	Scalability	The operational coverage of FANETs can be easily increased by adding new UAVs and adopting efficient dynamic routing schemes.
2	Cost	The deployment and maintenance cost of small UAVs is lower than the cost of a large UAV with complex hardware and heavy payload.
3	Survivability	In FANETs, if one UAV becomes non-operational (due to weather conditions or any failure in the UAV system), FANETs missions can still proceed with the rest of flying UAVs. Such flexibility does not exist in a single UAV system.

4.9 INTERNET OF THINGS (IoT)

The IoT refers to the billions of physical devices around the world that are now connected to the Internet, all collecting and sharing data (Sharma and Garg, 2019). With the availability of super-computer chips and the ubiquity of wireless networks, it is now possible to connect anything into a part of the IoT, ranging from a chip to an airplane. Connecting all these different objects and sensors adds a level of digital intelligence to devices, enabling them to communicate the data in real-time without involving human beings. The IoT, therefore, represents a heterogeneous network scenario with virtually unlimited uses: smart homes, smart cities, industry, smart grids, etc. (Giyenko and Cho, 2016).

It is a common practice in IoT to use small (and low-cost) sensor devices to capture data from multiples sources. These data are usually sent by means of wireless technology to a gateway that provides Internet connectivity to the cloud, where network servers are located. Network servers are responsible for collecting and processing the data and also making decisions or defining specific actions to be carried out. Consequently, these small sensors used in IoT, together with their communications capabilities, could be easily embedded in UAVs. By doing so, the UAVs provide a new framework to deploy IoT-based services using the WSN (Martinez-Caro and Cano, 2019). With the massive growth of the IoT, more UAVs will be performing important tasks, especially where it is expensive, dangerous, or impossible for humans to travel.

The UAVs play an important role in the IoT because they are critically dependent on sensors, antennas, and embedded software to provide two-way communications for remote control and monitoring. The latest advances in electronics allow putting more and more processing power onboard the UAV, which allows offloading the processing power from the ground stations. This will reduce the required bandwidth, which allows creating more intelligent UAVs that use Machine-to-Machine (M2M) principles to communicate directly with different devices and each other to find the optimal way for performing the tasks. To solve the problem of designing the smart UAV platform, there is a need to address and integrate its four elements (Giyenko and Cho, 2016): data, connectivity, device, and service. A smart UAV platform can be achieved only by the convergence of these four elements, as shown in Figure 4.13.

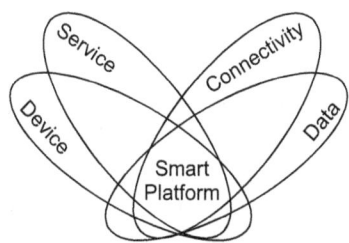

FIGURE 4.13 Elements of a smart UAV platform (Giyenko and Cho, 2016).

UAV flight monitoring is a key element provided by the integration of IoT in UAVs. Using IoT technology, it is possible to keep an updated flight log records available online. This is a powerful tool for monitoring the flights by operators and storing a flight record database that can be used for extracting system reliability data or controlling the productivity of a company. The main advantage of using M2M and IoT technology in UAVs is the capability to perform UAV operations from any PC connected to the Internet. In this way, the operators can not only control the flying of UAV, but at the same time, it is possible to monitor and control UAV swarm flights from the main control center. The M2M multipoint network structures permit to have as many nodes in the network as required, including UAVs and control stations. The swarm systems flying simultaneously can share information for enabling advanced flight modes. Formation flights and collaborative sense-and-avoid technology become a reality with this interconnection.

UAVs are an emerging form of new IoT devices, flying in the sky with full network connectivity capabilities (GSMA, 2018). Intelligent UAVs with cognitive computing skills need the capability to automatically recognize and track objects to free the users from the tedious task of controlling them. The UAVs are used in various applications, such as telecommunications, defense, traffic management, surveillance, mapping, emergency services, weather monitoring, resources exploration, and environmental analysis. As UAV collects huge data of an area, so the time required for processing this data quickly is an important consideration in these applications for near-real-time decision making. The UAVs can be used as signal boosters and mobile nodes for IoT and sensor networks, dispersal of pesticides in agricultural areas, or emergency surveillance in case of fire, earthquake, or other disaster events. This will allow reducing the deployment time and costing compared to traditional aircraft and transport (Giyenko and Cho, 2016). Furthermore, the rollout of 5G technology is expected to enhance the ability of UAVs to react to commands in real-time, requiring instant feedback.

The IoT-enabled UAVs are used to carry out comprehensive monitoring of crops (Sharma and Garg, 2019). Based on the crop data, the UAVs will be able to provide recommendations to the farmers on optimal productivity conditions and interventions that need to be made. This will lead to increased productivity in farming and consequently increased yields. Mines, power grids, tunnels, and industrial sections, such as boilers and furnaces, are hard to access and pose health hazards to inspectors. However, with sensors and cameras, the smart UAVs can gather data that is then relayed in real-time to an inspector to undertake maintenance in a safer manner. This reduces the human-machine intervention resulting in reduced costs and reduced safety hazards (D'mello, 2019). The UAVs are also helpful in giving a first-hand view of an upcoming structure as well as monitoring its progress. It also reduces the time and effort used in making regular site visits. Inspection of roofs and utilities will also be enhanced with UAVs, which will reduce accidents and thus increase safety at construction sites.

Intelligent UAVs are also changing the goods delivery mode, as they can easily reach places where other modes of transport are not viable. The UAVs are also expected to reduce response time to emergencies and will reduce challenges associated with delivering relief and medical supplies in disasters and emergency situations. In addition, UAVs are also going to be used to help firefighters determine the exact locations of a fire as well as locating the affected people. Police can use intelligent UAVs to locate violent spots and release tear gas or pepper spray to disperse the crowds. It is expected to increase the capabilities and performance of UAVs further. With technological advances, IoT-based UAVs will continue to discover new uses (D'mello, 2019).

The IoT devices often have small transmitting power and may not communicate with UAVs over a long-range (Ding *et al.*, 2018). In fact, the UAVs can be optimally placed based on the locations of IoT devices, enabling those devices to connect to the network using a minimum transmit power successfully. The UAVs can also serve massive IoT systems by dynamically updating their locations based on the activation pattern of IoT devices. Therefore, IoT networks' connectivity and energy efficiency can be significantly improved by exploiting this unique feature of UAVs.

Mozaffari *et al.* (2017) proposed a framework for UAVs' dynamic deployment and mobility to enable reliable and energy-efficient IoT communications. Figure 4.14 shows results about the UAVs' locations and their associated IoT devices (indicated by the same color). Here, five UAVs are efficiently deployed to serve and collect data from 100 active IoT devices, which are uniformly distributed in the area of 1km × 1km. Using tools from optimization theory and facility location problems, optimal 3D positions of these UAVs are derived as well as the device-UAV associations, such that the total uplink transmit power of IoT devices is minimized while ensuring reliable communication. The results show that UAVs can be optimally deployed to enable reliable and energy-efficient uplink communications in IoT networks.

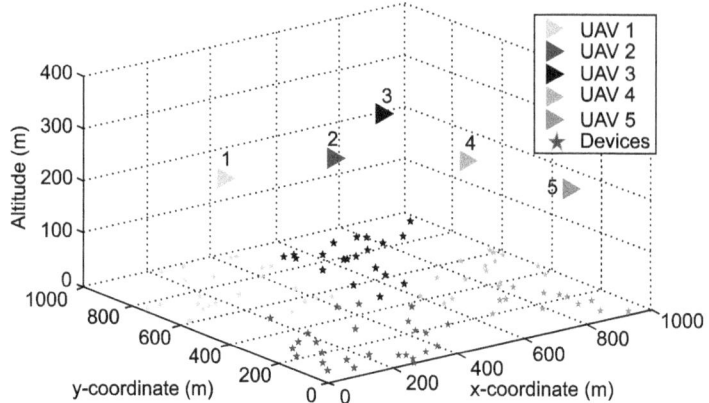

FIGURE 4.14 UAVs' locations and associations (Mozaffari et al., 2017) [See Plate 2 for Color].

Figure 4.15 shows the total transmit power needed by the IoT devices for reliable uplink communications with the number of UAVs in the interference scenario. It is seen that the total transmit power of the IoT devices can be reduced by deploying more UAVs. For instance, considering 100 active devices and 20 available channels, the total transmitted power decreases from 2.4 W to 0.2 W by increasing UAVs from 5 to 10 using the proposed approach. Furthermore, the total transmitted power of the devices decreases by 45% (as an average) compared to the stationary case. Clearly, for a lower number of UAVs, the proposed approach in the study leads to higher power reduction compared to the stationary UAVs. In other words, intelligently optimizing the locations of UAVs provides more power reduction gains when the number of UAVs is low.

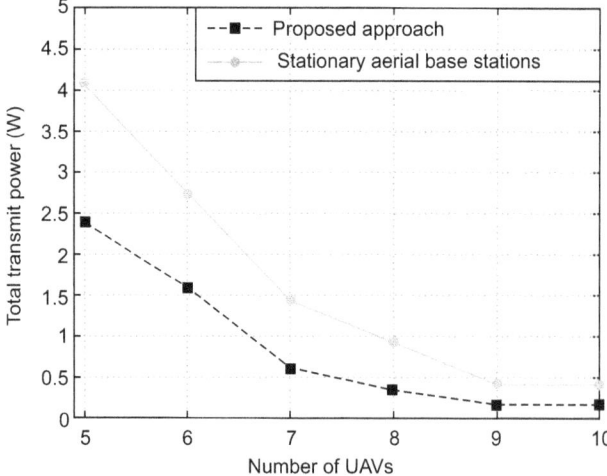

FIGURE 4.15 Total transmit power of devices and number of UAVs in the presence of interference (Mozaffari et al., 2017).

The UAVs can serve as wireless relays for improving the connectivity and coverage of ground wireless devices. A key application of this is improved surveillance scenarios. The real-world example that employs UAVs for wireless connectivity includes Google's Loon project from an industry perspective. The work AT&T (2016) provides a comprehensive survey on the potential use of UAVs for supporting IoT services.

Wireless networking technologies are rapidly evolving into a massive IoT environment that must integrate a heterogeneous mix of devices ranging from conventional smartphones and tablets to vehicles, sensors, wearables, and UAVs. Applications, such as smart cities, infrastructure management, healthcare, transportation, and energy management (Bamburry, 2015) require effective wireless connectivity among a number of IoT devices that can reliably deliver their data, typically at high data rates. The massive use of IoT-based devices requires a major rethinking of the way in which conventional wireless networks (e.g., cellular systems) operate. In an IoT-based UAV environment, energy efficiency, ultra-low latency, reliability, and high-speed uplink communications become major challenges that are not typically critical in conventional cellular networks (Wang *et al.*, 2017).

4.10 SUMMARY

The communication poses many challenges, including the need for good coordination and collaboration among UAVs; effective control mechanisms; reliable communication among UAVs and between UAVs & GCS; safe actions; and good scheduling of tasks. The UAV-to-UAV and UAV-to-Infrastructure communication is an essential component, which is vital to enable these devices to perform many collaborative tasks and services. These challenges present the main obstacles in effectively utilizing the UAV technology for future potential applications (Zang *et al.*, 2019).

Main design considerations for UAV deployment include (i) shorten average communication distance, (ii) collision and terrain avoidance, (iii) path optimization, (iv) ubiquitous coverage (aerial BS), (v) tradeoff in UAV altitude (higher altitude, larger path loss), (vi) energy-efficient operations, (vii) energy-efficient mobility, for example, avoid unnecessary vehicle maneuvering, and (viii) size, weight, and power limitations of UAVs (Chandrasekharan, 2106). The UAVs have promising applications in future wireless communication, such as (i) on-demand deployment, fast response, cost-effective, (ii) flexible in deployment and reconfiguration,

(iii) short-distance LOS channels, (iv) UAV-aided ubiquitous coverage, and (v) UAV-aided relaying.

Future research directions may include UAV swarm operation and communications, aerial BS, for example, mobile LTE BS/relay, UAV communication with limited buffer size/energy storage, UAV deployment/movement optimization, MIMO communication in UAV, and energy-efficient UAV communications. The UAV-aided wireless communication offers one promising solution to provide wireless connectivity for devices without infrastructure coverage due to, for example, severe shadowing by urban or mountainous terrain or damage to the communication infrastructure caused by natural disasters. These benefits will make UAV-aided wireless communication a promising integral component of future wireless systems.

REVIEW QUESTIONS

Q.1. Discuss in brief the role of various components of a UAV communication system. Draw a diagram.

Q.2. Explain various types of possible communications with UAVs, along with their utility.

Q.3. What do you understand by WSN? Describe various wireless technologies used in UAVs along with their advantages and disadvantages.

Q.4. Explain the utility of the UAV-based FSO technique in wireless communications with the help of a sketch.

Q.5. Write short notes on (i) FPV technology, (ii) Cybersecurity threats, and (iii) the Utility of IoT in UAVs.

Q.6. Explain the term swarm. What is the utility of swarm in UAVs-based work?

Q.7. Explain, with the help of diagrams, the differences between MANET, VANET, and FANET in Swarm technology.

CHAPTER 5

UAV Data Collection and Processing Methods

5.1 INTRODUCTION

The UAVs collect huge data from various sensors and deliver the collected data to ground base stations. The UAVs equipped with WSN or swarm UAVs can also be used as aerial sensor network for collecting and transmitting huge data for various applications. Many UAVs consist of onboard computational capabilities that have greatly improved UAV technology in recent years. However, the computationally intensive nature of real-time image processing has been a great concern for the uses of UAVs, especially with the availability of increased data from high resolution and multimodal imaging systems. Today, UAV data analysis software provides the opportunity to seamlessly export point cloud formats into common CAD tools or Geographic Information System (GIS) or open-source software (Garg, 2019).

For all applications that involve measuring and mapping in real-world, georeferencing and geometric correction of the images is imperative. However, a highly accurate geometric correction process requires time, effort, a DEM or good ground control points (GCPs). Some of the UAV photogrammetry mapping software used for various purposes in different applications include DroneDeploy 3D Mapping Solutions, Pix4D Mapper Photogrammetry software, AutoDesk ReCap Photogrammetry software, Maps Made Easy-orthophotos and 3D models, 3DF Zephyr Photogrammetry software, Agisoft PhotoScan Photogrammetry software, PrecisionHawk Precision Mapper/Viewer, Open Drone Map, and ESRI Drone2Map for ArcGIS. The typical output from these software includes orthomosaic images, map layers (Google maps, mapbox, etc.), DEM/Geotiffs, Normalized Difference

Vegetation Index (NDVI) images, 3D point clouds, meshed 3D models, and many more. In addition, UAV collects live video images for many applications, such as journalism and media (Molina, 2017). A good quality video images processing software may also be required for the post-processing of such images.

For the last decade, feature extraction is a very prominent technique to process the multi-images for classification, detection, objects retrieval, object recognition, change monitoring, pattern study and many more. Image processing methods can be used in autonomous UAV systems to locate potential targets in support of search and rescue operations. Moreover, location information can be augmented to aerial images of target objects (Vergouw *et al.*, 2016).

Advanced RTK positioning methods, such as those using network-based architectures, have great potential to benefit data collection for all integrated sensors (Garg, 2019). The sensor fusion technology, which intelligently combines data from several different sensors, such as a thermal camera and a regular RGB camera sensor, can be used for improving application or system performance. The UAV data is also driving advancements in AI and predictive analytics. In future, many technologies will change the world, the way we collect the data and analyze them. These technologies include intelligent flights, IoT, and cloud computing, among others (Corrigan, 2020).

5.2 DATA PRODUCTS

The UAV data typically starts with lots of images that are collected according to a set of specifications suited for a particular application (Garg, 2019). Many UAV missions require dealing with large amounts of data. Improvements are being made in methods and systems that handle the data in its various stages: acquisition, storage, and transmission. Some UAVs transmit data in real-time to ground stations. Once the data is obtained, it is usually processed with the software. For example, data is often georeferenced and corrected for various distortions and errors. Many new models, methods and techniques are being developed for processing the UAV data. To make these data meaningful for an application, the following steps may be performed:

- *Processing:* Several images are identified, arranged, georeferenced and/or stitched (mosaicked) together to create an organized, usable, and cohesive data set.

- *Measurement*: These data are often processed into data products that can be accurately geolocated and used for measurement, similar to a map or computer model. For example, thermal imagery enables temperature measurements.
- *Analysis*: Data are anlayzed and classified, like the severity of damage after an earthquake.
- *Monitoring*: Collecting data over time can be used to predict the changes, e.g., progress at a construction site.
- *Comparing*: Various scenarios may be compared to assess the best one. For example, 3D models produced by different algorithms may be compared to find the accurate one.

Various types of data/products to be collected and employed in UAVs are summarized below.

5.2.1 Aerial Remote Sensing Data

There are two broad types of remote sensing systems: active and passive remote sensing systems. In an active remote sensing system, the sensors have their own source of energy required to detect the objects. The sensor transmits radiations toward the objects to be investigated, and the reflected radiations from the objects are detected and measured by the active sensor. Most active sensors used in remote sensing applications operate in the microwave portion of the electromagnetic spectrum, which is able to propagate through the atmosphere (Garg, 2019). The active remote sensing systems include laser altimeter, LiDAR, radar, ranging instrument, scatterometer and sounder.

On the other hand, in a passive remote sensing system, the sensor detects the natural radiations emitted or reflected by the objects to be investigated. The majority of passive sensors operate in the visible, infrared, and thermal infrared portions of the electromagnetic spectrum (Whitehead and Hugenholtz, 2019). Such systems include accelerometer, hyperspectral radiometer, imaging radiometer, radiometer, sounder, spectrometer and spectro-radiometer.

A UAV can carry various sensors, such as digital camera, multispectral camera, hyperspectral, LiDAR, thermal, and many more. Datasets acquired from UAVs remote sensing sensors have a broad range of applications: crop monitoring, yield estimates, city mapping, forestry survey, drought monitoring, water quality monitoring, tree species, disease detection, etc. (Watts *et al.*, 2012). The UAVs can be integrated with thermal sensors and vision cameras useful for target detection. Thermal IR cameras can be used to detect the heat profile to identify problems in buildings, electrical components, mechanics, and pipes and waterproofing systems. Moreover, the IR camera is used to acquire images, specifically differentiating between water and land (Garg, 2019).

Table 5.1 presents a comparison among the UAV remote sensing systems based on their operating frequency and applications.

TABLE 5.1 The UAV Aerial Sensing Systems (Shakhatreh et al., 2018).

Types	Operating Frequency	Type of Sensor	Applications
Active	Microwave portion	Laser altimeter	It measures the height of a UAV with respect to the mean Earth's surface to determine the topography of the underlying surface.
		LiDAR	It determines the distance to the object by recording the time between transmitted and backscattered light pulses.
		RADAR	It produces a two-dimensional image of the surface by recording the range and magnitude of the energy reflected from all objects.
		Ranging Instrument	It determines the distance between identical microwave instruments on a pair of platforms.
		Scatterometer	It derives maps of surface wind speed and direction by measuring backscattered radiation in the microwave spectral region.
		Sounder	It measures the vertical distribution of precipitation, temperature, humidity, and cloud composition.

Types	Operating Frequency	Type of Sensor	Applications
Passive	Visible, IR, thermal IR, and microwave portions	Accelerometer	It measures two general types of accelerometers: (i) The translational accelerations (changes in linear motions), and (ii) The angular accelerations (changes in rotation rate per unit time).
		Hyperspectral radiometer	It discriminates between different targets based on their spectral response in each of the narrow bands.
		Imaging radiometer	It provides a 2D array of pixels from which an image may be produced.
		Radiometer	It measures the intensity of electromagnetic radiation in some bands within the spectrum.
		Sounder	It measures vertical distributions of atmospheric parameters such as temperature, pressure, and composition from multispectral information.
		Spectrometer	It designs to detect, measure, and analyze the spectral content of incident electromagnetic radiation.
		Spectro-radiometer	It measures the intensity of radiation in multiple wavelength bands. It designs for remotely sensing specific geophysical parameters.

5.2.2 LiDAR Data

A LiDAR is an active system that directs laser beam on the Earth's surface and determines the distance to the object by recording the time between transmitted and backscattered light pulses (Garg, 2019). The combination of UAV and laser sensor has become popular for effectively mapping the

Earth's surface since it can capture real-time 3D data (Lin *et al.*, 2011). LiDAR sensing is one of the most powerful technologies to acquire spatial data. The LiDAR point clouds can yield up to 500 points/m^2 with vertical elevation accuracy of 2–3 cm. The denser the data, the greater the flexibility to generate high-fidelity maps of densely vegetated and populated terrain.

There are numerous applications of LiDAR UAV data in different fields, such as civil engineering, cultural heritage, ecology, hydrology, risk assessment, early warning systems, and post-event phase of natural hazards, such as landslides, earthquakes, volcanic eruptions, and meteorological events. Airborne LiDAR is widely used to generate high-resolution and accurate contours, DEM, Digital Terrain Model (DTM), and detailed terrain analysis (Nagai *et al.*, 2009). When combined with aerial images of the ground surface, it offers various advantages, such as orthophoto generation, Digital Surface Model (DSM), and 3D terrain visualization, providing more detailed information of the ground. The elevation data from airborne LiDAR can be used as an alternative for georeferencing UAV-based photogrammetry images. It provides an alternative source for GCPs and photogrammetry of airborne imagery.

5.2.3 Ground Control Points (GCPs)

The GCPs are those points whose coordinates are measured precisely with the GPS or any other devices. These GCPs have to be well distributed around the area to be mapped for the best results. More GCPs generally permit more accurate results (Garg, 2019). Photogrammetry mapping with UAV needs very precise georeferencing that can be achieved with the help of GCPs, already established using a GNSS with very high accuracy (Jiang and Jiang, 2017). The GCPs typically are marked before the UAV flight takes place, using easy to see aerial targets that can later be identified through data processing software (Tomaštík *et al.*, 2019). Another way could be collecting the GCPs coordinates with onboard GPS on UAV, which is stored in the flight log generated by the flight controller. The time stamp of each photograph is then matched with the corresponding location of the UAV.

The GCPs can be used to find the coordinates of remaining other objects present in the image of the area. Image processing/GIS software is used for georeferencing, which requires the real-world coordinates of the sufficient number of visibly identifiable GCPs in the collected UAV aerial imagery. Accurate GCPs are very important in UAVs-related work. These are required to carry out georeferencing to create the mosaic and orthophotos and generate the DEM and DSM.

5.3 DATA PROCESSING APPROACHES

Data processing constitutes an important step in deriving any useful information or analyzing the data for an application. There are many approaches available, but their use will depend on the purpose of output required and specific application in mind. This section describes some data processing approaches.

5.3.1 Georeferencing Approach

Georeferencing is an essential and important process so that the UAV data corresponds to real-world coordinates (Garg, 2019). In simple terms, georeferencing is the "process of assigning geographical coordinates to the acquired spatial data" (Cramer and Stallmann, 2002). While collecting data and creating the maps without any georeferencing process, these data/maps cannot be used for actual measurements but for viewing the terrain only. Georeferencing is also important to determine the accuracy of UAV data collected for each mission. Georeferenced UAV data/maps are also much easier to work with, as they can be overlaid on other existing thematic maps in GIS/image processing software, like Open Street map and Google maps. For GIS-based applications, a large number of thematic maps and data may be required to integrate together, and georeferencing of these maps and data to a common reference system and datum is a pre-requisite for their uses (Cramer and Stallmann, 2002). The QGIS is an open-source GIS software useful for the extraction of information from multiple data. It is therefore important to understand the importance of georeferencing process and the procedure to collect the GCPs.

There are two possibilities for georeferencing the photographs using a given coordinate system. The first possibility is to identify GCPs on the photographs such that their locations in a reference system are known. Polynomials are used to carry out accurate georeferencing, and *rmse* is determined to know the accuracy of georeferenced photos. The second possibility is to create georeferenced maps/images without the GCPs, using more sophisticated onboard GPS units on UAVs. The GPS receivers and IMU in UAVs are able to provide input to create accurate georeferenced maps/images (Mian *et al.*, 2015).

In the second case, most photogrammetry software packages, such as Agisoft PhotoScan and Pix4D, can be used. This software will use GPS data collected by a GPS logger or by a GPS-enabled camera to create reasonably geographically accurate images/maps. The software first carries out the automatic process of alignment of photographs by searching common GCPs on photographs and matching them. It deduces the position of the camera for each photo so that it can refine its camera calibration parameters. Once photo alignment is completed, the software generates a sparse point cloud with a set of associated camera positions and internal camera parameters. A point cloud will have a set of points in 3D space, where each point, in addition to its coordinates, may have additional information, such as color intensity. The software requires this set of camera positions and an optimized sparse point cloud to advance the process of producing a dense point cloud, which can often take long hours on a reasonably high-powered laptop. The software builds a 3D polygonal mesh based on the dense point cloud, representing the 3D surface of the object (Qu *et al.*, 2018). A proper projection system is selected based on the requirements.

In fact, the second approach determines the exterior orientation-positions (x, y, z) and rotational parameters (yaw, pitch and roll of the platform or κ, φ and ω of the image) of the images during the UAV flight. Processing of such UAV images has its own challenges. This recorded data may not be precise enough to enable direct georeferencing as compared to image scales and resolution, but the system uses these data as initial values for approximate orientation. This is an iterative step that continues until the desired accuracy is achieved. The exterior orientation values are taken from the log files recorded during the UAV survey flight for initial direct georeferencing of the images (Cramer and Stallmann, 2002).

The locations of UAV can be improved by feeding IMU and camera data into a Kalman filter (Mian *et al.*, 2015). Some recently developed UAV platforms utilize multiple sets of IMU and GPS receivers to improve the localization redundancy. However, many simple UAVs do not log their GPS coordinates, but merely use the onboard GPS to feed data into the autopilot system (Hugenholtz *et al.*, 2016). GPS loggers, such as the Flytrex Core 2 Flight Tracker, collect longitude, latitude, and altitude values during UAV flight, using the same GPS chip as used for navigation. This data is available in formats that can be directly used to georeference the images/maps. Some UAVs, such as the MAVinci SIRIUS Pro and the SenseFlyeBee RTK model, including specialized mapping UAVs, use direct georeferencing techniques without the use of GCPs (Forlani *et al.*, 2018).

5.3.2 Terrain Modeling Approaches

Terrain modeling involves creating 3D models (e.g., DEM, DSM, DTM) from 2D images and point cloud data. It also requires creation of 3D visualization of terrain using VR, AR and other visualization techniques. Terrain modeling is required in many applications where elevation of ground, slope of ground and height of various objects on ground are needed. Some approaches are discussed below.

5.3.2.1 Digital Elevation Model (DEM) Creation

A DEM is a bare Earth raster grid referenced to a vertical datum. It is a digital data file consisting of elevation information in a color code (raster) or grid form (Figure 5.1a). A DEM will not include the built features (e.g., power lines, buildings, bridges, roads and towers) and natural features (e.g., trees). The elevation values and their spatial variations are of importance in many applications. The DEMs with high spatial resolution (<1 m) and height accuracy (< 0.1 m) in elevation measurements can be generated with LiDAR data. Various studies showed it was possible to estimate the elevation with errors ranging from 0.5 m to 0.05 m (Manousakis *et al.*, 2016; Forlani *et al.*, 2018). The DEMs can be used to estimate the slope and the aspect in each grid which are useful in hydrology, infrastructure and land use planning (Gajdamowicz *et al.*, 2007). The DEM data are also used for calculation of volume, finding out higher elevation regions, lower elevation regions, flat areas and flow of the drainage. Table 5.2 presents a comparison of height modeling from photogrammetry and LiDAR data.

TABLE 5.2 A Comparison of Height Modeling from Photogrammetry UAV and LiDAR UAV.

	Photogrammetry	**LiDAR**
Platform	Aerial and space-borne.	Aerial
Sensor	Passive optical sensor. Images acquired by area or line sensors.	Active near-infrared sensor. Typical wavelength 900 nm.
Geometry	Perspective geometry, requiring at minimum two images of overlapping area from different view directions.	Near-vertical geometry with polar coordinate determination. Narrow scan widths of a km or less typical incidence angle ±20°.

(continued)

	Photogrammetry	**LiDAR**
Flight plan	Weather and illumination dependent, large range of flying height and speed. Sub-meter pixel size satellite imagery can capture up to 300 km² scenes.	Weather dependent, flying height 300–2000 m, flying speed 200 km/h, ground strip width depends on flying height.
Elevation measures	Image matching from stereo pairs is used to find differences in parallax to calculate surface elevation.	Distance to a feature is a function of the time it takes for the emitted signal to return.
Vertical accuracy	Vertical accuracy is a function of a photoscale; a rough indication for aerial photography is that vertical accuracy is around 1:8000 of flying height. Satellite imagery can achieve height accuracy to 2–3 pixels.	*rmse* 0.1 – 0.3 m.
Factors influencing vertical accuracy	Photo scale, Image contrast, Image geometry, Pixel size, Terrain steepness, and Image matching algorithm.	Flying height and speed, Scan angle, Point density, Terrain steepness, and Interpolation algorithm
DEM post-spacing (horizontal distance between data points)	Manually collected information can derive DEM and DSM points and lines. Automatic image matching algorithms result in DSM grid points; processing is then required to reduce the DSM to a DEM.	10,000–100,000 points/s can be produced for dense elevation data. Irregular mass points are derived from laser footprints of 0.40–2m, separated 1–5 m apart (depending on flying height). Interpolation is required to produce a regular grid. First return (surface data) and last return (hard surface) data are available; however, processing is required to generate DEMs from DSMs. Typical outputs will range from 0.5 to 5m post-spacing.

	Photogrammetry	**LiDAR**
Advantages	• Large area coverage • Breaklines can be easily applied such that elevation over complex terrain surfaces is calculated • Using additional images with differing view angles allows redundant elevation measurements, potentially increasing accuracy.	• Active sensor, therefore able to collect data at any time irrespective of light • High point density, typically 1 point/m^2 • Penetration through vegetation • Fast data-to-information processing and high turnaround rates in production
Disadvantages	• Relies on complex software systems • DEM extraction by automated image matching algorithms still requires interactive stereo editing and can be labor intensive, thus increases the cost of data extraction.	• Surface discontinuities are not well preserved due to footprint sizes • Multipath and vertical profile strikes may be misleading • No data is available for water surfaces • Does not penetrate in rain or fog

In photogrammetry, high quality and a high-resolution camera can yield DEMs with horizontal (x-y) accuracies in the range of 1 cm (0.4 in) and elevation (z) accuracies in the range of 2 to 3 cm (0.8 to 1.2 in), making them ideal for use in construction planning, hydrology, route alignment, and precise volumetric analysis (Westoby *et al.*, 2012). While taking high-resolution images with a camera mounted on UAV, it is important to have an angular separation not exceeding 30 degrees between the images (considering the camera's position during the capture) and a minimum of 60–80% overlapping images for efficient construction of the 3D surface. The location and number of GCPs strongly influence the accuracy of the 3D model used to georeference the model, the flight altitude, the inclination of the camera axis, the image network geometry, image matching performance, surface texture and lighting conditions.

The DEM can be generated either from LiDAR 3D data or photogrammetry images. High-resolution imagery from UAV can be processed through software, such as Agisoft PhotoScan or Pix4D to obtain DEMs. Dataset fusion of RGB images obtained from UAVs and elevation information provides a more holistic representation for the construction of accurate maps for planning purposes. Various sources of data using different techniques and methods used to generate DEM at different resolutions are given in Table 5.3. Limitations of each method have also been explained. It is clear from this table that no method is perfect for creating a DEM, but the data and approach would entirely depend on the application for which DEM will be used.

TABLE 5.3 Various Sources, Techniques, Methods and Limitations for Generating DEM (Lakshmi and Yarrakula, 2018).

Methods	Data Formats	Interpolations Methods	Availability	Limitations
Ground surveys	Ground elevation point	Triangular Irregular Network (TIN) to interpolate	Expensive and time-consuming to collect data for large areas. Suitable for small areas only.	Not suitable for large areas. GPS does not provide reliable height under the canopy. Problem with TIN interpolations.
Airborne photogrammetric surveys with manual interpretation	Contours and measured points	Kriging	Require aerial photography and skilled operators.	Problem with vegetation and measurement frequency.
Airborne photogrammetric surveys using automatic interpretation method	Using correlated points	Kriging	Require aerial photography.	Problem with vegetation and non-ground points with medium frequency resolution.
Existing topographic map data	Primarily contours	Kriging	Readily available and can be done relatively cheap.	Problem with vegetation and measurement frequency. Added errors with digitization.

Methods	Data Formats	Interpolations Methods	Availability	Limitations
Airborne laser scanning	Point data	Inverse Distance Weights	Low cost. High-resolution DEM, DSM and DTM are the products.	Problems may occur with steep slopes and heavy vegetated areas.
Automated stereoscopic based satellite imagery	Point data	Use correlation for surface points and filtering is required.	Can do at a fraction of costs in Photogrammetry. Resolution is much lower. 30m SRTM DEM is freely available.	Problem with clouds, non-ground points and vegetation with medium frequency resolution.
Radar based satellite imagery	Raster DEM	None	Cost is lower than Photogrammetry. High resolution is possible but lower than LiDAR. 30 m SRTM DEM is freely available.	Problems with vegetation and steep slopes.
High resolution satellite data		Contour interpolation	Available at the cheaper cost. Higher-resolution DEM is possible. Some data are available free of cost.	Requires a cloud-free view to generate a good quality and high-resolution DEM.
LiDAR based satellite imagery	Point values: LAS format or ASCII	Kriging	Provide high-resolution DEM with good accuracy. Low cost than photogrammetric methods. Cover large area.	Suffers from an inability to penetrate in the dense canopy. Difficult to interpret and process large datasets.

5.3.2.2 Digital Surface Model (DSM) Creation

A DSM represents the elevations of the natural surfaces and built features above the Earth's surface. In a LiDAR system, pulses of light travel to the ground, bounce off their target and return to the sensor to give a variable distance to the Earth. The LiDAR delivers a massive point cloud of varying elevation values from natural terrain features on the Earth's surface, including the top of buildings, bridges, towers, tree canopy, power lines, etc. A DSM, therefore, represents the height of various objects present on the ground (Figure 5.1b). The DSM generates a dense point cloud by matching features across multiple image pairs (Manousakis *et al.*, 2016). The DSM is especially useful in 3D modeling, smart city planning, telecommunications, aviation and forestry, archeology, hydrology, engineering or agriculture (Al-Najjar *et al.*, 2019).

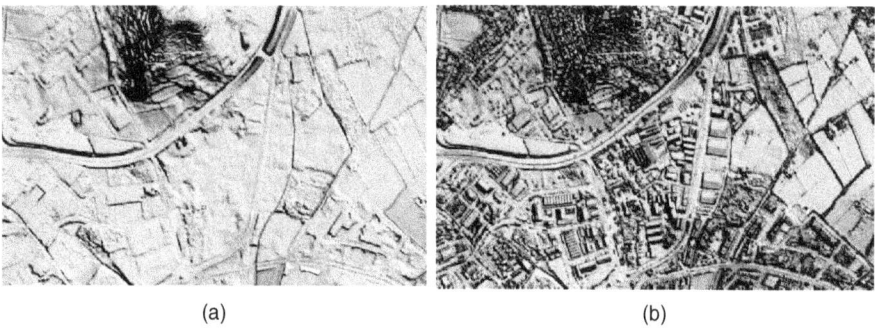

(a) (b)

FIGURE 5.1 (*a*) DEM from LiDAR data (*b*) DSM of terrain [*See Plate 3 for Color*].

Photogrammetry UAV also allows the production of DSMs with a high spatial resolution. There are several steps to be performed before using UAV images to create a DSM, as shown in Figure 5.2. The DSM can also be considered an additional feature with remote sensing images, which can improve the classification results by image segmentation.

Nex and Remondino (2014) presented the details of time taken in each activity of a UAV-based field campaign considering the automation in automated tie point extraction and DSM generation, as shown in Table 5.4. It is seen that a high percentage of time is spent on image acquisition, image orientation, GCPs measurements, and DSM generation, in particular, if direct georeferencing cannot be performed. However, the time required for the feature extraction would depend on the types of feature to be extracted.

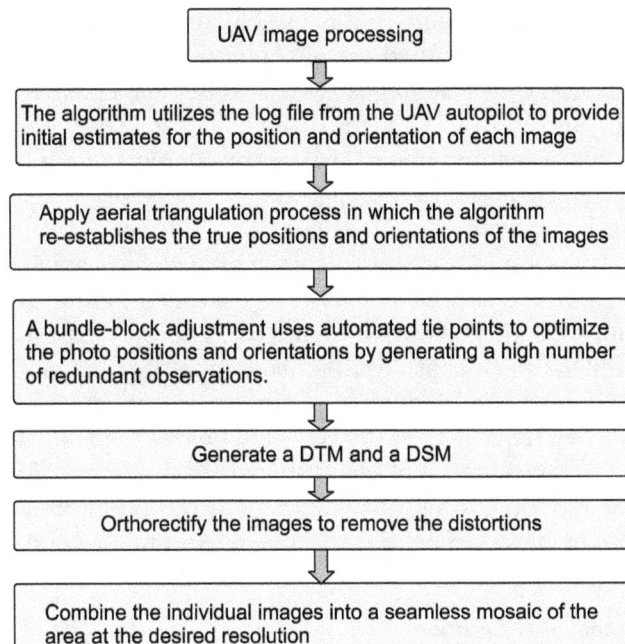

FIGURE 5.2 Steps for orthophoto generation (Shakhatreh et al., 2018).

TABLE 5.4 Time Efforts in Each Activity of Photogrammetry UAV.

Activity	Time Required (%)
Flight planning	5
Image acquisition	20
GCPs field measurements	15
Image triangulation	15
DSM generation	25
Orthomosaic	10
Feature extraction	10

5.3.2.3 Digital Terrain Model (DTM) Creation

The DTM is a grid of 3D points (x, y, z) that does not consider objects above the ground. The density of the grid in a DTM varies dependent upon the terrain, which is stored in digital form. It is a vector data set composed of regularly spaced points and natural features, such as ridges and breaklines. The DTM represents distinctive terrain features much better because of its 3D breaklines and regularly spaced 3D mass points. The DTMs are normally derived from DSM by digitally filtering vegetation, building or other man-made features. The DTM is considered to be a more useful product than a DSM, because the high-frequency noise associated with vegetation cover is removed. The DTM can be used to model water flow, documentation, transportation system planning or geological applications.

There are several steps to be performed before using UAV images to create a DTM, as shown in Figure 5.2. The UAV-based methodologies have benefited from recent advances in the structure from motion (SfM) approach, which, much like traditional photogrammetry, utilizes image overlap from multiple perspectives of the collected images to identify key points and generate an accurate point cloud model the surface (Kaminsky *et al.*, 2009). Once georeferenced, SfM point clouds can be filtered and interpolated to create highly accurate, site-specific DTMs and/or DSMs upon which image mosaics are georectified (Westoby *et al.*, 2012).

5.3.2.4 Orthophotos Creation

Orthophotos are common products of photogrammetric applications, as they combine geometry and photorealism in order to provide a metric visualization of the area. The quality of orthophotos depends on image resolution, camera calibration and orientation accuracy, and DTM accuracy (Kraus, 2007). Orthophotos, thus generated from high-quality UAV images, show great details which can be used for area, distance and other planimetric measurements (Hugenholtz *et al.*, 2016). In addition, the applications of Google Earth-Google Maps, Bing Maps, etc., have increased the demand for orthophotos. The DSMs and orthophotos are also used for object detection or quantification of topographic changes.

Orthomosaic maps are generated from a large number of image dataset collected using a UVAs camera. An aerial orthophoto mosaic can be created following the standard photogrammetric procedure that consists of (i) image orientation, (ii) re-projection with a DTM, and (iii) image mosaicking. Image orientation is carried out from a set of GCPs and tie points using collinearity equations (bundle block adjustment). An orthophoto map is created by the vertical parallel projection of a surface and consists of the geometric accuracy of an image's map and visual characteristics. These image datasets are then processed for georeferencing and mosaicing into a single image layer. The UAV photogrammetric data processing workflow to create a DSM, DTM and an orthophoto is shown in six steps in Figure 5.2.

First, the algorithm utilizes the log file from the UAV autopilot to provide initial estimates for the position and orientation of each image. The algorithm then applies an aerial triangulation process in which the algorithm re-establishes the true positions and orientations of the images from an aerial mission. During this process, the algorithm generates a large number of automated tie points for conjugate points identified across multiple images. A bundle block adjustment then uses these automated tie points to optimize the photo positions and orientations by generating a high number of redundant observations, which are used to derive an efficient solution through a rigorous least-squares adjustment (Benasse *et al.*, 2017). To provide an independent check on the accuracy of the adjustment, the algorithm includes a number of check points. Thereafter, the oriented images are used to create a DSM, which provides a detailed representation of the terrain surface, including the heights of objects, such as towers, trees, bridges and buildings.

The images taken from the UAV are processed with a SfM and multi-view stereo (MVS) approaches. These approaches allow the geometric constraints of camera position, orientation and GCPs from many overlapping images to be solved simultaneously. Computations are performed to generate the DSM and orthophotos.

Figure 5.3a shows the original image collected by UAV photogrammetry, and its orthophoto is shown in Figure 5.3b. The orthophotos can be used to create 3D models of the area. After combining these datasets, a virtual 3D dynamic model is generated, which can be visualized for forest, mountains, heritage monuments, temples, palaces, archeological sites, statues, etc. The DSM and orthophoto generation have been accomplished by SfM methods which combine photogrammetry and computer vision methods (Westoby *et al.*, 2012).

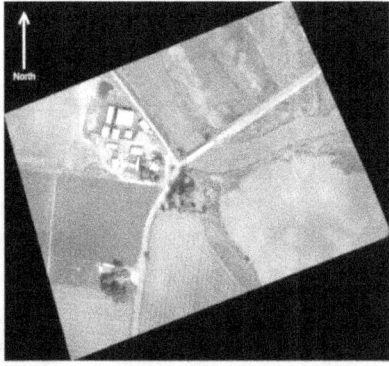

(a)   (b)

FIGURE 5.3 (*a*) Image, and (*b*) Orthophoto of an area.

5.3.2.5 Multi-View Stereopsis (MVS) Approach

The UAV photogrammetry using MVS, known as UAV-MVS, is a technique that combines photogrammetry and computer vision. It is becoming increasingly popular for 3D reconstruction surveys (Harwin, 2015). If the positions and orientations of the UAV cameras are known, the UAV-MVS algorithm can reconstruct the 3D structure of a scene by using multiple-view images. One of the most representative methods was proposed by Furukawa and Hernández (2015).

This method estimates the 3D coordinates of the initial points by matching the difference of corner points between different images, followed by patch expansion, point filtering, and other processing (Qu *et al.*, 2018). The patch-based matching method is used to match other pixels between the multi-images. Thereafter, a dense point data cloud and mesh data cloud can be obtained. Using the concept of Furukawa and Hernández (2015), Koci *et al.* (2017) have proposed several 3D reconstruction algorithms which are based on depth-map fusion. These algorithms can obtain reconstruction results with an even higher density and accuracy.

The method proposed by Shen (2013) is one of the best approaches. The important difference between this method and the method used by Furukawa and Hernández (2015) is that it uses the position and orientation information of the cameras as well as the coordinates of the sparse feature points generated from the structure calculation. The estimated depth-maps are obtained from the mesh data generated by the sparse feature points. After depth-map refinement and depth-map fusion, a dense 3D point data cloud can be obtained. This method can easily and rapidly obtain a dense point cloud. An implementation of this method can be found in the open-source software openMVS (Shen, 2013).

5.3.2.6 Structure from Motion (SfM) Approach

The dominant photogrammetry technique adopted to derive 3D point clouds from the sequence of overlapping UAV images is SfM (Turner *et al.*, 2012). The SfM is a photogrammetric technique that is based on two principles: (i) the binocular vision and (ii) the changing vision of an object that is moving or observed from a moving point. The 3D model now can be computed from overlapping images without the need for pre-requisite information on camera location and orientation, camera calibration and/or surveyed reference points in the scene.

The SfM differs from the traditional photogrammetry mainly in three aspects: (a) features can be automatically identified and matched in images at differing scales, viewing angles and orientations, which is of particular benefit when small unstable platforms are considered for image collection; (b) the equations used in the algorithm can be solved without information of camera positions or GCPs, although both can be added and used; and (c) camera calibration parameters can be automatically solved or refined during the process. Thus, the SfM can automatically deliver photogrammetric models without requiring rigorous homogeneity in overlapping images, camera positions, and calibrations.

The general 3D reconstruction algorithm without a priori position and orientation information can be roughly divided into two steps (Dellaert *et al.*, 2000). The first step involves recovering the 3D structure of the scene and the camera motion from the images. This is generally referred to as the SfM problem. The second step involves obtaining the 3D topography of the scene by generating a dense point data cloud or mesh data cloud from multiple images. This is generally referred to as the MVS problem (Harwin, 2015). The entire process can thus be referred to as SfM-MVS (Iglhaut *et al.*, 2019). Figure 5.4 depicts a schematic workflow of the SfM-MVS process.

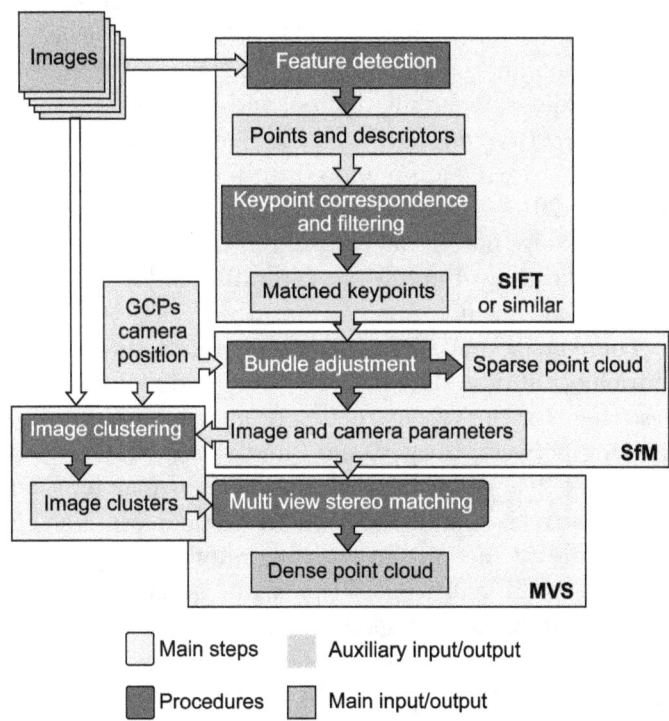

FIGURE 5.4 Workflow of the SfM-MVS process (Iglhaut *et al.*, 2019).

The SfM-MVS process starts with the automatic extraction of keypoints (i.e., points or sets of pixels with distinctive contrast or texture) in the images. The keypoints are identified in all images and then tied (matched) across images where they appear. The Scale Invariant Feature Transform and its variations are the most common algorithms for keypoint identification and matching in SfM. Given a sufficient number of images and keypoint matches, SfM performs bundle adjustments (Jiang and Jiang, 2017) to simultaneously compute the camera positions and parameters and a sparse 3D point cloud of the scene (consisting of the position of keypoints matched in different images). The outputs of SfM are then scaled and georeferenced based on GCPs and/or data from navigation devices from the camera or its platform. The camera positions and parameters obtained from SfM are then applied to generate a densified point cloud using MVS algorithm. Prior to the MVS densification and for computational efficiency or even viability, images are clustered based on their locations. In this way, the dense point cloud of each cluster (i.e., group of images) is computed separately.

The SfM techniques have revolutionized 3D topographic surveys in many fields, such as physical geography, map landslides and assess glacier movement. The orthophotos and DSM/DTM can be utilized for topographic map creation and updation. Using SfM, it is possible to acquire images of entire buildings, including the roof and other parts inaccessible to the scanner (Micheletti *et al.*, 2015). In addition, 3D modeling of cultural heritage sites has led to scientific and cost-effective documentation and archiving (Wu *et al.*, 2018). However, the SfM algorithm has limitation in many applications because of the time-consuming calculation (Fonstad *et al.*, 2013). In many applications, the SfM algorithm has higher requirements of computing speed and accuracy.

There are several improved SfM methods, such as the incremental SfM, hierarchical SfM, and global SfM. The incremental SfM is the most popular strategy for the reconstruction of unordered images using feature point matching between images, and bundle adjustment. As the resolution and number of images increase, the number of matching points and parameters optimized by bundle adjustment will increase dramatically. This results in a significant increase in the algorithm's computational complexity and will make it difficult to use it in many applications (Turner *et al.*, 2012). Micheletti *et al.* (2015) highlighted the potential of SfM applications in geomorphological research and suggested freely available SfM tools to process high-resolution topographic and terrain data acquired through a smartphone.

5.3.2.7 Visualization Approaches

Data visualization is the use of human natural skills to enhance data processing and organization efficiency (Ham *et al.*, 2016). It refers to techniques used to communicate relationships of the data with images. With the rise of big data, it is important to interpret increasingly larger batches of data. Machine learning makes it easier to conduct analyses, such as predictive analysis, which can then serve as helpful visualizations to present. Data visualization is going to change the way we work with data, as they are going to be expected to respond to issues more rapidly. Geospatial or spatial data visualizations relate to real-life physical locations, overlaying familiar maps with different data points. These data visualizations are commonly used to display spatial and temporal trends and patterns and can be most useful for proper planning. Many software, such as R language, ggplot2, Tableau, FineReport, Power BI, etc., are used for data visualization (Zollhöfer *et al.*, 2018).

The UAV-based computer vision includes mechanical vision, optical recognition, and comprehension of images. In computer vision, the UAVs act as a mobile sensor; 3-dimensionally scanning the environment (Pflanz *et al.*, 2018). A flight system uses various onboard sensors to collect information from the area around it which can be used to detect the defects, control layers and objects, inspect shapes and sizes, examine surfaces, assess the condition of objects, check and evaluate inventories, and many more. A UAV can also collect visible information and automate to process it. Image handling and processing systems have shown considerable improvement and provided the opportunity for increasing interest in the use of UAVs in many sectors. Such large interest in the use of UAVs has grown because of the improvements in technology and the increasing demand for timely and accurate information to cope up with the changing environment.

UAVs are revolutionizing the applications by data collection in terms of ease, accuracy, and cost. The emerging technologies, such as AR, VR, AI, 3D modeling, and IoT, can further enhance the information and results (Sharma and Garg, 2019) drawn from UAV's data (Qu *et al.*, 2018). There is a synergy between these technologies, which complement each other when coupled with UAV to derive a meaningful output from the data. For example, images from a camera mounted on a UAV can be feed over the Internet to a VR headset in real-time to offer an enhanced experience to the users. In the long run, UAV computer vision will assure quality in many business areas.

5.3.3 Image Processing Approaches

Image processing methods can be used to process the multi-images for classification, detection, and objects retrieval. The movement of autonomous UAV is mostly based on self-monitoring and self-targeting on the basis of real-time object detection and classification techniques. Vision cameras can help in the detection process of objects and persons. Thermal IR cameras can also be used to detect the heat profile to locate pipeline leakage, forest fire etc. The UAVs can be integrated with thermal and vision cameras for target detection.

In search and rescue operations, image processing can be done either at the GCS for post target identification or at the UAV itself, using onboard processors with real-time image processing capabilities (Shakhatreh *et al.*, 2018). The image post-processing approaches using 3D point clouds, DTMs, and orthomosaics can be used to analyze a large number of images with higher accuracy (Harwin, 2015). Artificial intelligence (AI) is being used in image processing to find accurate information in the image and extract the desired information.

Visockiene *et al.* (2014) compared two popular photogrammetic software, Pix4D Mapper (Switzerland) and PhotoMod (Russia) software, in a study to process the UAV images. The Pix4D Mapper is specialized for UAV images processing, while PhotoMod is designed for common photogrammetry processing. Orthomosaic may be saved into the following formats: BMP, DGN, IMG, GeoTIFF, JPEG, NITF, PCIDSK, PNG, TIFF. Table 5.5 presents a comparison of this two software (PhotoMod and Pix4D Mapper) with regard to performing various photogrammetric steps.

TABLE 5.5 Comparison of Features of Two Popular Software.

Photogrammetric Steps	PhotoMod	Pix4D Mapper
Images deployment at the world map	No	Automatically
Photo camera calibration	No	Yes
Images correction depending on photo cameras lens disorientations	Automatically	Automatically
Image orientation: (interior orientation, relative orientation, triangulation)	(Manually, Semi-automatically, Automatically)	Automatically

Photogrammetric Steps	PhotoMod	Pix4D Mapper
Orthomosaic generation	Semi-automatically by Mosaic module	Automatically
Terrain model (DTM) creation	Semi-automatically by DTM module	Automatically
Topographic map creation and updation	Manually by Stereo Draw module	Manually

5.3.3.1 Data mining approaches

Today's digital world is surrounded by big data forecasted to grow 40% per year into the next decade. Data mining is defined as the process of applying specific algorithms for extracting patterns and new knowledge from large banks of data (Austin, 2010). Today's UAVs are small, simple and economical to operate, allowing survey data to be collected in a continuous stream throughout the flight and instantly uploaded to a server for near real-time analysis. All this data also creates noise which requires powerful tools or techniques to mine useful data. Relying on techniques from the intersection of database management, statistics and machine learning, data mining techniques can process and draw conclusions from a vast amount of information (Sharma and Garg, 2019).

The UAVs create massive amounts of data during the survey of landscapes, habitations, forests etc., that need to be managed, stored and analyzed efficiently. The data streams collected on board UAV sensors, such as ultraviolet, infrared and thermal imaging cameras, hyperspectral sensors and LiDAR scanners, may be required to integrate with other collateral data and information for complete analysis (Fahlstrom and Gleason, 2012). In many applications, the surveyed data are required to be uploaded instantly to the cloud, processed, and then downloaded to both mobile devices of the users and to a central computer in the office. So a high-performance analysis of the database becomes very important.

Data mining allows users to look beyond simple factual information and consider the consequences of each decision in a wider context. It can highlight implicit, previously unknown and potentially useful knowledge from large data sets (Sowmya and Subramani, 2018). For example, if localized adverse weather occurs, data mining will identify similar weather patterns from previous years' data, and predict how market prices of agricultural produce will change. It is expected that in future, information and knowledge that can be drawn (mined) from big data will be one of the greatest assets available to users.

5.3.3.2 Fuzzy Logic Approaches

Fuzzy logic was proposed by Zadeh in 1965 as a way to more accurately capture the real-world. It is a form of first-order logic that can determine the value of a statement as being true or false in the form of a number between 0 and 1, where true is 1, and false is 0. In fuzzy logic applications, a number of fuzzy sets and linguistic variables are defined (Sabo and Cohen, 2012). Its control can deal effectively with highly non-linear systems over conventional control methods. Modularity and adaptability are the key cornerstones of fuzzy logic control.

Fuzzy logic has benefits of adaptability to uncertainty, near-optimal results, ability to compute online, etc. It allows the user to capture much more information about the environment in a more efficient manner. Fuzzy logic excels in circumstances where the environment is continually changing, and incomplete information is available. In recent times, due to the advantages of fuzzy logic systems, they are also being used in expert systems (Garcia-Aunon, et al., 2017). The fuzzy logic approach can lead to lower development costs, superior features, and better end-product performance.

Landing a UAV on a moving platform is a very difficult problem scientifically and from an engineering perspective. Fuzzy logic provides the solution to a challenging environment when the UAV tries to land on a moving platform, such as a delivery van or a warship pitching in high seas (Sabo, and Cohen, 2012). The UAV has to land within a designated area with a small margin of error. Similarly, fuzzy logic helps the UAV make good navigational decisions amid a sea of statistical noise called "genetic-fuzzy" because the system evolves over time and continuously discards the lesser solutions. Compared to other state-of-the-art adaptive thinking and deep learning techniques, genetic fuzzy is scalable, adaptable, and very robust (Ross, 2010). The reduced failure rate is a distinguishing attribute of fuzzy logic.

The benefits of fuzzy logic combined with other intelligent control techniques, like genetic algorithms, artificial potential fields, and neural networks, are the real-time computations, and therefore can effectively be used for path planning of UAVs. In addition to artificial potential fields as an intelligent control method, fuzzy logic has shown excellent results for path tracking and collision avoidance for both mobile robots and UAVs (Hui et al., 2006). Compared with other path planning methods, the fuzzy logic control gets good results, and it can be obtained online with any type of environment. When introducing stationary obstacles and exploiting information about these obstacles (i.e., distance and angle), path tracking of UAVs using fuzzy logic has shown promising results (Pérez-Rastelli and Santos, 2015).

Fuzzy logic, based on multi-valued logic, provides a unique method for encoding knowledge about continuous variables. Juan *et al.* (2018) proposed four methods to carry out the path planning: attraction, fuzzy logic, adaptive-network-based fuzzy inference system, and a particle swarm optimization algorithm. All of them calculated the location of the waypoints to be followed by the UAVs to minimize the distance and the risk exposed to the people.

As fuzzy logic offers several advantages in a challenging environment which is continually changing and where the information available is incomplete, the fuzzy logic controller can be used for motion planning of UAV, as shown in Figure 5.5. Four inputs into the system are (Sabo and Cohen, 2012) (i) distance from the UAV to the obstacle, (ii) angle between the UAV and the obstacle, (iii) the distance to the target, and (iv) the error between the current heading angle of the UAV and the angle of the target in relation to the inertial reference frame. Fuzzy logic manipulates the input using heuristic knowledge and typically converts them to outputs in three stages: fuzzification, decision-making logic, and defuzzification. Fuzzification will take analog inputs and converts them to a continuous value between 0 and 1 based on their degree of membership in each function. A knowledge base is then used to form a set of rules relating to the inputs and outputs in the form of if-then statements. The outputs are finally converted back to crisp numbers using a defuzzification method.

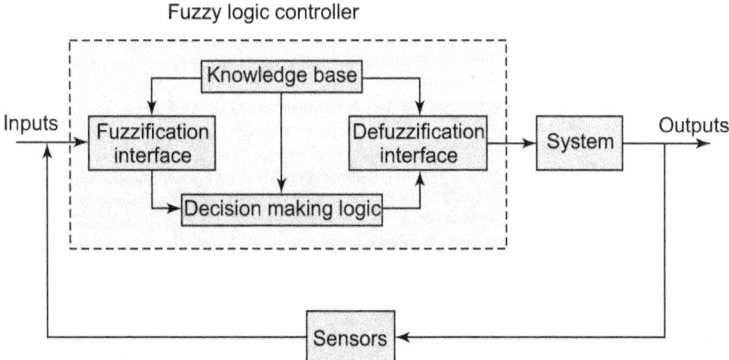

FIGURE 5.5 Fuzzy logic control system.

5.3.3.3 Machine Learning Approaches

Machine learning is about learning from data. Machine learning techniques can be applied to images captured by UAVs. The first step of machine learning is data cleansing, which includes cleansing up the image and/or textual-based data and making the data manageable. This step sometimes might include reducing the number of variables associated with a record. The next step is selecting the right algorithm, which includes getting acquainted with the application in hand. Machine learning uses three popular algorithms in remote sensing (Ma *et al.*, 2017): (i) random forest; (ii) support vector machine (SVM); and (iii) artificial neural network (ANN). In some scenarios, where there are multiple features but limited records, the SVM algorithm might work better. If there are a lot of records but less features, ANN might yield better prediction/classification accuracy. Machine learning has also played an important role to perform automatic land classification from remote sensing images (Gibril *et al.*, 2018).

Normally, several algorithms are applied on a dataset, and the one that works best is selected. Bur in machine learning, a combination of multiple algorithms can also be employed, which is referred to as ensemble, to achieve higher accuracy of the results (Wang and Wang, 2019). Similarly, multiple ensembles may be applied on a dataset in order to select the best ensemble. It is practical to choose a subset of algorithms based on the problem and then narrow down the algorithm that performs best.

A simple block diagram to use a machine learning algorithm is shown in Figure 5.6. The first challenge in machine learning is that the training segment of the dataset, should have an unbiased representation of the whole dataset, and should not be too small as compared to the testing segment of the dataset. From the input data, the system would attempt to extract data, so-called "feature extraction," to reduce more resources. After finishing this process, the acquired model is compared with the accurate model to ensure the selection of the most suitable model. If the acquired model is suitable, it must be kept to predict or test data accuracy with other data sets. This step is called cross-validation. The second challenge is overfitting, which can happen when the dataset used for algorithm training is also used to evaluate the model (Duro *et al.*, 2012). This will result in very high prediction/classification accuracy. However, if a simple modification is performed, then the prediction/classification accuracy may reduce.

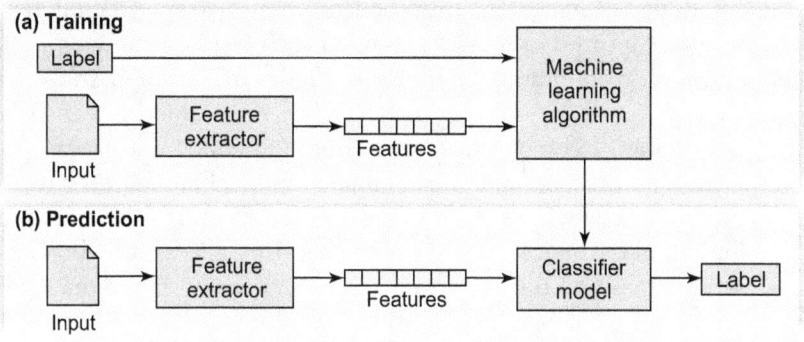

FIGURE 5.6 A block diagram to use a machine learning algorithm.

Machine learning algorithms can be simply classified as (i) supervised, (ii) unsupervised, and (iii) reinforcement learning. The supervised learning algorithms utilize known models and enable the estimation of unknown parameters. They can also be utilized in higher-layer applications, such as estimating the mobile users' locations and behaviors, which can help the UAV operators enhance their services' quality. The unsupervised learning algorithms utilize the input data itself in a heuristic manner. They can be used for heterogeneous base station clustering in wireless networks and in fault/intrusion detections. The reinforcement learning algorithms utilize the dynamic iterative learning and decision-making process. They can be used for estimating the mobile users' decision making under unknown scenarios (Jiang *et al.*, 2017).

The techniques used for image analysis and classification are generally either object-based or pixel-based techniques. Both have been integrated into different machine learning models for semi-automated feature extraction (Duro *et al.*, 2012). A number of object-based and pixel-based semi-automated approaches have already been developed for mass movement detection, using different machine learning models (Tsourdos *et al.*, 2011). According to Shakhatreh *et al.* (2018), five machine learning steps can be utilized by UAV remote sensing, which includes (i) gathering data from various sources, (ii) cleansing data to have homogeneity, (iii) selecting the right machine learning algorithm, (iv) gaining insights from the results, and (v) data visualization of results.

Bejiga *et al.* (2017) proposed machine learning techniques applied to images captured by UAVs equipped with vision cameras to help in search and rescue operations, as shown in Figure 5.7. In their study, a pre-trained Convolutional Neural Network (CNN) with trained linear SVM is used to determine the exact image/video frame in which a lost person is detected. Search and rescue operations employing machine learning technology may face many challenges. As UAVs are battery-powered, so there is a significant limitation on their onboard processing capability. Protection against adversarial attacks on the employed machine learning techniques poses another important challenge (Shakhatreh *et al.*, 2018).

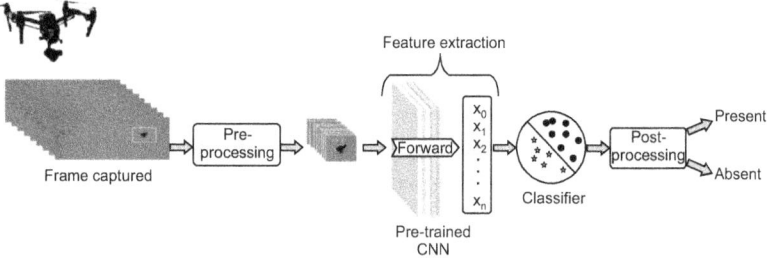

FIGURE 5.7 Use of UAVs with machine learning technology (Bejiga *et al.*, 2017).

5.3.3.4 Deep Learning Approaches

Over the last decade, deep learning models, and in particular CNNs, have been successfully used for the classification and segmentation of remotely sensed imagery for a broad range of object detection applications (Hui *et al.*, 2006). The CNNs are mostly supervised machine learning in which masses of labeled sample patches are used to feed learnable filters to minimize the applied loss function (Mazumdar and Padhi, 2011). The performance of these approaches is strongly dependent on their network architecture, the sample patches selected for input, and on graphics processing unit (GPU) speed (Kim and Bhattacharya, 2007). While CNNs have shown superior capabilities for extracting the features from remote sensing images, only a few studies have been carried out to date using CNN. Deep learning techniques are also least used because of the limitations of onboard processing capabilities and power resources on UAVs. Therefore, there is a need to design and implement onboard, low power, and efficient deep learning solutions using UAVs (Carrio *et al.*, 2017).

Al-Najjar *et al.* (2019) used UAV data to produce an RGB orthomosaic photo and a DSM of the study area. The workflow and the procedure are shown in Figure 5.8. A deep learning classifier (CNN) was used to analyze two different datasets to map the existing land cover classes into bare land, buildings, dense vegetation (trees), grassland, paved roads, shadows, and water bodies. The CNN was trained using only the RGB image. Then, the RGB image and the DSM data were fused to study the effectiveness of a combined dataset. The two networks were trained to assess whether the fusion of RGB and DSM would yield superior results. This study compared the capability of CNN to accurately classify UAV images of lower resolution with very-high-resolution aerial photos and found promising results with the fusion of datasets.

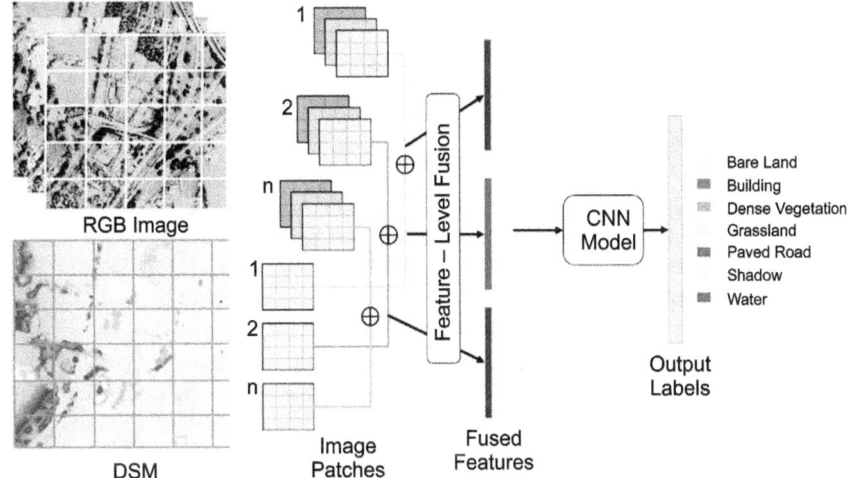

FIGURE 5.8 Flowchart of the proposed CNN based classification model (Al-Najjar *et al.*, 2019) [*See Plate 4 for Color*].

Figure 5.9 shows the CNN based classification of the above two datasets. The overall accuracy of land cover classification into seven classes increased in RGB data from 96.5% (without DSM) to 99.1% in RGB plus DSM fused image. A similar improvement in the average accuracy was observed from 93.3% (without DSM) to 98.9% when considering the DSM data. Thus, the classification that did not include the DSM feature had a lower performance than the fused datasets using the testing & training datasets.

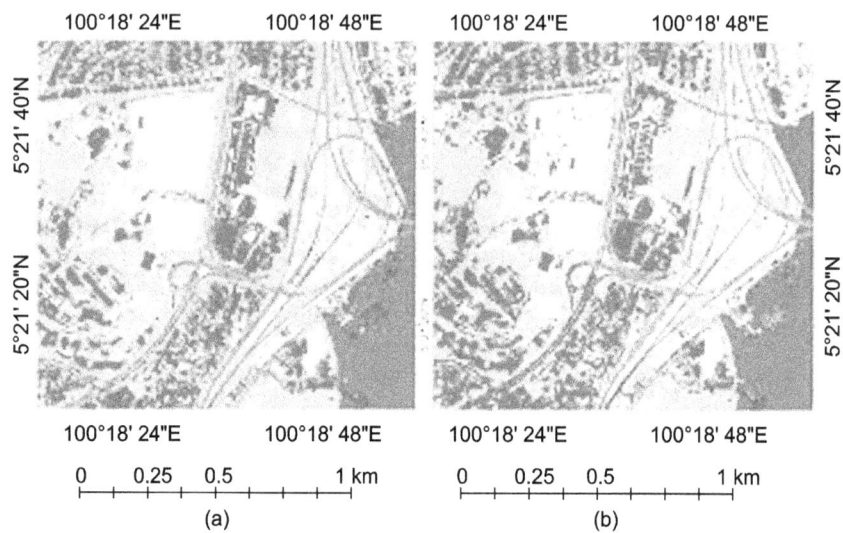

FIGURE 5.9 Classification map produced by the CNN based on the two datasets: (*a*) RGB only dataset and (*b*) RGB + DSM dataset (Al-Najjar *et al.*, 2019) [*See Plate 5 for Color*].

5.3.4 Cloud Computing Approaches

Cloud computing is defined as a computing service from anywhere, anytime, using any computing device through the Internet (Sharma and Garg, 2019). It is provided through a type of parallel and distributed system on virtual inter-connected computers, which can be dynamically provisioned and presented as one or more unified computing resources based on "service level agreements" established between the service provider and the users (Paloz-Sanchez *et al.*, 2017). Cloud computing technology does not require servers, hardware and software in the offices, as the processes are carried out elsewhere over the Internet. All of this occurs in a dynamic and scalable way, on-demand. In addition, the resources required can be increased or decreased according to the operation needs. Compared to traditional computing, cloud computing provides systematic information that directly impacts the final recipient of the process in terms of effectiveness and analysis process. It also provides decision support for an activity.

In this approach, the data captured by UAV sensors is automatically sent to the cloud computing servers, where it is immediately processed with the power of the servers and cloud processing. The processed data is analyzed and optimized so that the results can be shared automatically. For example, multispectral imaging technology can help with land management, crop fertilization or early detection of biotic stress, monitor irrigation by identifying areas where water stress is suspected to estimate crop yields and explore agronomic indices (Saura *et al.*, 2019).

Some companies now also offer UAV mapping and processing software that can carry out real-time image processing on their servers, such as Drone Deploy, Drone Mapper, and Airware. Outsourcing the computing work to process several UAV images would not only shorten the long processing time required by sophisticated photogrammetry software, but also it can provide quick output while the survey team is still in the field. The disadvantage is that the secrecy of the data may be lost while sharing for outsourcing. Using these services requires access to mobile data or the Internet, which may not be available in remote areas or during disasters events. Some services, such as Drone Deploy, require the purchase of a separate unit that is mounted on the UAV to function for the transfer of data.

5.3.5 UAV Photogrammetry Software

The camera on UAV takes a large number of photographs which are stitched together using specialized photogrammetry software to create the 3D images. The photogrammetry software is required to process the images into maps and 3D models. The typical output from photogrammetry UAVs software includes orthomosaic images/map layers (Google maps, Mapbox, etc.), 3D point clouds, meshed 3D models, DEM, DSM, DTM, NDVI images and classified outputs. There are many photogrammetry software for creating 3D maps and models, which would require a lot of computer processing power for a large dataset.

Aicardi *et al.* (2016) carried out a detailed study and compared the results of different photogrammetric software packages. This type of comparison is very beneficial to understand the capability of software to be able to best handle UAV images with reference to the point density of the model for 3D visualization. Table 5.6 presents a summary of the number of millions of points created using 190 UAV images in each software. The acquired data is processed using the SfM approach implemented in different software tools: Agisoft Photoscan Professional, Pix4D, 3Df Zephyr, SURE, MicMac, VisualSfM, and ContextCapture (Federman *et al.*, 2017). This software are commercial and proprietary solutions, except MicMac and VisualSfM, which are open-source software. It is observed from the table that the Context

Capture software obtains the denser cloud (19 million points). The other software allows extracting dense clouds (between around 4–7 million points), except for 3Df Zephyr and VisualSfM where the obtained point clouds are <1 million points.

TABLE 5.6 Cloud Points Created Using Various Software (Aicardi *et al.*, 2016).

	Photoscan Professional	Pix4D	3Df Zephyr	MicMac	SURE	Context Capture	Visual SfM
Millions of cloud points	3.5	4.4	<1	4	7.5	19	<1

The use of open-source software is another possibility for post-processing for aerial imagery. Some of these include MapKnitter from Public Lab, OpenDroneMap, MicMac, VisualSfM, OSM Bundler, Microsoft Photosynth, Photosynth Toolkit, My3DScanner PhotoTourism and AutoDesk 123D Catch (Bartoš *et al.*, 2014). The Microsoft's Image Composite Editor may be a good choice for panoramic image stitching, although it does not create geometrically corrected orthophotos. Such open-source and free software packages can be more difficult to use, as they may have fewer features than commercial software, but they are nonetheless powerful enough to achieve useful results. The Flight Riot software (*http://www.flightriot.com*) provides a comprehensive overview of open-source mapping software and associated techniques, with instructions on the proper use and selection of cameras, UAV platforms, and processing software. Processing big bunches of high-definition aerial imagery can be slow and require a powerful computer processor.

Sometimes low-quality image processing software is required in the field just to check that an adequate number of images with overlaps have been taken, thereafter a higher quality model is created from software in the office. Additional care is taken to avoid systematic errors created by the processing software for applications requiring precision (Rothermal *et al.*, 2012). For instance, the combination of radial lens distortion and many images taken from near-parallel directions can introduce an effect called "doming," which makes a flat surface into a dome.

Some of the UAV photogrammetry mapping software is given in Table 5.7 and briefly explained below. The readers can get more details of software from its respective Website.

TABLE 5.7 The UAV Photogrammetric Software.

S. No.	UAV Photogrammetry Software	Link
1	Agisoft PhotoScan photogrammetry software	https://www.agisoft.com/features/professional-edition/
2	Agisoft Metashape 3D software	https://www.agisoft.com/
3	AutoDesk ReCap photogrammetry software	https://www.autodesk.com/recap/introducing-recap-photo/
4	AutoDesk 3D mapping software	https://www.autodesk.com/solutions/3d-mapping-software
5	Bentley Context Capture	https://www.acute3d.com/contextcapture/
6	3DF Zephyr Photogrammetry software	https://www.3dflow.net/3df-zephyr-feature-comparison/
7	DroneDeploy	https://www.dronedeploy.com/product/platform/
8	DroneDeploy 3D mapping mobile app	https://www.dronezon.com/drone-reviews/drone-deploy-app-review-with-tutorials-pricing-contact-details/
9	ESRI Drone2Map for ArcGIS	https://www.esri.com/en-us/arcgis/products/drone2map/overview
10	Maps Made Easy orthophoto and 3D models software	https://www.mapsmadeeasy.com/guide
11	MicMac open-source software	https://www.micmac.ensg.eu
12	Open Drone Map photogrammetry software	https://www.opendronemap.org/
13	Pix4D Mapper photogrammetry software	https://www.pix4d.com/product/pix4dmapper-photogrammetry-software
14	Precision Mapper 3D map software	https://www.precisionhawk.com/precisionmapper
15	SimActive Correlator3D software	https://www.simactive.com/
16	SURE software	http://www.nframes.com
17	VisualSfM open-source software	http://www.ccwu.me/vsfm/

5.3.5.1 Agisoft PhotoScan Photogrammetry Software

The Agisoft LLC, a Russian-based research company, was founded in 2006, focusing on computer vision technology. With time, Agisoft LLC has gained expertise in image processing algorithms with digital photogrammetry techniques setting the direction for the development of applied tools. Agisoft PhotoScan software performs photogrammetric processing of digital images and generates 3D spatial data. It is also widely used in GIS applications, cultural heritage documentation and visual effects production, and indirect measurements of objects at various scales. The professional version of Agisoft PhotoScan has the features for photogrammetric triangulation, editing and classification of dense point cloud, export of DSM/DTM and georeferenced orthomosaic, 3D model generation and texturing, 4D modeling for dynamic scenes, measurement of distances, areas, volumes, customize processing workflow with Python scripts, multispectral imagery processing, stitching panorama stitching and network processing.

5.3.5.2 Agisoft Metashape 3D Software

The Agisoft Metashape is a standalone software product that performs photogrammetric processing of digital images and generates 3D spatial data. It is useful for (i) aerial triangulation for dense point cloud and orthomosaic generation, (ii) creation of accurate DEMs for volume calculation in mining & quarrying, (iii) precision agriculture and customizable vegetation index calculation using panchromatic, multispectral and thermal imagers, (iv) archeology & documentation, (v) aerial and oblique imagery support for cultural heritage preservation, and (vi) game & video design models.

5.3.5.3 AutoDesk ReCap Photogrammetry Software

The ReCap (stands for Reality Capture) is AutoDesk's software for creating 3D models from photographs or laser scans. It is used to create point cloud or mesh ready for CAD and Building Information Modeling (BIM) tools. About 1,000 images can be processed on the cloud and then integrated into AutoDesk products, like AutoCAD Civil 3D and InfraWorks, for further design.

5.3.5.4 AutoDesk 3D Mapping Software

The AutoDesk has a vast range of products for creating 3D maps using UAVs or ground-based system. The 3D mapping products include (i) AutoCAD professional design software for 2D and 3D CAD, (ii) REVIT- a BIM software to plan, design, construct, and manage buildings and infrastructure,

(iii) AutoCAD Civil 3D for civil engineering design and construction, and (iv) AutoCAD Map 3D which is a model-based mapping software with access to CAD and GIS data.

5.3.5.5 ContextCapture Software

The Bentley's ContextCapture, which is the evolution of the previous Acute3D, can quickly produce the 3D models derived from photographs and/or point clouds. It can create highly detailed 3D meshes to provide precise real-world context for design and construction. Hybrid processing in ContextCapture enables the creation of engineering-ready meshes that incorporate the versatility and convenience of high-resolution photography supplemented, where needed, by additional accuracy of point clouds from laser scanning. It can use up to 300 gigapixels of photos and/or 3 billion points from a laser scanner, resulting in fine details, sharp edges, and high geometric accuracy 3D models. It reduces processing time with the ability to run two ContextCapture instances in parallel on a single project.

The ContextCapture Editor enables fast and easy manipulation of meshes at any scale as well as the generation of cross-sections, extraction of ground and breaklines, and production of orthophotos. These data can be integrated with GIS and engineering data to enable the intuitive search, navigation, visualization, and animation of information within the visual context of mesh to quickly and efficiently support the design process.

5.3.5.6 3DF Zephyr Photogrammetry Software

The 3DFlown, also known as 3DF Zephyr, is a complete photogrammetry software package produced by the Italian company 3Dflow. It is a perfect tool to capture reality automatically and reconstructs 3D models from photos. It has a user-friendly interface and the capability to export in many common 3D formats. It has tools for laser scan support, point cloud/mesh support, statistical analysis, orthophoto and orthomosaic, contour lines generation, DEM and multispectral viewer. It can even generate high definition videos without the need for external tools.

5.3.5.7 DroneDeploy Enterprise 3D Map Software

It is the leading cloud software platform for drone mapping and 3D modeling. DroneDeploy can be used to create accurate, high-resolution maps, reports and 3D models, as well as real-time 2D live maps for analysis. One of the interesting features of DroneDeploy is the thermal live map which allows visualizing the temperature range variability and generating thermal maps over large areas. DroneDeploy can be extended with an ecosystem of industry-leading tools and customized services. It can be used to install apps for construction, inspection, compliance, agriculture and more. For other details, see Chapter 3.

The DroneDeploy Enterprise Mapping software is a fully featured aerial information platform that allows managing several UAVs, with each UAV creating a 3D photogrammetry map. The processing of all the data into 3D maps takes place in the cloud, making it available to teams to collaborate and analyze the data.

5.3.5.8 DroneDeploy Mobile App

The DroneDeploy mobile app can be used to fly drone to make interactive 3D maps and models. It is very easy to use with both beginners and professionals. It is the best 3D mapping apps to produce orthomosaic, volumetric computation, 3D mapping and agricultural surveying. It provides a specialized solution for the agriculture sector. The software works with most of the latest drones, such as Phantom 3, Phantom 4, Inspire 1 and 2, Mavic Pro, Matrice 100, 200 and 600 models.

For autonomous flight, the app is used for auto-flight and camera settings, automated pre-flight safety checks, auto-takeoff, flight and landing, live stream FPV, smart auto-exposure adjustments, disable auto-flight and resume control with a single tap, multi-flight mission support and starting waypoint selection for continued missions, offline flight capabilities, and custom parameters (altitude, front and side image overlap, camera settings). In-field data analysis can explore interactive maps, measure distance, area and volume, analyze elevation and NDVI, share maps, annotations and messages.

5.3.5.9 ESRI Drone2Map for ArcGIS Photogrammetry Software

The ESRI Drone2Map photogrammetry software turns the drone into an enterprise GIS productivity tool, allowing creating orthomosaics, and 3D meshes. It works with the ArcGIS products; ArcGIS Online, ArcGIS Pro, and ArcGIS Enterprise. It streamlines the creation of professional imagery products from UAV captured still imagery for visualization and analysis in ArcGIS.

It performs in-field image processing to verify the coverage and quality of imagery collection. The Drone2Map streamlines the ground control workflow by using Collector for ArcGIS to capture high-accuracy points that are ready for integration into Drone2Map for the ArcGIS project. It automates image processing routine and produces contours and 3D products (DSM) that can be used further for analysis and visualization. It can be used to visualize imagery from multiple points of view. The software can be used to create base-maps, perform image analysis and measure features. 3D point clouds enable an analysis of natural and built-up features, including volumetric measurements, change detection, lines of sight and obstructions.

5.3.5.10 Maps Made Easy for 3D Maps

The Maps Made Easy processes the aerial images taken from any platform into orthophoto maps and 3D models. It stitches the photos into 3D maps using its own 3D photogrammetry software. It combines multiple looks of an area to create a 3D model from which accurate orthophoto maps are produced. The textured 3D model can be downloaded and previewed online. Its web-based volume measurements make measuring and tracking stockpile volumes quick and easy. It creates and stitches NDVI images. Multiple georeferenced visits to the same mapping site can be combined and arranged to view history for every point on the map. This feature is especially useful for construction monitoring, agriculture, pipeline documentation and solar array inspection. It provides full online processing of data. Any map created on this site can easily be embedded in your own website or blog. The service of the imagery is backed by the Maps Made Easy reliable servers and gives a better control over the map.

5.3.5.11 MicMac Software

The MicMac has been developed at the National Institute of Geographic and Forestry Information (IGN), France, since 2003. It is a free open-source photogrammetric suite that can be used in a variety of 3D reconstruction scenarios. One of MicMac strengths is its high degree of versatility. It can be used in various fields: cartography, environment, industry, forestry, heritage, and archeology, to create 3D models and ortho-imagery. The software is suitable for every type of objects at any scale. The tools can be used for the georeferencing of the end products in the local/global/absolute coordinates system. Some complementary tools open the fields of metrology and site surveying.

5.3.5.12 OpenDroneMap Photogrammetry Software

The OpenDroneMap is an open-source toolkit for processing aerial imagery. For other details, see Chapter 3.

5.3.5.13 Pix4D Mapper Photogrammetry Software

The Pix4D, from the well-known Swiss software house, is especially dedicated to process the UAV images. It is used to generate highly precise, georeferenced 2D maps & 3D models and professional maps from images. The 3D maps are customizable and complement a wide range of applications and software. The software is used for creating various outputs in large number of output formats, which include contour lines, DSM and DTM, 3D textured model, orthomosaic, volume calculation, assess reflectance based on the pixel value in multispectral or thermal imagery, create NDVI, aggregate and visualize the values derived from index maps.

5.3.5.14 Precision Mapper & Viewer Photogrammetry Software

The PrecisionMapper works online to automatically process the aerial images into 2D or 3D products. It features a continuously expanding library of on-demand analysis tools and makes sharing or collaborating very easy. The software can be used to (i) collect aerial data with UAV, (ii) upload data and process 2D or 3D products, (iii) manage, collaborate, and share data, and (iv) analyze data with a library of on-demand analysis tools for creating orthomosaics and 3D models, crop health analysis, and volume measurement.

PrecisionViewer is a desktop software that allows users to easily view flight path coverage, add GCPs, and attach flight logs and flight bounds to surveys.

5.3.5.15 SimActive Correlator3D™ Software

The SimActive is the developer of Correlator3D™ software. The Correlator3D™ software is an end-to-end photogrammetry solution for the generation of high quality geospatial data from UAVs, including satellite and aerial imagery. SimActive first introduced the aerial triangulation and DSM generation engines in the industry, enabling multiple-fold increase in processing speed. It also produces DTM, point clouds, orthomosaics, 3D models, and vectorized 3D features. The Correlator3D™ ensures matchless processing speed to support the rapid production of large datasets being powered by GPU technology and multicore CPUs.

5.3.5.16 SURE Photogrammetry Software

The SURE photogrammetry software is developed by nFrames GmbH, a Germany based company, for 3D reconstruction from images with unique performance and flexibility. The SURE software transforms imagery from aerial cameras, multi-head oblique systems, UAV cameras as well as most consumer-grade terrestrial cameras into 2.5D or 3D data, such as point clouds, photo-realistic textured meshes and true orthophotos in a streamlined full automatic integrated process with image processing techniques. It has multiple interfaces and modular (SURE Aerial, SURE Aerial Mesh Add-on, and SURE Editor) design for quality production. The software accepts many different image and orientation files (Rothermal *et al.*, 2012).

5.3.5.17 VisualSfM

The VisualSfM is a GUI application for 3D reconstruction using SfM. It is open-source which runs very fast by exploiting multicore parallelism in feature detection, feature matching and bundle adjustment. In addition, VisualSfM (64-bit) provides the interfaces to run with several additional tools, including SURE software for dense reconstruction.

5.3.6 LiDAR UAVs Software

For LiDAR and photogrammetry mapping, the UAV can be programmed to fly over an area autonomously, using waypoint navigation. Various software and tools are available, either open-source or commercially, for processing the LiDAR point cloud data collected by UAVs. These are presented in Table 5.8. For further details about this software, readers are advised to go through the respective website.

TABLE 5.8 Various Software and Tools Available for Processing the LiDAR Point Cloud Data.

Software	Basic Functions	Link
AutoCAD MAP 3D	Creates DEM from LiDAR data. It uses the indexing method on the point cloud data, which make it easy to load and perform analysis on it.	*http://www.autodesk.com/products/autocad-map-3d/features/all/gallery-view*
BCAL LiDAR tools	Open-source tools used for processing, analyzing and visualizing LiDAR data. Written in IDL programming language and intended to be used as an add-on in the ENVI remote sensing software.	*http://bcal.geology.isu.edu/Envitools.html*

(continued)

Software	Basic Functions	Link
Bentley Points tools V8i	A stand online software to process point cloud data. It creates point clouds quickly for distribution and reuse.	*http://www.bentley.com/en-US/Promo/Pointools/pointools.htm?skid=CT_PRT_POINTOOLS_B*
CloudCompare	Open-source software for processing point cloud data. Can build TIN and raster surfaces and then generate a DEM and DSM.	*http://www.cloudcompare.org*
Computree	Open-source platform dedicated to LiDAR data processing.	*http://computree.onf.fr/?lang=en*
Dielmo3D products and lidar2dems	For LiDAR data processing.	*http://applied-geosolutions.github.io/lidar2dems/*
FME	Feature Manipulation Engine (FME) has some tools for the LiDAR point cloud data.	*https://www.safe.com/data-types/lidar-point-cloud/*
FullAnalyze	For handling, visualizing and processing LiDAR data.	*http://code.google.com/p/fullanalyze/*
FugroViewer	A freeware designed for use with LiDAR and other raster- and vector-based geospatial datasets, including data from photogrammetry.	*http://www.fugroviewer.com/*
Fusion/LDV	Free software which allows 3-D terrain and canopy surface models and LiDAR data to be fused with 2-D imagery (e.g., ortho-photographs, topographic maps, satellite imagery, GIS shapefiles).	*http://forsys.cfr.washington.edu/fusion/fusionlatest.html*
Global Mapper	The LiDAR module provides numerous advanced LiDAR processing tools, including automatic point cloud classification, feature extraction, cross-sectional viewing and editing, dramatically faster surface generation, and much more.	*http://www.bluemarblegeo.com/products/global-mapper-lidar-download.php*

Software	Basic Functions	Link
GRASS GIS	Open-source for LiDAR data processing.	Open-source *https://grasswiki.osgeo.org/wiki/LIDAR*
LAS tool	Used for processing small number of points.	*http://rapidlasso.com/LAStools/* and *http://www.cs.unc.edu/~isenburg/lastools/*
LiDAR Analyst	An extension for ESRI ArcGIS, it is used for feature extraction from airborne LiDAR data, allowing geospatial analysts to automatically extract 3D objects, such as bare earth, trees or buildings.	*http://www.vls-inc.com/login/downloads/LIDAR_Analyst_v5_1_2_ArcGIS.zip*
LiDAR360	Contains tools for generating DEM, DSM, editing DEM, and other spatial products.	*http://greenvalleyintl.com/product/lidar360/*
LiForest	Manipulates LiDAR point cloud and extract useful information, especially for forest studies, forest metrics, regression models, individual tree segmentation, and point cloud segmentation.	*http://www.liforest.com/*
Lizardtech–LiDAR Compression tool	It compresses the LiDAR data which reduces 25% of the total size (in Mr SID format) while retaining every return.	*http://www.lizardtech.com/products/lidar/*
MCC-LiDAR	A free software in C++ application for processing LiDAR data in forested environment.	*http://sourceforge.net/projects/mcclidar/*
Merrick's Mars Viewer	Free viewing application which supports basic LiDAR data navigation and 3D visualization.	*http://www.merrick.com/index.php/geospatial/services-gss/mars-software*
Point Cloud Library	Used for making your own tools and workflow.	*http://pointclouds.org*

(continued)

Software	Basic Functions	Link
Points2 Grid Utility (winP2G)	Tool used for generation of DEMs from LiDAR point cloud data.	*https://cloud.sdsc. edu:443/v1/AUTH_open topography/ww w/files% 2Fp2g%2FGEON-points2 grid-Utility-Setup-v1_3. exe*
Quick Terrain Reader	Free software for visualizing point clouds and pre-built DEMs in a fast and intuitive way.	*http://appliedimagery. com/download*
RIEGL RiALITY – IPAD App	It is an IPad application to visualize points cloud acquired through RIEGL laser scanners. It shows the true 3D color of the point cloud.	*http://www.riegl. com/media-events/ multimedia/3d-app/*
Saga GIS	Modules available for points cloud.	*http://www.saga-gis.org*
SPDLib	Open-source software for processing LiDAR data.	*http://www.spdlib.org/ doku.php*
Terrascan, Terra match and Terra Modeler	Used for LiDAR data processing software.	*www.terrasolid.fi*
VCG Library	Used for making your own tools and workflow.	*http://vcg.isti.cnr.it/vcglib/*
libLAS	Used for making your own tools and workflow.	*https://www.liblas.org/ start.html*

The software simplifies the flight planning, processing and analysis of UAV data to provide actionable intelligence to service providers so that inputs can be optimized and better decisions can be made with reduced cost. A good UAV software must include the automation of UAV flight plans, augmented view, georectification of images and 2D/3D models generation. Table 5.9 presents a comparison of various UAV software along with their primary uses. The information given in the table may help the users decide about the selection of a software.

TABLE 5.9 Comparison of UAV Data Processing Software (Singhal et al., 2018).

Software	Uses	Pros	Cons
Aero Points	Provides accurate GCPs for UAV survey.	Automatic data upload, fast and easy to setup	Lack of functions, like NDVI maps, 3D models creation
Agisoft PhotoScan	Generation of 3D models and photogrammetric processing	Cost-effective all-in-one software suite with a full range of image sensors, including NIR, RGB, thermal and multispectral.	Compare to Pix4D it is clunky and with less functionality. One license for one computer only.
Drone Deploy	Creates accurate, high-resolution maps, reports, and 3D models, as well as real-time 2D live maps for analysis.	More suited to agricultural applications and compatible with third-party UAV hardware and accessories	Surface detail for buildings is somewhat disappointing as compared to more specialized solutions from Agisoft and Pix4D.
Drone Logbook	Imports flight log to fill automatically, view GPS trace and replay it in 3D.	Compliance and custom reporting, mission planning & operations calendar	Not capable to make DEMs and DTMs
Drone Mapper Rapid	Ortho-mosaicing, DEM, and robust processing algorithm	Freely available for download and testing on a limited data set	Limited to processing of 150 images for an area
Drone Mapper Rapid Expert	Full functionality and allows input of up to 1000 geotagged JPEG images of 12 MP format or greater.	Can be used up to 1000 georectified images and can produce X8, X4 or X2 DEM	Does not have self-calibration facility
Field Agent	Replacement of Sentera's AgVault, allows to scout crops, capturing health and vegetation index data in near real-time.	IOS Mobile device enabled automated flight with unlimited NDVI images	Limited functionality for a mobile-free version, but pay USD 25 per month for full functionality

(continued)

Software	Uses	Pros	Cons
Live NDVI	Video technology-live stream NDVI video at the field's edge	Provides NDVI video in real-time while drone is still in air	Compatible only with Sentera Double 4K sensor
Pospac UAV	Direct georeferencing and post-processing software	Maximum accuracy and efficiency with compatible UAV	Compatible with APX-15 L-UAV
Pix4D	Plans UAV flight, creates orthomosaics, point clouds and professional 3D models	Photo camera self-calibration, and automatic DTM generation	Topographic maps can be created only manually.
Photo Mod	Provides all photogrammetric products, like DEM, DSM, 2D and 3D vectors, and orthomosaics	High performance, simplified user-friendly interface and automation of photogrammetric operations	Does not have the camera self-calibration facility, unlike Pix4D
Sensefly eMotion	UAV flight planning and data management	Connects wirelessly to UAVs, finger swipe flight planning and full 3D environment, multi-flight mission support	Very advanced software but no real-time NDVI processing.
Sense fly Survey 360	Complete aerial mapping system	Precise geodata provides accurate point cloud and surface model outputs in lesser time	Does not provide fully 3D environment
Virtual Surveyor	Use for 3D visualization photogrammetry, virtual surveying, flight planning and civil designing	Flood simulation capability, fast simulation and interoperable with CAD environment, can process huge amount of data files	Not responsible for reliability, accuracy and suitability of information.

5.4 SUMMARY

Having images and spatial data is not the same as having a map, and therefore the collected imagery/data can be processed through software to generate a map. Good software will stitch together and georectify the UAV images quickly for their further uses. Proper georeferencing of UAV collected images is traditionally based on highly accurate GCPs and DTM of the area. In a large number of overlapping photographs, common GCPs are identified using an image matching algorithm to create a mosaic. The 3D model of an object is built up from the images acquired during the UAV survey (Xu et al., 2014). Increasing levels of autonomy in both flight software and post-processing software allows the creation of 3D maps quickly with less direct human intervention and without having GCPs.

Technology keeps on changing with time. A large number of processing software is used for UAV mapping, and the demand is changing as applications of UAVs are growing. This software is regularly updated and improved upon, as the demand for UAV mapping and the market for photogrammetry software are expanding. The Pix4D and Agisoft PhotoScan are considered the preferred choice for processing and analyzing UAV aerial images, with relatively simple user interfaces. However, this photogrammetry software is expensive and require considerable processing power to operate. Choosing a software package is highly dependent on the funds available, the processing power available, and the goals defined.

REVIEW QUESTIONS

Q.1. Discuss various data products that can be collected with UAVs.

Q.2. Explain the role of GCPs. How do you get GCPs for processing UAV images?

Q.3. Discuss the georeferencing process of UAV images.

Q.4. Differentiate between a DEM, DTM and DSM. What are the different methods to collect elevation data?

Q.5. What is an ortho-image? What is its use? Explain the general procedure with a flow diagram to generate ortho-images.

Q.6. Write the short notes on (i) MVS approach, (ii) SfM approach, and (iii) visualization approach.

Q.7. What is data mining? Describe the data mining approaches to be used with UAV images.

Q.8. Describe in brief the function and working of (i) Fuzzy logic approaches, (ii) Machine learning approaches, (iii) Deep learning approaches, and (iv) Cloud computing approaches to be used with UAV data.

Q.9. Discuss the salient features of two popular UAV-based photogrammetric software.

Q.10. Describe the main characteristics of two popular UAV-based LiDAR software.

CHAPTER 6

REGULATORY SYSTEMS FOR UAVS

6.1   INTRODUCTION

Aviation rules emerged in the 20th century. The Chicago Convention (also known as the International Civil Aviation Convention) was signed in 1944 and established the International Civil Aviation Organization (ICAO), the specialized agency of the United Nations (UN), which is responsible for coordinating and regulating international air travel (ICAO, 2011). The Federal Aviation Administration (FAA) of the US granted authority to use the first UAV in the US national airspace in 1990 (FAA, 2010).

A UAV may be navigated by remote control or by an onboard computer that provides it a degree of autonomy (Dempsey, 2015). The degree to which a UAV can be autonomous for an application is a matter of debate. Civilian and commercial UAVs could be used for various purposes, including surveillance around malls or monitoring and control of road traffic, domestic policing, oil, gas, mineral exploration, and the delivery of goods. The UAVs can be flown in a park as a hobby for pleasure or used by a photographer to take stunning photographs. Additional uses might include crop monitoring and spraying, fire services, and forestry surveillance.

There is no doubt that UAVs are a game-changing technology in terms of security and privacy and have become a growing societal threat in recent years. The UAV, which was considered earlier an immature technology, has now started drawing policymakers and administrators' attention. With terrorism on the rise and the growing use of UAVs in society, these could become prime vehicles for terrorists and criminals. These devices can not only be used to crash into densely populated areas, injuring and killing people but also to harvest the information to aid criminals and terrorists in cyber/digital crimes that could range from theft of personal bank details to the loss of highly sensitive military recon footage (Andrews, 2020).

Right policies are essential for UAV technologies to develop and revolutionize the world. The UAV regulations vary widely from country to country or may not even exist in some country. The regulations vary from users to users, depending on where in the world they're using the technology (Floreano and Wood, 2015). Countries, worldwide, are working to integrate UAVs into their national airspace program. Regulatory bodies are working with government agencies, industries, and institutions to streamline the policies (Stöcker et al., 2017). Flexible rules/policies have been made by the United States, Europe, and China, the three largest potential markets in the world for commercial UAVs.

With the recent increase in UAVs in space, adequate measures are required to be taken to properly regulate and carry out safety measures with the movement of UAV's operation (Andrews, 2020). This process will require the coordination and involvement of academia, industry, government agencies, like the Regulatory authority and Department of Defense, local authorities, and private sectors in order to successfully implement an efficient and robust UAV system for safe commercial operations (Finn and Wright 2012). The socio-economic impact of UAVs will require not only creating reliable products through scientific and technological advances but also developing a regulatory framework that will enable public and private entities to operate in safety, privacy, and security.

6.2 ISSUES OF CONCERN FOR UAVs

There are several important issues to be considered while flying UAVs in an area/region for data/information gathering. Several issues are common throughout the globe, while flying regulations may vary from country to country (Dalamagkidis et al., 2012). Earlier, there were many restrictions that limited the use of UAVs. Still, now, operations BVLOS and at night are being considered, and unmanned traffic management is being tested in urban environments (Gajdamowicz et al., 2017). The growing number of incidents has highlighted the need to detect and disable UAVs that are maliciously used.

The UAVs are available in different shapes and sizes, having different ethical problems associated with them. There are some serious ethical and legal issues, in contrast to UAVs' benefits in gathering information. The ethical and legal issues (Finn and Wright 2012) about gathering geographical data by UAVs can be divided into four main areas: regulations, privacy, safety, and noise (Dempsey, 2015). The UAV should also avoid the restricted areas (i.e., airports, protected areas, or urban areas). Safety can be improved through compliance with operational rules, mass limitations, and other guidelines. With UAVs being allowed for commercial use, an entire industry has emerged, enabling businesses to enhance safety and delivery standards (Floreano and Wood, 2015).

There is very dense air traffic in the sky, and therefore it calls for the proper regulation of flying UAVs. Additionally, national laws and social concerns related to privacy and security are to be considered to adopt the UAV technology and its full utilization. It has also created new UAV research and development avenue for academia and industry focused on anti-drone methods. For example, commercial drones provide optimal physical attack or smuggling infrastructure due to their flight range, size, speed, and carrying capabilities.

Table 6.1 lists some of the threats anticipated from the UAVs. The anti-drone market is expected to reach US$1.85 billion by 2024 (Reportlinker, 2018).

TABLE 6.1 Threats Mapped to Types of UAVs (Nassi *et al.*, 2019).

UAVs	Privacy		Physical Attacks	Crime	Cyberattacks
	Video Streaming	Carrying Surveillance Equipment			
Nano	Yes	—	Targeted assassination	Targeting homes for burglaries	—
Micro	Yes	3D mapping using radio transceiver. MITM attacks against cellular networks. Tracking a person according to his/her device.	Carrying radioactive sand	Smuggling goods into prison yards	—

(continued)

UAVs	Privacy		Physical Attacks	Crime	Cyberattacks
	Video Streaming	Carrying Surveillance Equipment			
Mini	Yes	Yes	Carrying a bomb. Colliding with an airplane	Hijacking radio controlled devices. Smuggling goods between countries	Establishing a covert channel.
Swarm	Yes	Yes	Multiple casualty incidents	Cyber warfare	—

Some major issues of concern for uses of USVs, almost common to all countries, are briefly discussed below.

6.2.1 Air Traffic Control

Integration of UAVs into the airspace is a critical issue, which has been discussed in depth by Dalamagkidis *et al.* (2012). It has two parts: about aviation issues and regulations about social issues. Air traffic control is an essential condition for coordinating the large fleets of UAVs, where regulators do not permit large-scale UAV delivery missions (Business Insider, 2017). A UAV can determine, through GPS, if it is flying in "No-fly zone."

Amazon has designed an airspace model for the safe integration of UAV systems, as shown in Figure 6.1 (Kimchi, 2015). In this proposed model, the low-speed localized traffic (200 ft/61 m) will be reserved for (i) terminal non-transit operations, such as surveying, videography, and inspection and (ii) operations for lesser-equipped UAVs, for example, ones without sophisticated sense-and-avoid technology. The high-speed transit (400 ft/122 m) will be reserved for well-equipped UAVs as determined by the relevant performance standards and rules. The "No-fly zone" (500 ft/152m) will serve as a restricted area where UAV operators will not be allowed to fly, except in emergencies. Finally, the pre-defined low-risk locations will include areas, like the designated academy of model aeronautics airfields, where altitude and equipage restrictions in these locations will be established in advance by aviation authorities.

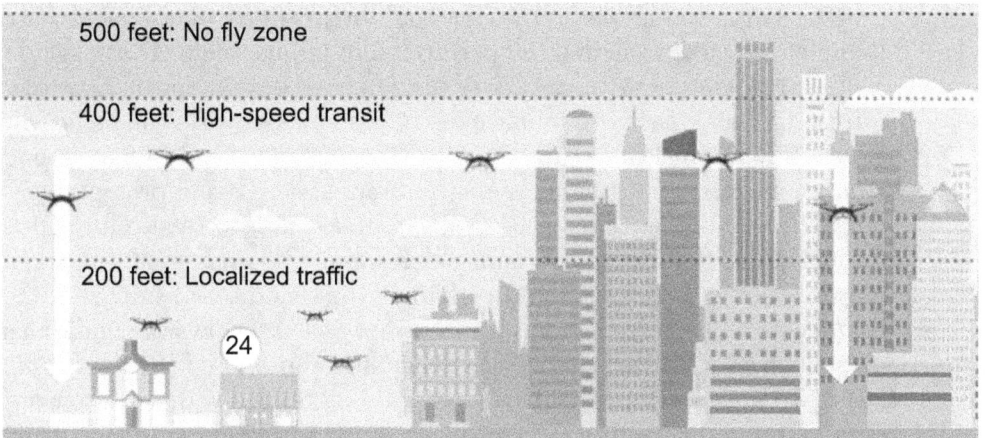

FIGURE 6.1 Amazon airspace model for the safe integration of UAV systems (Kimchi, 2015).

6.2.2 Wi-Fi/Bluetooth Security Issues

Some UAVs can be controlled by smart devices through Wi-Fi. Compared with the use of a remote control, there is no doubt that the use of a smart terminal brings much convenience. However, Wi-Fi brings security risks as well. Attackers can attach a small computer, such as Raspberry Pi to a UAV, fly it over places where they wouldn't normally be able or allowed to enter, and then use Wi-Fi, bluetooth, or RFID for vulnerability and hijacking information. The UAVs equipped with a radio transceiver could be used to hijack bluetooth mice. In addition, any other bluetooth-connected devices could be accessed, such as computers, from which attackers could obtain key information and figure out users' login credentials (Dempsey, 2015). The UAVs equipped with radio transceivers can be used for (i) locating and tracking people across a city by extracting unencrypted information from the lower layers of the "open systems interconnection" model of radio protocols (e.g., Wi-Fi and bluetooth) that reveal the "media access control" addresses of their device's owners (e.g., smartphones, smart watches), (ii) 3D through-wall imaging with UAVs, and (iii) performing a "man-in-the-middle" attack on telephony by downgrading 4G to 2G (Nassi *et al.*, 2019).

Cyber-security could be another big concern. The potential harm from the cyber-security threats can be amplified when a malicious operator is able to operate a swarm of UAVs simultaneously to perform the task, turning a targeted cyber-attack against an organization into cyber warfare and a targeted assassination into a massive terror attack (Nassi *et al.*, 2019).

The UAVs can be hacked, which could have serious consequences on the safety of the system and the perception of public safety (Dalamagkidis et al., 2012). These UAVs can now be used to hack servers, spy on networks, extract data, and block communications. Corporate networks can be heavily affected by the malicious use of UAVs, so companies need to have sound security measures in place to prevent unwanted access and protect themselves from cyber warfare attacks. Some attacks aim to steal information through security holes of communication links, while others aim to spoof sensors, such as GPS spoofing. In addition, the security and storage of the data collected by these systems are of concern as it could have additional consequences related to the privacy debate if compromised (Richard et al., 2014). To improve security of UAV, it is also equipped with extra sensors, such as infrared detectors, ultrasonic detectors, and vision sensors to avoid obstacles. Since the information obtained by these sensors will change the behavior of a UAV, it can be seen that these sensors have a strong influence on security (Dempsey, 2015).

6.2.3 Privacy Issues

The right to privacy is a civic right that is applicable to all citizens. Despite the promising applications of UAVs, there are several concerns regarding privacy, public safety, security, collision avoidance, and data protection. The biggest concern related to ethical and legal issues of UAVs is about the privacy. The UAVs have become a threat to privacy since they can be used as spying devices. Private UAV operators are invading the privacy of beachgoers and private property owners. There has always been a concern about the invasion of privacy, in particular private property, whether it is gathered by high-resolution satellite images, airplanes, or UAVs (Vattapparamban et al., 2016). The operations of low altitude flying UAV, mandated by the regulating authority, can also be noisy and, therefore, may adversely affect birds, wildlife, and local populations.

The DJI has detected a "third-party plug-in" (called JPush) for the Android platform's applications, collecting personal information by updating the application. The DJI believes that the available "plug-in" may also breach privacy. The JPush is accused of collecting the app names installed on the user's Android device and sending the entire information to the JPush server (Zhi et al., 2020).

6.2.4 Moral and Ethical Issues

Currently, the moral and ethical issues are far from being resolved. The ethical issues emerge as the result of how a UAV is used by a user in contrast with how those who experience the UAV could have been used by the users (Richard et al., 2104). Unless UAVs are fully autonomous, it can be assumed that the actions of UAVs are subject to ethical evaluation based upon (i) the actions of the person controlling the UAV, (ii) the intentions of that person, and (iii) the consequences produced by the UAV. Ultimately, it is the person who is controlling the UAV that is subject to moral evaluation.

There is an increasing ethical violation in military operations, and UAV operators need to be trained to respond in such ethical situations (Finn and Wright, 2012). It is claimed that autonomous systems would lead to an increase in ethical behavior in the battlefield. For defense uses, moral and ethical issues are similar to the issues of using long-range precision weapons. As UAVs can cause significant human life losses on the ground in combat zones, these systems are required to adapt the current ethical implications, and the operators or decision makers must be held responsible for both intentional and unintentional damages created by these systems. It should consider moral uncertainties to establish ethical standards as guidelines for behavior.

Among the most prominent concerns raised are the threat of civilian casualties, adherence to the principles of international humanitarian law, and the potential for overuse. The ethical concerns raised by the capabilities of UAVs can be mitigated with relative ease, first and foremost, by adapting protocols to reinforce the principles of international humanitarian law (Andrews, 2020). Ethically, there is a fear that UAVs will take away lots of jobs; create more unemployment and under-employment; drive down wages; and generally move us one step further away from being a caring society. Therefore, there is a need to develop ethical guidelines to set the boundaries about UAVs' proper and improper use.

6.2.5 Safety Issues

Since UAVs are flown with a low attitude, safety is also a primary concern. The regulations must prohibit flying UAVs near populated areas or over people for safety reasons. A common practice to prevent UAVs from causing harm to manned aircraft and infrastructure is to enforce "No-fly zones" around airports and sensitive areas (Urekwitz et al., 2016). For example, the famous Gatwick Airport, United Kingdom, stopped 1,000 flights from

December 19 to 21, 2018, which affected the travel plans of around 140,000 people when a drone was sighted over the airport.

In many countries, these regulations mandate that the UAV operator be in constant visual contact with the device. Several tech-savy companies and research institutes have proposed obstacle avoidance solutions for UAVs. The UAV threat to our physical safety ranges from accidental harm through miscalculations to targeted attacks. Accidental harm could be the legitimate owner losing the control, a hacker losing the control, or a hardware or software malfunction within the device (Helnarska *et al.*, 2018). A UAV crashing out of the sky and hitting the humans may cause significant damage, and the bigger the UAV, the greater the damage.

UAVs pose an unacceptable risk to jetliners. The heavy Li-Ion batteries in UAVs could puncture the skin of an aircraft wing or smash the blades in an engine. As soon as Amazon announced that it was planning to start delivering packages to consumers by automated drones, there was a concern about the drones colliding in mid-air; crashing into buildings; and creating havoc in aviation (Business Insider, 2017).

6.2.6 Legal Issues

Legal issues related to civilian use of UAVs include data protection, copyright law, and private property concepts (Demir *et al.*, 2015). Data protection focuses on the rules to prohibit the acquisition of personal information by UAVs. In most countries, there are no regulations prohibiting the use of UAVs for image capturing. Copyright laws regulate capturing images of people in public and private areas. Although there are several regulations available in some countries, there is still a need to develop more comprehensive regulations. Private property rights prevent capturing images of private properties under defined conditions. In many countries, this issue is under the regulation of personal rights.

To minimize the risks of UAV-triggered incidents or accidents, an increasing number of national and international authorities have introduced legal provisions that mandate "Go," "No go," or "How to go" statements that either allow, prohibit or restrict flight operations. Such regulations significantly impact how, where, and when data can be captured and the diffusion of the technology within a national context (Stöcker *et al.*, 2017). The biggest barrier to the adoption of UAVs in gathering geographic information is that while hobbyists are free to fly UAVs, several commercial UAV activities are currently not legal.

6.3 AVIATION REGULATIONS

The history of UAV regulations dates back to manned aviation and the emergence of airplanes during World War II. In 1944, the international community established the first globally acknowledged aviation principles- the Chicago Convention (FAA, 2016). After years of technological research and innovation, UAVs developed into a commercially workable system in 2000s. In 2006, the ICAO identified and declared the need for international harmonized principles of the civil uses of UAVs (ICAO, 2015). At this time, five countries had already established and enacted the UAV regulations. The United Kingdom and Australia were the first nations that promulgated regulations in 2002. Several countries then followed, and from 2012 onwards, the trend changed. An increasing number of countries established national UAV regulations between 2012 and 2016, and more than 80% of the 65 countries with national regulations enacted them.

In general, six main criteria are often considered when developing UAV regulations, as given in Table 6.2 (Stöcker *et al.*, 2017). These criteria are important from the operation point of view of UAVs.

TABLE 6.2 Six Main Criteria while Developing UAV Regulations.

S.No.	Criteria	Details
1.	Applicability	Pertains to determining the scope (considering type, weight, and role of UAVs) where UAV regulations are applied.
2.	Operational limitations	Related to restrictions on the locations of UAVs.
3.	Administrative procedures	Specific legal procedures could be needed to operate a UAV.
4.	Technical requirements	Includes communications, control, and mechanical capabilities of UAVs.
5.	Human resource requirements	Qualification of pilots.
6.	Implementation of ethical constraints	Related to privacy protection.

Aviation regulations determine the rules and minimum flight requirements for a UAV to be used for a specific purpose. Regulatory issues are important limiting factors for the deployment of UAVs, including communication systems. The UAV regulations are being continuously developed to control the safe operations of UAV, considering various factors, such as UAV type, altitude, and speed. A comparative analysis of 19 national UAV regulations is given in Stöcker et al. (2017).

Table 6.3 presents the characteristics of the regulatory framework in force in 16 countries in 2019 (Faraz, 2020). However, many countries have yet to develop regulations to ensure a legal environment for safe operations of UAVs. Since these regulations keep on changing with time, the users are advised to go through the latest regulations prevailing at that time from their official website.

TABLE 6.3 The Characteristics of the Regulatory Framework in 16 Countries.

Territory	Possibility of Commercial Flights	License Required to Fly	Training Required for Pilots in Order to Obtain Licenses	Insurance Required for Commercial Flights	Possibility to Perform BVLOS Flights	License Required for BVLOS Flights
Poland	✓	✓	✓	✓	✓	✓
United Kingdom	✓	✓	✓	✓	✓	✓
China	✓	✓	✓	✓	✓	✗
Canada	✓	✓	✗	✓	✓	✗
Germany	✓	✓	✓	✓	✗	✗
India	✓	✓	✓	✓	✗	✗
France	✓	✓	✓	✗	✓	✗
South Africa	✓	✓	✓	✗	✓	✗
Indonesia	✓	✓	✓	✓	✗	✗
Australia	✓	✓	✓	✓	✗	✗
Brazil	✓	✓	✗	✗	✓	✗
Mexico	✓	✓	✓	✗	✗	✗
United States	✓	✓	✗	✗	✗	✗
Japan	✓	✗	✗	✗	✗	✗
Russia	✗	✗	✓	✗	✗	✗
Argentina	✗	✗	✗	✗	✗	✗

6.3.1 The FAA Regulations

For maintaining the safety of UAVs and the public, the FAA in the United States has developed rules to regulate the use of small UAVs (FAA, 2016). In response to broad interest in the United States, the FAA initiated a registration program of small UAVs in December 2015 with certain regulations. These regulations in the United States imposed by the FAA have created a debate over how extensively UAVs should be used in the national airspace system. The FAA approved the civil use of UAVs in the United States, provided they are flown at a defined altitude (US Department of Transportation, 2013). For example, the FAA operational rules require small UAVs that weigh more than 0.55 lbs (250 g) and below 55 lbs (25 kg) to be registered in their system.

The FAA regulations are broadly organized into two categories: prescriptive regulations and performance-based regulations. The prescriptive regulations define what should not be done, whereas the performance-based regulations indicate what must be attained. Almost 20% of FAA regulations are performance-based regulations (FAA, 2016). These small UAVs require flying less than 400 feet (122 m) without obstacles in their vicinity. The operators maintain a line of sight with the operated UAV at all times. For small UAVs, the minimum flight visibility, as observed from the location of the control station, should not be <3 miles (4.8 km) (US Department of Transportation, 2013).

The FAA regulations allow the operations of UAVs only during daylight and during half-light if the UAV is equipped with anti-collision lights (FAA, 2016). The rapid development of commercial UAVs and potential benefits motivated the FAA to define a road map for the gradual integration of civil UAVs in the national airspace system that can fly beyond the operator's line of sight by 2028 (Shakhatreh et al., 2018). These regulations also initiate height and speed constraints and other operational limits. The ground speed of a small UAV may not exceed 100 mphh or 160 kmph. Although, there is a process through which users can apply to have some of these restrictions waived off (NCLS, 2016).

Other FAA regulations include that UAVs are not to be operated over or close to the public infrastructure (e.g., stadiums) or critical infrastructure as they may pose a danger to the public. The critical infrastructure includes petroleum refineries, chemical manufacturing facilities, pipelines, wastewater treatment facilities, power generation stations, electric utilities, chemical or rubber manufacturing facilities, and other similar facilities. For critical infrastructure, legislation has been developed to protect such infrastructure from UAV operators (FAA, 2016). The regulations should prohibit flights over unprotected people on the ground who are not directly participating in the UAV operation. They also require UAVs not to fly within 5 miles from an airport without permission from the airport and the control tower; thus, avoiding the endangerment of people and aircraft.

UAVs can be employed for public use, provided they are flown at a certain height. In Figure 6.2, the FAA categorizes the airspace for small UAVs in accordance with the FAA regulations (FAA, 2016). The FAA pays a lot of attention to Class A—all US airspace from 18,000 to 60,000 feet where commercial planes fly. This is followed by the airspaces around airports, called Class B (big airports), C, and D (smaller cities). Class G (700-1,200 feet), an unregulated airspace, is allocated to use UAVs for civilian use.

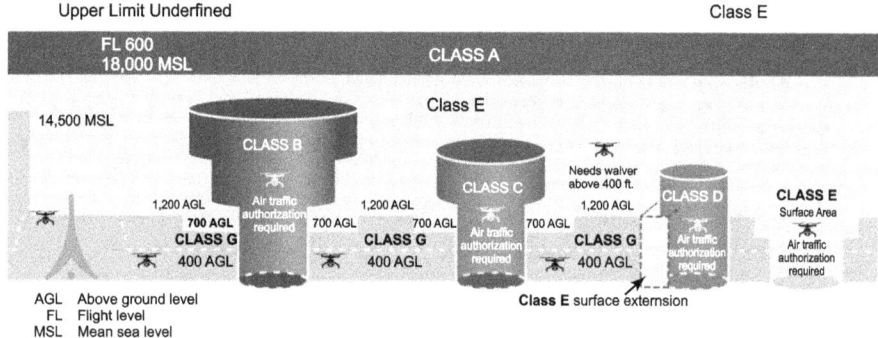

FIGURE 6.2 The FAA airspace classification (FAA, 2016).

In violation of the FAA regulations, owners of UAVs can face civil and criminal penalties. The FAA has also provided a smartphone application, "Know Before You Fly" (B4UFLY), which gives UAVs' users important information about the limitations that are location specific. The "Know Before You Fly" campaign by the FAA aims to educate the public about UAV safety and responsibilities (Shakhatreh et al., 2018). The FAA also enforces its regulations along with Law Enforcement Agencies to deter, detect, investigate, and stop unauthorized and unsafe UAV operations (Vattapparamban et al., 2016).

6.3.2 The European Union (EU) Regulations

In the EU, regulations are developed by the European Aviation Safety Agency (EASA) (EASA, 2017). These act as prototype regulations that are to be adapted by the member states to their nations. The regulations regulate the operation of UAVs within the Single European Sky (SES) airspace (SES, 2017). They do not regulate other operations related to the police, army, or firefighting. The prototype regulations do not introduce changes to the Standardized European Rules of the Air (SERA, 2017). As per EU regulations, the operator of a UAV shall be registered with the competent authority (AA, 2017) and is responsible for the safe operation of the system. The operator would comply with the EASA requirements and any other applicable regulations (i.e., security, privacy, data protection, environmental protection, and insurance). The UAV shall display the unique identification mark. The competent authorities may designate zones or airspace areas where UAV operations are restricted (Floreano and Wood, 2015).

The operation of large civilian drones (>150 kg) in the EU member states is regulated by EU laws and monitored by the EASA. Smaller drones are largely unregulated at the EU level and are regulated by the individual EU member states (European Commission, 2014). Some countries, including Belgium, Czech Republic, France, Ireland, Italy, Sweden, Germany, Switzerland, and the United Kingdom, have national regulations for unmanned aircraft, although these regulations differ between these countries.

In Belgium, the Federal Public Service Mobility & Transport and the corresponding ministry (i.e., BCAA) is the key regulator for safe UAV flights. Its main tasks are to authorize UAV flights with respect to Belgian Drone Law, organize theoretical drone exams, and issue pilot and drone licenses. The flexible and efficient regulations are proving to be a key contributor to the growth of the drone ecosystem in Belgium (Culus et al., 2018).

In France, the Directorate General of Civil Aviation is the regulatory body for UAVs. For "category A" UAVs (<25 kg), no prior authorization is required. The UAVs weighing less than 2 kg will have a ceiling of 50 m of maximum working height and 200 m of maximum working distance (UAV Coach, 2020). Autonomous UAVs cannot weigh >1 kg. Protection device/mechanism (e.g., parachute, emergency landing feature) should be available. An annual declaration of fitness to fly would be required for operators of UAVs weighing >25 kg (category B). The UAVs of 150 kg or more are subject to EU regulations, and their flight and use conditions are authorized on a case-by-case basis. The regulations allow UAVs to operate BVLOS and near cities under specific conditions (DHL, 2017). The operator should have a license with 100 h of flying experience and 20 h of UAV training for flying out of the line of sight. Cargo weight for commercial delivery UAVs are capped at 500 g. It is restricted to fly within 4 km from airport, heliport, or aerodrome, as well as in populated areas including events, concerts, festivals and firework shows, and sensitive airspace, including nuclear power plants, military courses, hospitals, prison, historical monuments, and national parks. The UAVs can fly within 150 m above the ground (Helnarska *et al.*, 2018).

In Israel, the country uses its civil aviation laws, including licensing, registration, and operational requirements, to regulate UAVs. Under the Civil Aviation Authority of Israel rules, UAVs cannot be flown in populated areas below 5,000 feet, except during take-off and landing or with prior approval. In the absence of prior approval, the rules also disallow the simultaneous remote operation of more than one UAV by the same operator from the same remote pilot station (Padmanabhan, 2017).

Poland currently excludes UAVs weighing <25 kg from registration requirements. For UAVs >25 kg, a permit to fly is required; however, operational restrictions may be applicable (LOC, 2019).

Switzerland allows autonomous operation of UAVs within the line of sight as long as the operator can regain control (Swiss Federal Office of Civil Aviation, 2014).

Swedish regulations establish several categories based on the weight and usage of UAVs. The UAVs used for commercial and research purposes and those flown outside the operator's sight generally require licenses from the Swedish Transport Agency.

The UK allows the operation of UAVs of <20 kg within the line of sight over congested areas with approval from the UK Civil Aviation Authority. For UAVs between 20-150 kg, operators of these UAVs are therefore required, among other things, to obtain a certificate of airworthiness, have a permit to fly, and have a licensed flight crew. Some common rules, such as flying in the line of sight and flying 50 m away from a person, would apply. For all commercial uses of UAVs, permission would be required prior to their operation (UK Civil Aviation Authority, 2012). In spring 2018, the UK government released a new set of regulations to increase the overall UAV safety while ensuring that Britain remains at the forefront of drone technology development. These measures also seek to expand their use with businesses and infrastructure, as given below (Patterson, 2018).

1. Drones over 250 gm will have to be formally registered.

2. Drone pilots will have to clear a safety awareness test before they are allowed to fly.

3. The police will have powers to ground drones if suspected of criminal activity or unsafe flying. Drone pilots will present their registration documents if required by the police.

4. Drone pilots will have to use apps to ensure their planned flights are safe and legal. Restricted areas will be easier for pilots to view, such as schools and military bases.

5. Drones may be completely banned from flying near airports at a height above 400 ft.

6. Use of geofencing is essential, which would help pilots comply with the guidelines.

In 2013, the European Commission defined a road map (EASA, 2017) to integrate remotely operated aircraft in the European aviation system using a staged approach, whose aims, timeline, and research priorities are similar to those defined in the US road map. However, the first stage of the European road map consists of extending European regulations for large UAVs to those below 150 kg by adapting national regulations, which may result in more liberal regulations than those in the United States (European Commission, 2014). Furthermore, the European Commission explicitly aims to improve UAVs' public perception and promote widespread use of these technologies for public and commercial uses.

The EU has classified the UAVs into three main categories; open, specific, and certified category, as below (González-Jorge *et al.*, 2017; Tsouros *et al.*, 2019):

6.3.2.1 Open Category

Operations in this category do not require authorization or a pilot license. This category is limited to operations in VLOS, up to 120 m height, and performed with UAVs compliant with some technical requirements defined. Safety is ensured through compliance with operational rules, mass limitations (25 kg maximum), and product safety requirements. Operation is allowed only during daylight hours, having a maximum range of 500 m.

6.3.2.2 Specific Category

This is the second level of operations, which covers operations that are not compliant to the restrictions of the open category, and are considered "medium risk" operations. Operators must perform a risk assessment (using a standardized method) and define mitigation measures. Operations involving UAVs of more than 25 kg or not operated in VLOS will typically fall under this category. The technical requirements for this category depend on the specific authorized operation.

6.3.2.3 Certified Category

This category is considered "high risk," including operations involving large UAVs in controlled airspace. Rules applicable to this category will be the same as for manned aviation.

6.3.3 The International Civil Aviation Organization (ICAO)

Regulatory impact is currently one of the most important factors affecting the adoption of UAV-based solutions by business and government officials. In recent years, the UAV regulations have changed from being treated as a niche hobby to becoming part of regular aviation operations to a point where national authorities have started developing special regulatory frameworks to address the most urgent issues. UAV regulations' elements are mainly focused on increasing the platforms' reliability, underlining the need for safety certifications for each platform, and ensuring public safety. As technical developments and safety standards govern them, rules and certifications are given equal weightages while flying UAVs (Nex and Remondino, 2014).

At the global level, the ICAO, a specialized agency of the United Nations (UN), prepares the standards and recommends national and international air navigation practices to ensure safe and orderly growth (ICAO, 2015). The ICAO has prepared a document to provide a fundamental international regulatory framework with supporting procedures for air navigation services and guidance material to outline routine operations of UAV throughout the world (Demir *et al.*, 2105). In addition, there are also several national regulations aiming to ensure safe operations of different UAV types in individual national airspaces. The UAV regulations vary between different countries and types of geographical areas (e.g., urban or rural). Table 6.4 lists important regulations to fly UAVs in some countries. Since these regulations keep on changing with time, the users are advised to go through the latest regulations prevailing from their official website.

TABLE 6.4 Important Regulations for the Deployment of UAVs.

Country	Maximum Altitude (m)	Minimum Distance to People (m)	Minimum Distance to Airport (km)
United States	122	8 (25 feet)	8.0
Australia	120	30	5.5
South Africa	46	50	10.0

Country	Maximum Altitude (m)	Minimum Distance to People (m)	Minimum Distance to Airport (km)
United Kingdom	122	50	5.0
Chile	130	36	2.0

The international community started to define the security criteria for UAVs some years ago. In particular, North Atlantic Treaty Organization and EuroControl in 1999 started to prepare regulations for UAV platforms and flights, but they could not lead to a common and international standard, especially for civilian applications (ICAO, 2011). The commercialization of new UAV systems has pushed several national and international associations to analyze the operational safety of UAVs (Demir *et al.*, 2015). The first country to implement all necessary sets of regulations was Poland in 2013 (Helnarska *et al.*, 2018). With the combined efforts of the civilian aviation authorities, the UAV community, and insurance companies, Poland allowed both VLOS and BVLOS operations of commercial UAVs in a secure and user-friendly way. It is observed that many countries have one or more authorities involved in the UAV regulations that operate independently. In Table 6.5, a schematic summary of the already existing regulations in several countries is presented.

TABLE 6.5 Regulations for UAV Use in Several Countries (Nex and Remondino, 2014).

Country	Civil Use of UAV (Laws and Regulations)
Australia	CASA Circular, July 2002 (CASA, 2002)
Belgium	Certification Specification, Rev. 00, 24.01.07
Canada	Approach to the Classification of Unmanned Aircraft, 19.10.10
Denmark	Regulations on Unmanned Aircraft not Weighing More Than 25 kg, Edition 3, 09.01.04
France	Decree Concerning the Design of Civil Aircraft Fly Without Anyone on Board, August 2010
Great Britain	CAP 722, 06.04.10 u. Joint Doctrine 2/11, 30.3.11
Norway	Operation of Unmanned Aircraft in Norway, 29.06.09
Sweden	Flying with UAVs in Airspace Involving Civil Aviation Activity, 25.03.03
Switzerland	Verordnung des UVEK über Luftfahrtzeuge Besonderer Kategorien, 01.04.11

(continued)

Country	Civil Use of UAV (Laws and Regulations)
Czech Republic	Czech Aviation Regulation L2—Rules of the Air, 25.08.11
United States	UAS Certification Status, 18.08.08; Fact Sheet-Unmanned Aircraft Systems, 15.7.10 and NJO7210.766, 28.3.11, 8.2.12 and FAA Bill

UAVs currently have different safety levels according to their dimension, weight, and onboard technology. For this reason, the rules applicable to each UAV could not be the same for all the platforms and categories. For example, in the United States, safety is defined according to its use (public or civic), while in some European countries, it is according to the weight, as this parameter is directly connected to the damage they can make in case of a crash (Helnarska *et al.*, 2018). Other restrictions are defined in terms of minimum and maximum altitude, maximum payload, area to be surveyed, GCS-vehicle connection (i.e., visual or radio), etc.

Regulations on operator and pilot licensing are in their infancy stage in many countries. The indirect control of an operator from the GCS may lead to increased accidents due to human errors. For this reason, in several countries, UAV operators need some basic qualifications, training, and certification. Several universities and organizations offer programs on UAV pilot certification (Stöcker *et al.*, 2017).

6.3.4 Australia

Australia's national aviation authority is the Civil Aviation Safety Authority (CASA). Newly adopted rules in Australia generally exempt commercial flights of UAV weighing <2 kg from having a remote pilot's license or operator's certificate. Operating UAV weighing between 2–25 kg will similarly be exempted if flown over a person's own land for certain purposes and in compliance with standard operating conditions. Operating UAV weighing between 25 and 150 kg under similar conditions will require a remote pilot's license (LOC, 2019). Operators of large UAVs, as well as smaller UAVs for other non-recreational purposes, will be required to obtain a remote pilot's license and operator's certificate. Large UAVs must also obtain airworthiness certifications.

In Australia, safety rules are to be followed when a UAV is flown for recreational use (CASA, 2002). According to the CASA regulations, the main safety rules that apply for recreational flying include (UAV Coach, 2020): (i) daytime flying only, (ii) flying in the visual line of sight, (iii) flying not closer than 30 m from other people, or in populous areas, (iv) in control airspace, its height must not exceed 120 m, (v) it should be kept 5.5 km away from airfield, aerodromes, and helicopter landing sites, and (vi) not to be flown in prohibited or restricted areas without permission.

6.3.5 Canada

In Canada, the national aviation authority is Transport Canada Civil Aviation (TCCA). There is a need to have a special flight operations certificate for UAV in Canada, depending upon whether the UAV is used for recreational or non-recreational purposes, and its weight. If the UAV weight is <1 kg and is used for recreational and not for profit-making purposes, permission from the aviation authority is not required, but the operator has to follow strict safety conditions (Transport Canada, 2017). The UAV weighs between 1 and 25 kg must be registered with the aviation authority (Terra Times, 2019). However, the users of UAV weighing 35 kg or more (irrespective of the purpose) will require a flight certificate (LOC, 2019). After receiving the certificate, the operation of the UAV will be subject to the rules defined by the TCCA.

Depending on the intended use of the system, different rules will apply (Demir *et al.*, 2015), but the restrictions may include locations, time of day, and other operating and safety parameters. The UAV can be flown below 122 m (400 feet) in the air, away from bystanders, at a minimum distance of 30 m for basic operations as well as <5.6 km from airports or <1.9 km from heliports. The UAV cannot fly at the site of emergency operations or advertised events, outdoor concerts, and parades.

6.3.6 China

In December 2015, the Civil Aviation Administration of China (CAAC) published UAV operation provisions to regulate the operation of UAVs. Since June 2017, any UAV weighing 250 g (0.55 pounds) or more must be registered with the CAAC. In China, all UAVs <7 kg are permitted to be flown, but in the case, they weigh between 7 and 116 kg (empty weight) or 150 kg (gross take-off weight) and airspeed of not greater than 100 kmph, a license and UAV certification from the CAAC is required (LOC, 2019). Flying beyond the visual line of sight and above 120 m is not permitted. It is prohibited to fly in densely populated areas, around airports, military installations, or other sensitive areas, such as police checkpoints or sub-stations. All UAVs are subject to China's "No-fly zones." There are further regulations in China that divide UAVs into seven categories according to their weights (Terra Times, 2019).

6.3.7 Germany

The Federal Aviation Office is the regulatory body in Germany. An individual license plate is required to identify the operator for UAVs weighing >0.5 kg. Flights >100 m height and BVLOS are prohibited (UAV Coach, 2020). In order to operate a UAV of >5 kg, the owner must obtain specific flight authorization from the concerned authority. For UAVs not more than 5 kg and without a combustion engine, an authorization to fly may be granted for a period of two years (LOC, 2019). Flights of UAVs weighing >25 kg are not allowed unless authorized by the regional state authorities. Commercial UAV operators are required to obtain a license after passing a test demonstrating the aeronautical skills and knowledge regarding the aviation laws (Terra Times, 2019). It is prohibited to fly within 1.5 km of airfield, above prison, military installations, the crowd of people, sense of accidents, disaster areas, and other sensitive areas affecting national security. Other "No-fly zones" include industrial facilities, power plants and power transmission networks, federal roads, and railroads.

6.3.8 Hong Kong

The regulatory body in Hong Kong is called the Hong Kong Civil Aviation Department (HKCAD). The UAVs <7 kg (without fuel) for recreational purposes can be classified as model aircraft flying, and therefore no permission is required (UAV Coach, 2020). If flying for commercial purposes and/or if flying a UAV that weighs >7 kg (15.4 lbs), permission is required from the HKCAD. The maximum working height of UAV is 300 feet which are flown in daylight only, having VLOS at all times (Stöcker et al., 207). Evidence of pilot competency is required when taking permission to operate UAVs. The UAV should not be flown within 5 km of any aerodrome, 50 m of any person, vessel, vehicle or structure not under the control of the operator, except that during take-off and landing, and 30 m of any person other than the operator.

6.3.9 Israel

In Israel, the operation of all drones is looked after the aviation national authority, the Israeli Civil Aviation Authority (CAAI). All drones are generally subject to extensive regulation, regardless of their weight or type of use (LOC, 2019). Two permits are needed to fly drones: one from the CAAI and the second from the ministry of communication. Commercial drone operations in Israel require a license. Drone pilots must maintain a full line of sight with their drones while flying. Drones cannot fly above 50 m (164 feet) from the ground and in "No-fly zones." It cannot fly within 2 km (1.2 miles) of any airport or airfield and within 250 m (820 feet) of people and buildings (UAV Coach, 2020).

6.3.10 Japan

The Japan Civil Aviation Bureau is the aviation authority in Japan. Japanese law exempts UAVs that weigh 200 gm or less from the licensing, as conditions given in the Aviation Act (AA, 2017). In case of commercial flying of UAVs, the CASA is to be notified, and a certificate is obtained. A different set of rules exist that is specific for the two categories of commercial flying; under 2 kg and over 2 kg. The Ministry of Land, Infrastructure, Transport, and Tourism in Japan has the authority to restrict the flying of UAVs above 150 m (492 feet) of ground level near airports or above densely inhabited areas for safety purposes. According to the Aviation Act and without the exception of approval, the UAV should (UAV Coach, 2020): (i) fly in day time only, (ii) fly in the visual line of sight, (iii) fly not closer than 30 m from other people, or in populated areas, (iv) not exceed 150 m height in control airspace, (v) be kept 9 km away from airfield, aerodromes, and helicopter landing sites, (vi) not fly within 30 m ($\approx$100 feet) of people or private property, (vii) not fly over crowds or sites where large groups of people are gathered, such as concerts or sports events, (viii) not carry specified dangerous items, and (ix) not drop objects while in flight, either intentionally or accidentally.

6.3.11 New Zealand

The Civil Aviation Authority of New Zealand (CAANZ), the aviation authority of New Zealand, differentiates among UAVs based on their weight. The UAVs weighing between 15 and 25 kg may not be operated unless UAV pilots ensure that their UAVs are safe to operate before flying, and also approval is taken (LOC, 2019). Such UAVs must be operated in compliance with a list of operating limits enumerated by the law. The operation of UAVs that weigh >25 kg, and the operation of any size UAV outside of the operating limits, requires an operator's certificate from the CAANZ.

The UAV pilots must maintain a direct visual line of sight with their UAV at all times, and should not fly >120 m (394 feet) above ground level (UAV Coach, 2020). The UAV pilots must have knowledge of airspace restrictions that apply in the area where they are operating. The UAV must not be flown within 4 km (2.5 miles) of any aerodrome, and near military operating areas or restricted areas. To fly in controlled airspace, pilots must obtain an air traffic control clearance issued by Airways. Pilots must at all times take all practicable steps to minimize hazards to persons, property, and other aircraft. The UAVs must fly only in daylight and give way to all manned aircraft. The UAV cannot be flown closer to marine life. No aircraft will operate at an altitude of <600 m (2,000 feet) above mean sea level and closer than 150 m (500 feet) horizontally from a point directly above any marine mammal.

6.3.12 Russia

The State Civil Aviation Authority of Russia (SCAA) is the national aviation authority in Russia. All UAVs weighing more than 250 gm (0.55 lbs) must be registered, while weighing <250 gm can be used without registration. The UAVs can fly only during the day with clear weather and visibility. The pilots will maintain a direct VLOS and will not fly over people or congested areas. Flying over sensitive areas, such as military installations, Moscow Kremlin, and Red Square are not permitted.

6.3.13 South Africa

South Africa's national aviation authority, the South African Civil Aviation Authority (SACAA), permits the use of what are known as Class 1 and Class 2 UAVs (LOC, 2019), which are further divided into private, commercial, corporate, and non-profit operations. The technical and operational requirements that apply to UAVs in South Africa differ based on their operation. For instance, a private operation may only be conducted using a Class 1A UAV (<1.5 kg weight) or Class-1B UAV (<7 kg weight). Private operations are exempted from the need to obtain approval, certification of registration, and UAV's operator certificate (AA, 2017).

The UAVs weighing >7 kg (15.4 lbs) may not be flown. Pilots must fly in daylight hours only and maintain a visual line of sight with their UAVs at all times while in flight. The UAVs are not permitted to fly within 10 km (6 miles) to an airport without special permission from the SACAA (UAV Coach, 2020). The UAVs may not be flown within 50 m (164 feet) of people or private property (without permission from the property owner). The level of restrictions applicable to a UAV operation is also related to the complexity of the operation. Additional restrictions apply to operations in controlled airspace and the BVLOS.

6.3.14 Ukraine

The national aviation authority is the State Aviation Administration of Ukraine. In Ukraine all aircrafts, including UAVs, that are owned by a legal entity incorporated in Ukraine or by a resident in Ukraine, or rented or leased by a Ukrainian operator from the non-resident owner, must be registered. The UAVs having a take-off weight <2 kg that is used for entertainment and sports activities are exempted (LOC, 2019). The UAVs weighing >2 kg must register and obtain a permit, but UAVs <2 kg do not require a permit (AA, 2017). Regardless of weight, UAVs should not fly over roads, central streets of cities and villages, industrial zones, train stations and railroads, seaports, fuel storages, prisons, places of crashes and emergencies, in the territories of anti-terrorist activities, and above objects of defense and other government and military organizations.

The pilot must not be >500 m away from the UAV and should not control more than one UAV at a time. It can be flown in the daytime only, and not above 50 m height, with a maximum flight speed of 160 kmph. Beyond 50 m height, it requires a permit. The UAV cannot be flown within 5 km from the borders of airport, and within 30 m of an individual, within 50 m of a group of <12 persons, or within 150 m of a group of >12 persons (UAV Coach, 2020).

6.3.15 India

In India, all aerial vehicles, manned or unmanned, are governed by the Directorate General of Civil Aviation (DGCA), under the Ministry of Civil Aviation (MoCA). The DGCA, in a public notification in October 2014, mentioned that due to safety and security issues, and because the ICAO has not issued its standards and recommended practices for drones/UAVs yet, civilian uses of drones/UAVs are banned until further notice. However, regulations must be in place for UAV/drone-based services to continue to thrive and grow.

In October 2017, the DGCA issued the draft guidelines on UAVs through the circular, "Requirements for Operation of Civil Remotely Piloted Aircraft System (RPAS)" (DGCA, 2018). The intent was to have a regulatory framework to encourage the commercial uses of drones/UAVs in areas from industrial monitoring to disaster management. The draft policy is under review with different ministries in the government and is expected to be formalized with further detailing by 2025 (FICCI Report, 2018). The policy would provide the existing UAV users the momentum to expand their scope of activities.

On August 27, 2018, India achieved a major milestone by releasing "Requirements for Operation of Civil Remotely Piloted Aircraft," also known as India's Drone Policy 1.0. The policy, which is applicable from December 2018, lifts the blanket ban on UAV-related activity and authorizes the use of drones/UAVs for commercial purposes. The current draft policy (2018) classifies UAVs into segments, as nano (<250 g), micro (250 g–2 kg), mini (2–25 kg), small (25–150 kg), and large (>150 kg) (FICCI Report, 2018). The micro and mini classifications cover a wide range of toys and mini drones, as this is a low-risk category. All drones except those in the nano category must be registered and issued a unique identification number (UIN). In India, the rules are fairly stringent, but once this policy comes into full effect, drones/UAVs are expected to be widely used in the country.

In September 2018, the MoCA legalized commercial flying drones under the Digital Sky rules. As per rules, the DGCA will grant a UIN to either an Indian citizen or a company, or a body corporate (DGCA, 2018). All drone operators will register their drones and take permission to fly for each flight through India's Digital Sky portal (*http://digitalsky.dgca.gov.in*). Before every flight, drone pilots must take permission to fly via a mobile app called "No Permission-No Take-off" (NPNT). A permit is required for commercial drone operations (except nano category flown below 50 feet and micro category flown below 200 feet). The owner/operator will be held responsible for the UAVs' safe custody, security, and access control. All UAVs can be flown by trained operators only. Drone pilots must maintain a VLOS at all times while flying. Drones cannot be flown more than 400 feet vertically, and in areas specified as "no-fly zones," which include areas near airports, international borders, secretariat complex in states, strategic locations, and military installations.

The DGCA released the draft note for *Drone Regulations 2.0* in 2019, which had proposed regulations for the operation of drones in public spaces, particularly the ones implemented at a commercial scale. Other salient features are (Barik, 2019):

Drone corridors: Creating space in the sky with dynamic red, yellow and green zones. For example, a 5 km radius around airports and areas around Rashtrapati Bhavan will be a red zone. There will be plenty of green zones, and some yellow zones.

Drone service providers: For registering, licensing, and regulation for flying drones

Drone ports: To enable drone corridors for different types of drones, and to operate drones BVLOS with payloads, and to enable automation of flight, each operational drone has to be registered.
Automatic air traffic management: Bi-modal control will ensure that both the Digital Sky Service Providers and air traffic management have control of the drone
Drone Directorate: To set up a Drone Directorate under the DGCA.

On January 13, 2020, the MoCA, through a public notice, directed drone operators to register their drones on the Digital Sky portal by January 31. The MoCA has registered 19,553 UAVs/drones till March 2020 since it was made mandatory for operators to register. However, this did not give rights to operators to fly drones in India without permission. The latest guidelines of 2019 with regards to drones may allow certain businesses and organizations to train pilots for these unmanned aerial drones. At present, the guidelines allow only 34 flight training organizations to provide training and certification of drone pilots in India. It is expected that government may allow airlines, state, and central government agencies, universities, and drone manufacturers to train pilots for drones.

Drones startups in India are weighed down by high costs and compliance regulations, which have negatively impacted investor confidence, though the interest remains high. According to global market intelligence and advisory firm BIS Research (2018), the Indian drone market is expected to be valued at US$885.7 million by 2021, while the global opportunity is estimated to touch US$21.47 billion by 2021.

Table 6.6 presents a comparison of salient features of India's draft policy with policies existing in the United States and China (FICCI Report, 2018). As seen in the table, the market regulations in both United States and China are conducive toward leveraging the UAV applications to maximize the benefits to organizations. For example, more area can be scanned in a shorter flight by allowing higher altitude flights. Further, organizations can carry out their operations for a longer duration with the desired payloads, if heavier payloads are allowed without much restriction. The Indian policy 2018 proposes that the operator shall submit a flight plan and obtain necessary clearances for operations at or above 200 ft AGL in uncontrolled airspace. The drawback in the policy is that a 60 m AGL ceiling may restrict many applications for UAVs in India, for example, in power and utilities, transmission towers have a height of up to 350 feet, whereas other countries also have adopted a limit of 400 feet AGL as a threshold value (FICCI Report, 2018).

TABLE 6.6 Salient Features of India's Draft Policy with Policies Existing in United States and China (FICCI Report, 2018).

	India's RPA Policy	US's Small UAV Policy	China's UAV Policy
Applicability for now	All RPAs	Only small drones weighing under 55 lbs (including attached systems, payloads and cargo)	For small drones weighing between 1.5 and 150 kg (including attached systems, payloads and cargo) and calibrated air speed not greater than 100 km/h
Categorization	Based on Maximum Take-off Weight (MTOW)	Based on MTOW	Based on MTOW
Security or Safety Aspect	DGCA should be informed prior to sale or disposing of the RPA or in case of damage	License holder is required to cancel registration through the FAA's online registration system in case of sale, loss or transfer of UAV	Manufacturer and owner to register through CAAC's online portal to obtain license. Account holder is required to cancel registration through the CAAC registration system in case of sale, loss, damage, discarded, stolen or transfer of UAV
Training requirement for remote pilots	– Minimum age requirement-18 years – Practical training with a proportion of simulated flight training – No-fly zone awareness – Safe recovery mechanisms – Not applicable for nano and micro category RPAs	– Minimum age requirement- 16 years – Remote pilot airman certificate with a small UAV rating or under the direct supervision of a person who holds a certificate	– Minimum age requirement-16 years (14 years with adult supervision) – Practical and theoretical knowledge training – Safety operation skill assessment method – Training on aircraft aerodynamics and flight principles – Training on aviation regulations and flight around the airport-No-fly zone awareness – Universal emergency operation procedures – Training on UAV system features and components – Not applicable for drones weighing less than 1.5 kg

	India's RPA Policy	US's Small UAV Policy	China's UAV Policy
Requirements for the operation of RPA	– Nano and micro RPA operative up to 50 feet and 200 feet respectively above ground level exempted from filing the flight plan and obtaining Air Defense clearance – Only during daylight (between sunrise and sunset) – All RPA operations to be day operations and within VLOS only – Minimum ground visibility of 5 km – No RPA shall be flown from a mobile platform – RPA shall not discharge or drop objects unless specially cleared via Unmanned Aircraft Operator Permit. – Only nano RPA exempted from informing local police authority prior to commencing operations – No specifications yet on the speed of the RPA	– Maximum altitude of 400 feet above ground level or higher (only if it remains within 400 feet of a structure) – Operations in class G airspace allowed without any air traffic control permissions – Daylight only operations/ civil twilight (30 minutes before sunrise and 20 minutes after sunset) – With the visual line of sight only of the remote pilot in command/ visual observer – Minimum weather visibility of 3 miles from the control station – No operation from a moving vehicle unless the operations are over a sparsely populated area – Not faster than 100 miles per hour	– Maximum altitude of 120 m (~400 feet) above ground level or higher – Real-time online systems to record flight data and prevent flight into prohibited areas – UAV flying within visual line of sight must be operated in the daytime – UAV flying BVLOS is not restricted with operating times, but a certain regulatory framework for addressing emergencies applies to such flights – Both UAV flying within VLOS and BVLOS must give way to manned aircraft – Air speed not greater than 100 km per hour – UAV operators to buy insurance for UAV covering liability for third parties on the ground

(continued)

	India's RPA Policy	**US's Small UAV Policy**	**China's UAV Policy**
Enforcement action	Cancellation of operator permit if the performance is no longer acceptable or if there is a breach of the compliance that may attract penalties	FAA may impose civil sanctions in instances of fraud and falsification for matters within its jurisdiction	The CAAC will impose limitations with respect to the use of UAVs and punishment by regulatory authorities in accordance with relevant regulations

Flying UAVs safely in India would require further R&D in a market-oriented environment if civilian uses of UAVs are permitted. The Government of India is considering broadening UAVs' classification and making the regulations less stringent to exploit the potential of UAVs. Companies are preparing themselves to adopt the UAVs-based solutions much faster and provide the training to pilots. Legislators need to assess the impact of UAV technology on other existing legal frameworks, such as IPR, security and privacy, and make suitable amendments in regulations accordingly (FICCI Report, 2018). Properly framed regulations on civilian UAVs in India will lead to developing several new applications, as UAVs are expected to stay in the market for a longer time. The drones/UAVs are expected to transform various operations in India by their innovative applications.

From UAVs, the main harms identified are; malfunction, mid-air collisions, and consequent damages to people & property on the ground. To address these harms, the UAV regulations focus upon three key aspects (Stöcker *et al.*, 2017): (i) targeting the regulated use of airspace by UAVs, (ii) imposing operational limitations, and (iii) tackling the administrative procedures of flight permissions, pilot licenses, and data collection authorization. Since the early 2000s, several countries have gradually established national legal frameworks for UAVs. Although all UAV regulations aim at one common goal, minimizing the risks for other airspace users and for both people & property on the ground yet a distinct heterogeneity of regulations exists worldwide. However, common issues, such as mandatory platform registration, obligatory insurance coverage, and standard pilot licensing procedures, indicate trends toward mature UAV regulations.

The past, present, and future trends of UAV regulations can be linked with developments in the law, the market, or information, as shown in Figure 6.3 (Stöcker *et al.*, 2017). Presently, legal frameworks are the driving force that regulates the use of UAVs worldwide. However, due to the growing UAV market, it is expected that the market mechanism will soon gain the most importance. Furthermore, information and education as enabling tools for public acceptance as well as integration of UAVs into society are expected to receive increased attention. Political will and clear-cut laws are also important to further develop UAV technology. Although UAV regulations are in place in about one-third of all countries (Jeanneret and Rambaldi, 2016), gaps and lack of capacity can be seen in both enforcement and implementation.

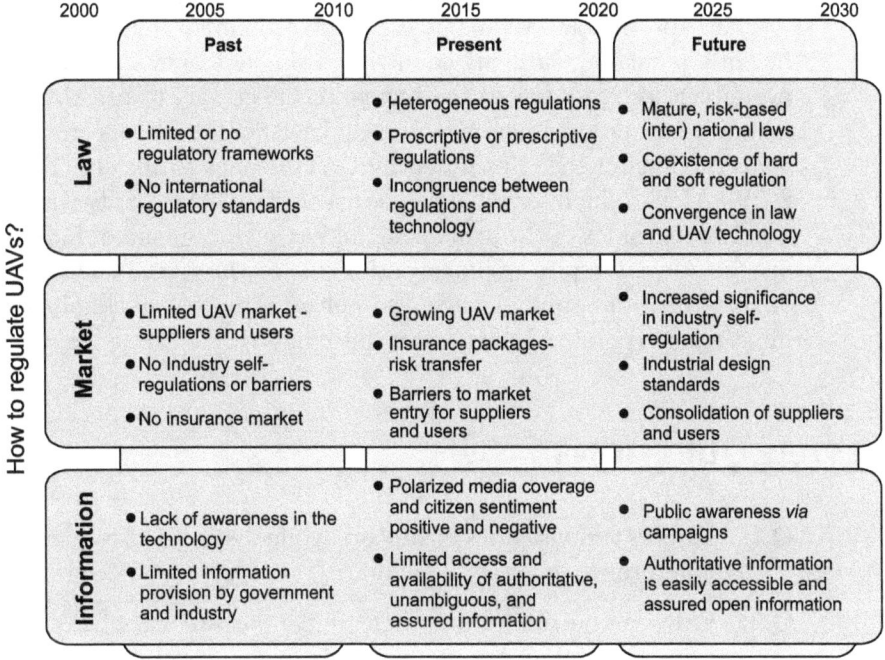

FIGURE 6.3 Past, present, and future developments with respect to regulatory laws, market, and information (Stöcker et al., 2017).

6.4 SUMMARY

The UAV regulations are under development/revision in several countries worldwide to propose some technical specifications and areas where these devices can be used (e.g., over urban settlements), with an increasing range of applications. Currently, the lack of precise rule frameworks and the time-consumed for taking flight permissions are considered the biggest limitations for UAV applications (Uerkwitz et al., 2016). To resolve potential regulatory hurdles, interested stakeholders as a partner with global bodies, such as ICAO, help shape transnational guidelines that could trickle down to individual countries. Advocacy in specific target countries may also be needed to speed up the adoption of favorable regulations that support the implementation of UAV solutions. The capabilities of UAVs will expand over the next few years, and society and law enforcement need to be aware of the threats this technology could deliver (Vattapparamban et al., 2016).

Perhaps, worldwide the biggest single problem is that UAVs are not sufficiently regulated. In many countries, a single agency has not yet claimed overall authority to deliver the regulations necessary to use UAVs in a safe manner. Law enforcement agencies and aviation regulators are increasingly concerned about the risks posed by UAVs/drones to society (Terra Times, 2019). As with the internet and cyber-security, UAVs' positive and negative uses are two sides of the same coin, and as such, we cannot have one without the other. Despite technical and regulatory hurdles, many scientific and engineering communities have also delved into UAV technology development or incorporated UAVs/drones into their respective fields.

REVIEW QUESTIONS

Q.1. Discuss various issues of concern while flying UAVs for data collection in reference to flying regulations.

Q.2. Why do you think that strict regulations are essentially needed for flying the UAVs? Are there any disadvantages also due to these strict regulations?

Q.3. Discuss six main criteria while developing the UAVs regulations.

Q.4. Write a short note on the (i) FAA regulations, (ii) European Union regulations, and (iii) the ICAO regulations for flying UAVs.

Q.5. Explain the common features among the regulations of various countries globally. Which country has more liberal regulations?

CHAPTER 7

VARIOUS APPLICATIONS OF UAVS

7.1 INTRODUCTION

Originally, UAVs were developed and used for military applications, such as reconnaissance and surveillance (Garg, 2019). From this perspective, the focus has primarily been related to their use in military operations (Blom, 2011). Therefore the UAV technology is much more advanced in the military sector than in other sectors. Before World War II, radio-controlled aircraft were used by the British military for target practice. During World War II, US and German militaries used such aircraft to fly into heavily fortified targets. Outside the military sector, UAVs have been used for decades by public entities, such as police and fire departments.

In the early 2000s, the public began to show greater interest in UAV technologies due to their reduction in costs and high-resolution images/data availability. Any application that could utilize a highly mobile data collection or communication platform could conceivably incorporate UAVs as a primary data collection component. In particular, UAVs have begun to emerge as an essential data collection tool for applications involving natural hazards (e.g., earthquakes and landslides) or snow-covered mountains where site accessibility can be challenging, and the need to collect the latest data is urgent. In such applications, UAV-based photogrammetry provides advantages over other remote sensing platforms such as satellites (Colomina and Molina 2014). Satellites are limited by temporal resolution, cloud coverage, and spatial resolution, whereas the UAVs, can be deployed on-demand, and flight parameters can be adjusted to acquire the desired image resolution and perspective.

More recently, the commercial sector has explored UAVs for their businesses (Boucher, 2015). Many UAVs originally developed for military purposes are now being used for various civilian applications (Watts *et al.*, 2012). For example, UAVs are used now in journalism, videography, search and rescue, disaster response, asset protection, wildlife monitoring, healthcare, and agriculture. Over the last decade, UAVs and other robotic systems have shown great potential in a wide variety of applications, such as environmental pollution assessment and monitoring, fire-fighting management, border security monitoring, fisheries, forest, logistics, oceanography, and communication relays (Jeanneret and Rambaldi, 2016). They are being used in many civilian applications due to their ease of deployment, low maintenance cost, high mobility, ability to hover for a long duration, and the possibility to reach difficult or impossible places (Crommelinck *et al.*, 2017).

There are three broad uses for UAVs: (i) to reach beyond cameras, (ii) to reveal new information by collating data from multiple sensors, and (iii) to transport materials. To date, the main applications of UAVs globally have been in the first two categories, as relatively inexpensive UAV-based cameras have been replaced with other forms of aerial imaging. Imaging flights have been successfully carried out in humanitarian missions, including surveillance, natural disasters, and vector research for malaria. The use of UAVs to transport materials has also begun. Militaries have created a suite of UAV systems with large-scale transportation capabilities, and commercial entities are investing heavily in drone/UAV delivery services. Business Insider (2017) predicts the total global shipments to reach 2.4 million in 2023, increasing at a 66.8% CAGR. The UAV/drone growth is expected to occur across the four main segments of the enterprise industry: agriculture, construction and mining, insurance, and media and telecommunications.

7.2   OVERVIEW OF UAVS/DRONES APPLICATIONS

The UAVs can perform both outdoor and indoor missions in challenging environments. These UAVs are equipped with various sensors and cameras for doing intelligence, surveillance, and reconnaissance missions. The applications of UAV/drones can be categorized based on the type of missions (military/civil), type of the flight zones (outdoor/indoor), and type of the environments (underwater/on the water/ground/air/space). Figure 7.1 presents a flowchart showing different types of drones/UAVs (Hassanalian and Abdelkefi, 2017). It is expected that drones/UAVs may have more than 200 applications in the future according to their types.

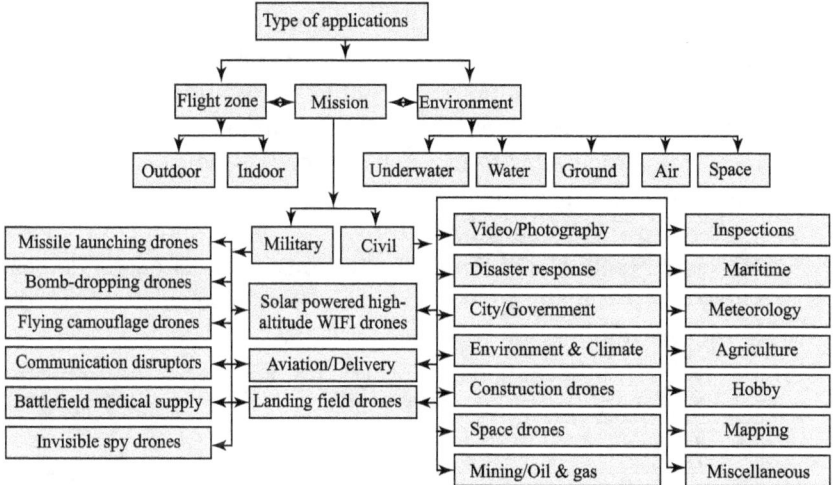

FIGURE 7.1 Applications of Drones/UAVs (Hassanalian and Abdelkefi, 2017).

Industries ranging from agriculture to entertainment and media are also taking full benefits of the UAVs/drones they offer. Figure 7.2 shows various industries adopting commercial UAVs and the market they control. It is seen that one of the most rapidly growing sectors is infrastructure development which includes construction.

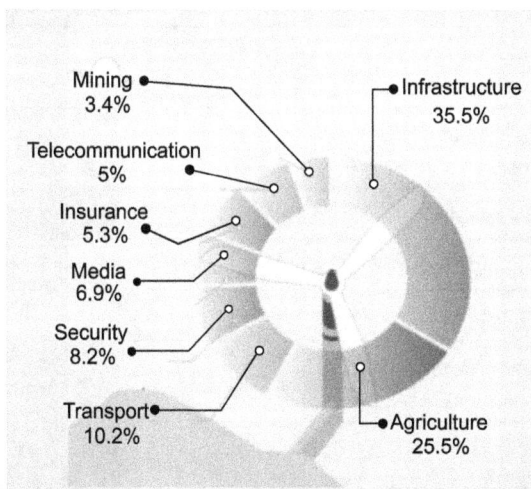

FIGURE 7.2 Various industries adopting the commercial UAVs/Drones (Patterson, 2018).

The UAVs carry photographic sensor types for data collection, such as RGB cameras, multispectral cameras, hyperspectral imaging, and thermal imaging sensors. In addition, biological sensors may be used in civil and environmental engineering fields to identify specific airborne contaminants or pathogens. These sensors are categorically similar to gas and radiation detection because they require contact between the sensor and the target. Magnetometers can probe beneath the ground surface useful for geotechnical infrastructure systems, such as pipelines, levees, earth dams, landfills, and sub-surface hazard detection.

Figure 7.3 presents the maturity level of common UAV applications and associated technology. Some applications, such as goods delivery, are still early, whereas many applications areas have reached maturity level, such as border patrol.

	Early Stage	Middle Stage	Late Stage
Applications			
Aerial photography			■
Border patrol			■
Construction and real estate images and monitoring		■	
Emergency management		■	
Infrastructure monitoring		■	
Mail and small package delivery	■		
Film-making and other media uses		■	
Oil and gas exploration			
Precision agriculture			■
Public safety			■
Weather forecasting and meteorological research		■	
Wildlife and environmental monitoring		■	

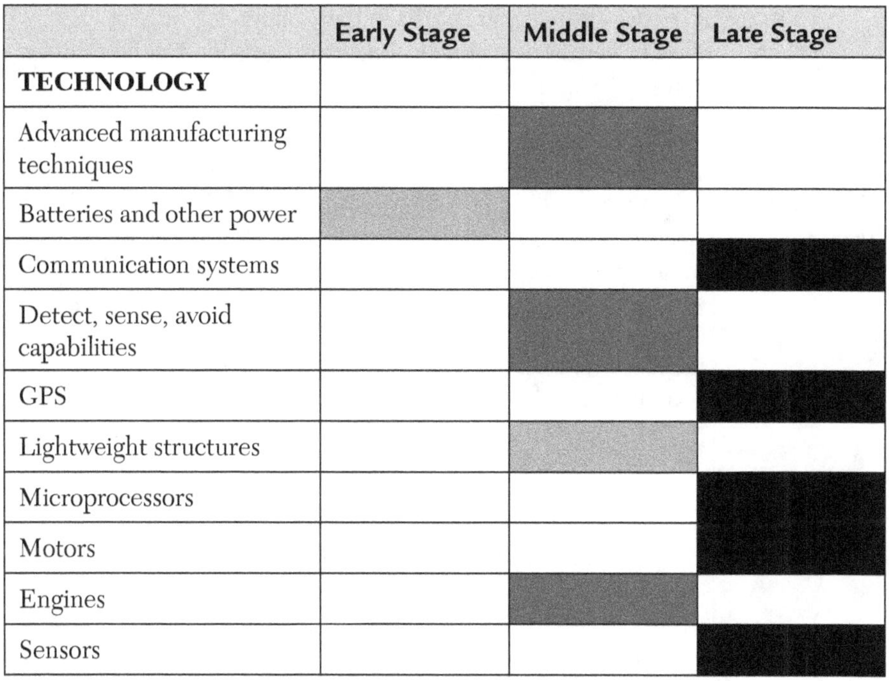

	Early Stage	Middle Stage	Late Stage
TECHNOLOGY			
Advanced manufacturing techniques		■	
Batteries and other power	■		
Communication systems			■
Detect, sense, avoid capabilities		■	
GPS			■
Lightweight structures		■	■
Microprocessors			■
Motors			■
Engines		■	
Sensors			■

FIGURE 7.3 Maturity levels of applications and technology of UAV/Drone (Uerkwitz, 2016).

Automated UAVs can contribute to a wide range of applications, such as oil & gas facilities, security, surveillance, emergency response, and infrastructure inspection. UAVs can perform applications in sea ports, such as surveying and mapping, port monitoring, and traffic control. Automated UAVs can be utilized in stockpile management, tailings dams, inspections, and more in mining operations. The UAVs allow mining companies to obtain precise volumetric data of excavations. Energy and infrastructure companies are able to survey pipelines, roads, and cables exhaustively. Humanitarian organizations could immediately assess and adapt aid efforts in continuously changing refugee camps (Tasevski, 2018). Coordinated teams of autonomous UAVs (i.e., swarm) will enable missions that last longer than the flight time of a single UAV by allowing some UAVs to leave the team for battery replacement temporarily.

Leveraging other emerging technologies, such as AR, VR, AI, 3D modeling, and IoT can enhance the output from the UAV data. There is a natural synergy between these technologies, which complement each other when coupled with UAV data (FICCI Report, 2018). Table 7.1 illustrates some of the potential applications of these technologies when coupled with UAVs across five key industries; power and utility, agriculture, highways, mining, and railways.

TABLE 7.1 Potential Applications of UAV Across Five Key Industries (FICCI Report, 2018).

Applications	Various uses				
Power and utility	Real-time information overlaid on an AR headset when scanning gas pipes through UAVs	Remotely inspecting a tower from the top using UAV and a VR headset	Automatically detecting corrosion on equipment on towers using a UAV feed	Creating 3D models of the site from the UAV data before construction begins to optimize layouts	Transferring information from UAV to a handheld device through IoT
Agriculture	Enhanced video output augmented with information about crop health overlaid on a UAV feed	Videos and images captured through UAVs can be used for training farmers	Automatically identifying insects and crop using NDVI and AI from a UAV feed	Planning for contour trenches and estimating the scope of work through a 3D model created from UAV feed	Collecting information from sensors and communicating with UAV remote operating center
Highways	Information for engineers on-site highlighting areas identified for maintenance through UAVs	Remote virtual tour of bridge construction site for higher management through UAV feed	Automatically monitoring the progress by calculating length of the road created through a UAV video	UAV-based information to create 3D models to estimate the scope of work while building a tunnel	Real-time visibility to remote engineers to assess road conditions through a UAV feed
Mining	Using 3D modeling and AR; engineers can run simulations for blasts to view through AR headsets to identify risks	3D models created with the help of UAV feeds can be used by engineers to access and monitor sections of the mine	UAVs and AI can help detect if workers on ground are complying with regulations, such as wearing hard hats etc.	Using 3D models created through UAV data for planning blast in a mine	UAVs can act as first responders to an accident site and transmit real-time information to rescue agencies

Applications	Various uses				
Railways	Site engineers can use AR headsets to consume information captured by UAVs, such as broken bolts etc.	Managers in railways can remotely visit a site through VR and inspect the site from the information captured by UAVs	Images and videos captured from a UAV can be used by algorithm to find the number of sleepers laid on track	3D models prepared through UAV can help engineers determine the volume of excavation required to be cut to construct a railway line	UAVs can transmit information through Internet in case of an accident and can help rescue teams to plan the best discourse
Emerging Technologies	**Augmented Reality:** AR is used to present enhanced natural environments overlaid with digital information to offer perceptually enriched experiences which can be digitally manipulated.	**Virtual Reality:** VR is used to present completely immersive digital environments or situations to offer perceptually enriched experiences in a digital environment	**Artificial Intelligence:** Intelligence is demonstrated by machines that perceive and learn from Big data to mimic and improve upon human cognitive functions, such as learning and problem solving	**3D Modeling:** Creating a digital model of the object in a 3D system by mathematically representing data points and rendering them in a 3D space	**Internet of Things:** Network of connected devices; robots, device and instruments, embedded with sensors to communicate with other objects to exchange data and information over the Internet
Problem classes	Enhanced experience, remote accessibility, and improved productivity	Virtual experience, remote accessibility, and improved productivity	Human perception, classification, and prediction	Digital model, volumetric estimation, and remote visibility	Real-time interaction, visibility, and gathering information
Industry examples	IKEA company used AR to showcase consumers how furniture would look in their homes	Audi company is using VR to allow customers to experience new car interiors	Technology giants, such as Google and Tesla are building self driving cars using AI	Lafarge Holcim company is using 3D models for blast planning and computing inventory volumes	GE company is using IoT to monitor and collect data on its aircraft engines

Aircraft present a compromise between satellites and UAVs, providing higher-resolution imagery than a satellite and covering much larger areas than UAVs. Due to battery life or regulatory restrictions on BVLOS operations, most UAVs have a comparatively shorter range. Average prices per area for all three technologies are presented in Figure 7.4. The prices are comparable between the two (aircraft and UAVs), but while UAVs can achieve that rate on just a few hectares, the high hourly price of aerial photography means those rates can only be performed on large projects. In the United States, some companies are providing services as little as US$ 0.73/acre to US$ 1.80/acre (DronaApps, 2018). For an area, it is unlikely that UAVs will compete with airplanes and satellites until their endurance has been drastically improved. But for smaller projects, high starting prices for aircraft and satellites can make UAVs an attractive option. The real advantage of UAVs lies not in their price per area but their ability to fly close to the ground so they can achieve few centimeters resolutions and capture imagery at oblique angles, which the other two technologies can't.

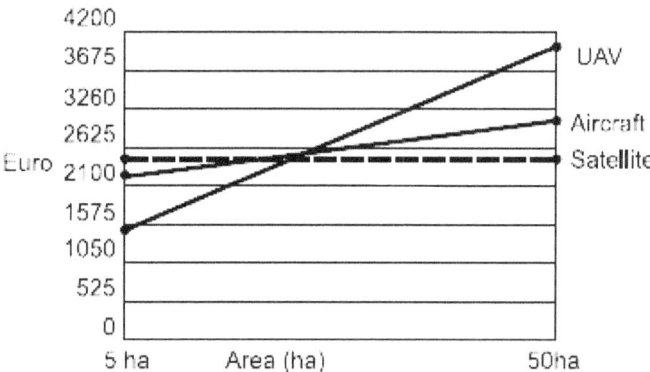

FIGURE 7.4 Cost for satellite, aircraft, and UAV-based mapping (DroneApps, 2019).

The UAVs are cost-effective and a good alternative to human-crewed helicopters. The UAV teams permit rescue organizations to deploy dedicated communication networks for ground operators quickly. Telecommunication companies can also leverage UAV networks to supplement or replace points of service temporarily. These days, drones are being used to create spectacular lighting effects in concerts, such as the super bowl by artists, Lady Gaga, or sports broadcasting (Ayranci, 2017). In the future, police will use drones for search and rescue missions and also track down criminals. Various police drones would serve different tasks, such as Drug Sniffing Drones (Turkmen and Kuloglu, 2018).

7.3   VARIOUS APPLICATIONS

Various applications of UAV technology are enumerated below.

7.3.1   Surveying and Mapping

For land surveying and mapping, UAVs offer high-quality output and remarkable locational accuracy. It offers an efficient supplement method to traditional survey methods, such as the total station and GPS survey (Garg, 2019). Surveying and mapping professionals routinely face difficult and often dangerous terrain for mapping. The UAVs provide the advantage of data collection in such environment/terrain with safety and security. Autonomous UAVs would allow surveyors and mappers to collect high-resolution aerial data and take precise measurements, thus saving time, money, and human resources. Eliminating the logistics of UAV operations, UAV can provide scheduled and on-demand aerial data collection and premium processing and analytics capabilities to support critical processes, such as stockpile volume measurements, terrain mapping, site planning, and more (Tatum and Liu, 2017).

Equipped with appropriate sensors and camera technologies, UAVs provide an economic platform for obtaining topographical data. Using photographic systems and high-resolution images, overlapping photos taken by the UAVs are put together in a mosaic that can be used for topographic mapping. The UAVs equipped with thermal imaging can be used to discover energy leaks by conducting building envelope surveys. Combining photogrammetry with LiDAR technology allows the production of ortho-photos, point clouds, DEMs to assist 3D building models, contour maps, volumetric calculations, and various other products (Garg, 2019).

RGB cameras on UAVs are dominantly used in mapping applications (Jordan *et al.* 2018). Due to the excellent spatial resolution of products from UAVs, they are an excellent choice for cadastral mapping (Crommelinck *et al.*, 2017). However, many other types of sensors, including hyperspectral imaging and thermal imaging (Nishar *et al.*, 2016), also play a role in civil engineering for multi-scale data collection. They can save considerable time, money, and human resources.

7.3.2   Industrial Applications

It is the most popular category for UAVs applications. Automated UAVs have emerged as incredibly powerful and versatile tools capable of having a wide range of industrial applications. In addition, the near real-time data gathering and analyzing capabilities make automated UAVs valuable for several industry sectors. Industries are increasingly using UAVs/drones to optimize industrial processes and improve their operational efficiencies (Uerkwitz et al., 2016).

The UAV's maneuverability makes it possible to inspect difficult or dangerous areas for humans to access. Such sites can provide valuable visual surveillance, assisting engineers in ascertaining structural integrity or general safety (Lee and Choi, 2016). Remote stretches of natural gas and petroleum pipelines are the best examples, as UAVs can travel and collect data faster than any human. They provide value addition by significantly reducing the costs. These data can be stored in the cloud or uploaded to facility management software to alert the technicians to maintenance issues, thereby increasing efficiency and accountability.

Many industrial inspections have involved ladders, ropes, and rigs to scale large machinery and towers, oversee processes and locate bottlenecks. These processes pose a great risk to managers and require shutting down of machinery, which may cause significant financial losses (Shakhatreh et al., 2018). An automated UAV platform can eliminate the logistics of its operations, providing on-demand aerial data and faster processing and analytics capabilities (Katukam, 2015). The UAVs can help managers with critical and ongoing industrial inspections in the following ways (Shakhatreh et al., 2018):

- Enable more efficient, cost-effective, and safe inspection.
- Offer managers a professional tool for viewing difficult-to-access areas.
- Provide non-disruptive technology that does not require ceasing operations.
- Positioning of the UAV at the required spot to get high-resolution and panoramic images for detailed topographic models, and 3D models for various purposes.

Some of the industrial applications where UAVs can potentially be used are discussed below.

7.3.2.1 Construction sector

The use of UAVs in construction has already led to many changes in the way buildings are made. The UAVs are changing the way construction companies do business, helping them coordinate teams more efficiently, track progress more regularly, and complete projects faster and economically. Scanning the sites with UAVs and add-on equipment, such as infrared, geo-locating, and thermal sensors, assist in mapping construction sites and modeling construction projects (Jones, 2019). Technology advances in recent years have resulted in the design of UAVs that are more reliable, less expensive, and easier to control. Such platforms enable construction companies to measure work progress in real-time. Using UAVs, construction owners are able to see what is happening on the site from the comfort of their offices.

The application of UAVs in construction management is developing in the construction industry (Greenwood *et al.*, 2019). The UAVs can do much more than produce high-definition images. The real estate industry has begun to take advantage of UAVs to conduct aerial surveys and mapping of planned developments. The development of 3D models at construction sites over time to monitor the progress has been the most common application of UAVs in construction management. Construction companies have primarily been using UAVs to provide real-time reconnaissance of their sites to provide high-definition video and still images for publicity and documentation of progress (Intelligence, 2018).

Construction management is best achieved through creating a Building Information Model (BIM). The BIM collects all data and any other info needed throughout a construction project from start to finish. It's the entire database of information for a project. With the proper information, a construction firm can build a 3D model of a project using BIM, which can significantly streamline the preparation and ongoing monitor the project progress, and reduce costs (Vacanasa *et al.*, 2015). The UAVs linked with BIM can potentially improve construction productivity by monitoring the construction process and material delivery, and equipment usage (Fan and Saadeghvaziri, 2019).

Construction monitoring can be achieved by using UAVs/drones to create 3D models and compare this model to BIM models, as shown in Figure 7.5. The UAV takes images of the project site, which are sent to the cloud and incorporated into BIM software to create 3D images. The 3D images are overlaid on the original CAD designs to look for problems and inconsistencies. As the model is created, the team members constantly refine and adjust their portions according to project specifications and change design to ensure the model is as accurate as possible before the project physically starts. As the project moves forward, construction

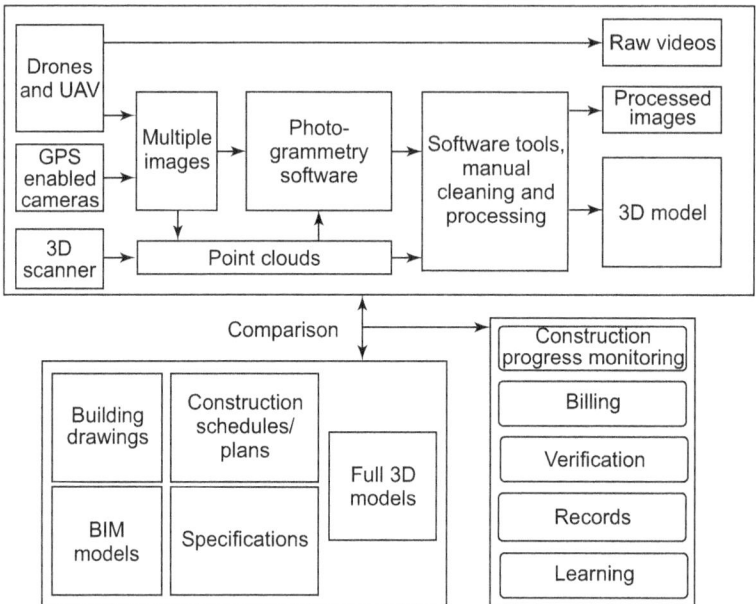

FIGURE 7.5 Overall concept of using drones in BIM to monitor the construction (Anwar *et al.*, 2018).

engineers and managers can locate the mistakes quickly, such as pipes or trenches that were not dug correctly (Anwar *et al.*, 2018). Identifying these inconsistencies upfront saves money and timeline setbacks, offering advantages for both contractors and their clients.

The UAV-based photogrammetric images with the LiDAR data can create 3D building models, contour maps, volumetric analysis, and various other products. The UAVs equipped with thermal imaging can be used to discover energy leaks by conducting building envelope surveys. The UAVs and GPS are being used with pre-determined waypoints to capture images from the same aerial perspective over time to monitor the actual construction progress against the planned progress. Routes are pre-programmed so that the UAV can follow pre-determined routes independently (Puri, 2015). Route planning software would allow UAVs to follow specific routes, speeds, altitudes, and targets to capture images. Aerial images produced by UAVs can be taken daily to plan the placement of stored materials, the flow of workers and vehicles in and around the site, and to identify potential issues with installed construction or the constructability of planned installations.

The UAVs have also been explored as safety inspection tools at construction sites. Images and video captured by UAVs can cover a larger area of the site in the shortest time which can be valuable information of the site conditions where accidents usually occur. Security of construction sites can also be enhanced using UAVs integrated into the security alarm system. A UAV could dock on a rooftop station, and when required, the UAV is deployed to maneuver above the construction site to capture real-time video (Heys, 2019). A high-definition camera at a low altitude could have the capability to identify small individual objects within its view, and the video could be streamed to a smartphone or other real-time device. The UAV could also be automatically programmed to conduct periodic security sweeps and returning to the rooftop station automatically.

In the construction industry, UAVs provide easy access to large or difficult sites and complex or tall structures (Patterson, 2018). Specifically, the UAVs have impacted the construction industry by:

1. Enabling better construction site monitoring,
2. Quick inspection of the buildings,
3. Monitoring on-site activities,
4. Reduction in or elimination of equipment downtime,
5. Higher precision and accuracy,
6. Increase in safety,
7. Reduction in regulatory demands for health and safety check-ups,
8. Reduction of time in data collection,
9. Security surveillance,
10. Detailed site analysis,
11. Computing the stockpile volume and material types for inventory,
12. Computing the length, width, and elevation for roads and structures,
13. Computing overburden to plan for an efficient removal,
14. Providing thermal images that can help examine pipelines, solar panels, power grids, and roofs for leaks, overheating, failure, and proper insulation, and
15. Photo documentation of before and after inspection.

The use of UAVs in various construction activities can result in cost-saving, as shown in Figure 7.6 (PwC, 2018a). The maximum saving is observed in accurate measurements as well as activities where improved communication and collaboration is required.

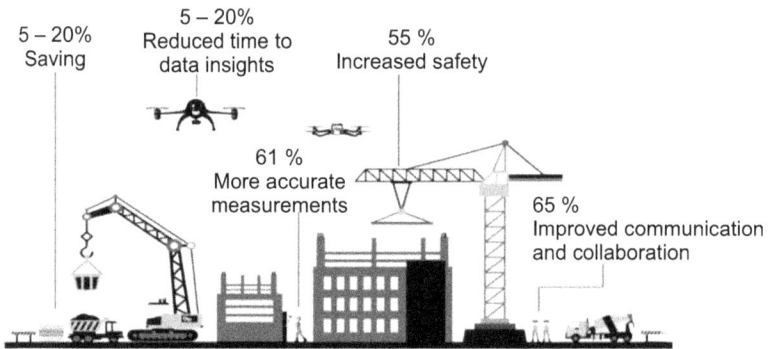

FIGURE 7.6 Advantages of using UAVs in construction (PwC, 2018a).

Innovative technologies, such as wireless sensor networks, and IoT are expected to help in better construction and safety management. Incorporating UAVs at construction sites introduces additional safety concerns, such as personnel distraction and increased collision risk with equipment or personnel. Future application of UAVs in the construction industry may include (Tatum and Liu, 2017): automated employee check-in/check-out, automated safety checks, scanning RFID tags on materials in laydown areas for inventory, material delivery, parts delivery, remote job walks, preview views from a building prior to construction, thermal scanning of utility-scale photo-voltaic (PV) plants, and interior missions. In the coming years, the construction industry is expected to adopt more expanding innovations. Construction is one of the world's biggest industries, worth about US$ 8.5 trillion a year. The Goldman Sachs report estimated in January 2017 that the businesses will invest US$ 13 billion in commercial drones by 2020, with more than US$ 11 billion in the construction industry alone (PwC, 2018a).

7.3.2.2 Oil and gas sector

The UAVs open up new opportunities in industries, such as mining, sea ports, oil & gas, and other large industries (Klochkov and Nikitova, 2008). Some of these applications include inspection of critical linear infrastructures, such as oil and gas pipelines, water pipelines, or electrical transmission lines (Silvangi, 2016), and inspection of wind turbine blades with higher-resolution imaging. The UAVs offer managers a professional tool for viewing difficult-to-access areas, giving them a safer and more cost-efficient way of gaining greater insight into operation-critical processes. The UAV can be autonomously deployed and landed, with pre-defined missions and applications to collect aerial data, which the system analyzes, providing a clear picture of sites.

Pipelines represent an infrastructure component of the oil and gas industry with significant failure. Monitoring pipelines over a large spatial range is critical and can be too costly to assess the optimum performance (Jawhar *et al.*, 2014). The UAV provides the solution to eliminate the need for logistics involved and provide reliable, on-demand aerial data, processing, and analytics capabilities while simplifying the inspection processes (Katukam, 2015). Automated gas inspection can be a significant safety feature to industrial facilities, such as chemical plants, refineries, and manufacturing sites (Culus *et al.*, 2018).

Several oil-field companies word-wide are making use of UAVs to provide customers with a value-added service. For example, British Petroleum (BP), a multinational oil and gas company, routinely uses UAVs to patrol their Alaskan oil field, the first authorized commercial UAV operation in the United States. Their UAVs will be used to monitor specific maintenance activities on roads, oil pipelines, and other infrastructure in northern Alaska's vast and potentially dangerous artic environment. Figure 7.7 shows the utilization of UAVs in oilfields.

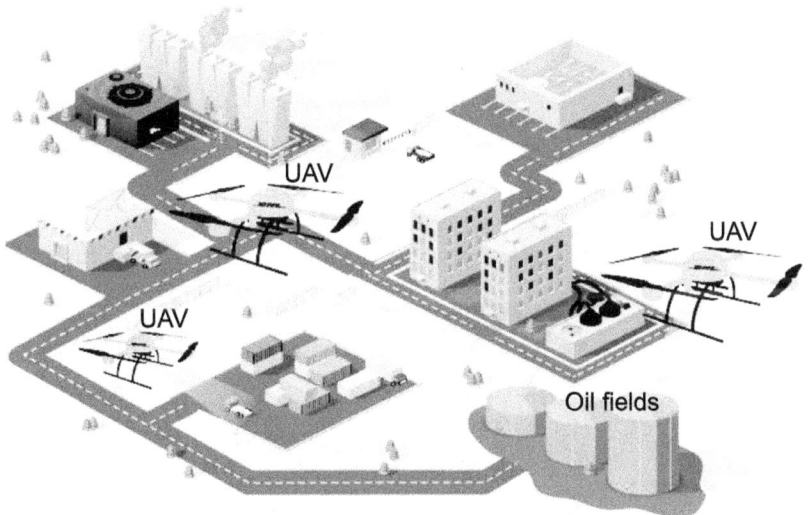

FIGURE 7.7 The UAVs for oil fields monitoring (Heutger and Kückelhaus, 2014).

7.3.2.3 Power sector

The UAVs can be utilized for high voltage inspection of the power transmission lines (Figure 7.8). They can be deployed to detect, inspect and diagnose the defects in power line infrastructure (Miller, 2016). Multiple images and data from UAVs can be processed to identify the locations of trees and buildings near the power lines and calculate the distance between trees, buildings, and power lines. A thermal IR camera can be employed for bad conductivity detection in the power lines (Jones, 2015). Jordan *et al.* (2018) provided a contextual review of UAV-based inspection of power facilities and structures.

FIGURE 7.8 The UAV operations of power lines.

Thermal imaging can be used to inspect inaccessible buildings, electric power lines (Heutger and Kückelhaus, 2014), PV systems, fire-fighting, and law enforcement. Thermographic inspection from the UAV can be particularly useful for monitoring different industrial facilities (Figure 7.9).

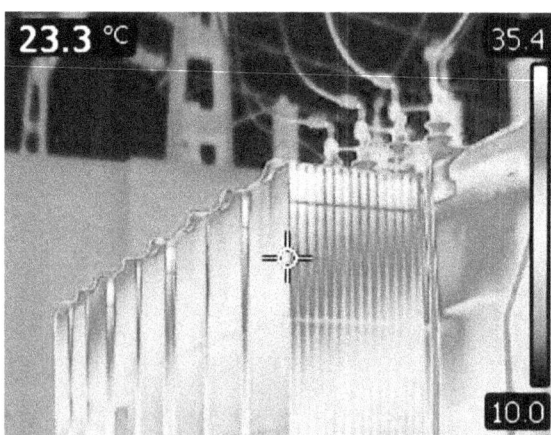

FIGURE 7.9 Thermal operations to evaluate energy efficiency [*See Plate 6 for Color*].

7.3.2.4 Mining sector

The UAVs are very useful in the mining industry, as their maneuverability is an asset that contributes to better work environments for workers (Katukam, 2015). Mining operators who previously chartered planes or helicopters to conduct inspections can now deploy UAVs (Figure 7.10) to collect information in far-flung or hazardous zones, bringing added safety and efficiency to site surveys, mapping, stockpile inventory management, road condition monitoring, and blast planning (Lee and Choi, 2015). The UAVs are being used to inspect mining plants, accurately manage tailings dams and view hard-to-reach structures, such as rooftops, towers, and conveyors (Lee and Choi, 2016).

FIGURE 7.10 Use of UAV in mining activities.

The UAV generates a dense point cloud of a quarry which makes it easy to frequently monitor the progress at a much higher speed and level of safety. Using RTK technology and/or GCPs, highly accurate results are achieved. Additionally, an initial calibration of permanent GCPs will help all recurring flights to reference the aerial data correctly. The use of a UAV leads to an 80% reduction in on-site time, and thus a saving up to 90% is possible by streamlining the overall process in mining activities.

7.3.3 Inspection and Monitoring

The UAV can be useful as a low-cost tool for visual inspection and monitoring infrastructure systems (Greenwood *et al.*, 2019). As structures are subject to deterioration due to increasing loads, weather conditions, and aging processes, UAVs offer a unique opportunity for their monitoring. Conventional inspections are based on time-consuming visual investigations, require technical experts, and are therefore expensive. Expert team on the ground (walking or using ground-based vehicles) can be challenged by difficult terrain, physical obstacles, or dangerous site conditions. Wind turbine inspections have traditionally been accomplished through the use of binoculars or by technicians that are required to climb to great heights. The introduction of UAVs has provided a much safer option for these inspections with higher resolution than that offered by binoculars (Tatum and Liu, 2017).

The UAV-based aerial data aids in visual inspection for monitoring the critical infrastructure (Shakhatreh *et al.*, 2018), or in some cases completely replace the field-based procedure. The UAV usually eliminates the need to work with scaffolds, cranes, or helicopters during the production or construction of high objects. It provides a more efficient, cost-effective, and safe inspection process that gives better, closer monitoring and control while business operations may continue as usual. As a result, UAVs offer an economic advantage, including reducing costs associated with personnel, travel, and site logistics.

Construction and infrastructure inspection using UAVs, equipped with onboard cameras and sensors, can efficiently employ image processing techniques. Thermal images using IR sensors from a UAV can be used for bridge deck inspection and geothermal field mapping. The LiDAR is the most popular non-imaging sensor used to develop a 3D mapping of infrastructure system (Nagai *et al.*, 2009). Combining RGB images with 3D point cloud data produces images that are valuable for analyzing various infrastructure system components. The use of image processing techniques allows for monitoring and assessing the construction projects, as well as performing inspection of the infrastructure, such as work progress monitoring, an inspection of bridges, railtrack, highways, irrigation structures monitoring, detection of construction damage, and surfaces degradation (Ham *et al.*, 2016).

Figure 7.11 presents an example that demonstrates the complexities involved in the process of UAV-based structural health monitoring (SHM) and to achieve improved performance. The two main parameters in the proposed model are environmental complexity and task complexity used to evaluate the operational scenario of a UAV (Vidyadharan *et al.*, 2017). The environment complexity is further sub-divided based on the meteorological conditions and the infrastructure types for which UAV will be collecting the data. The meteorological conditions will be based on factors, such as temperature, precipitation, wind speeds, and humidity, surrounding the infrastructure type where the UAV will operate. The weather conditions are incorporated as important attributes contributing to the infrastructure type's complexity.

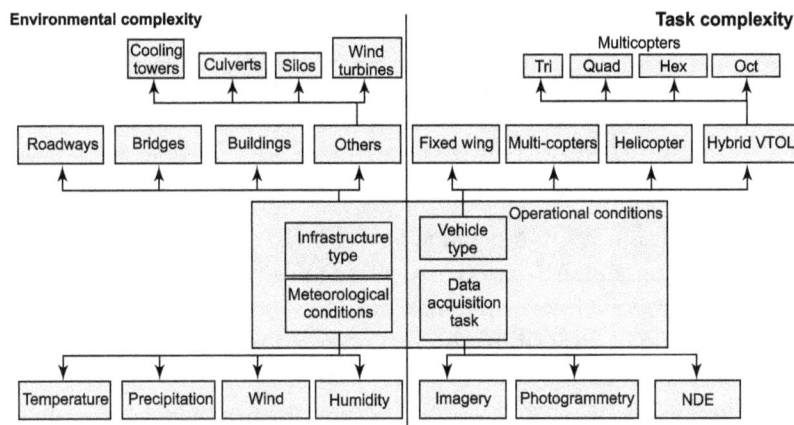

FIGURE 7.11 The SHM model representing the composition of the environment and task complexities (Vidyadharan et al., 2017).

The task complexity is sub-divided into data acquisition type and vehicle type. A combined analysis of each of these parameters can be used to understand the effect of operational scenarios on the UAV while performing SHM tasks. The data acquisition types from UAVs are broadly divided into three different categories: images (2D), 3D models (stereo-photogrammetry), and non-destructive evaluation (NDE). The cameras attached to the UAV will collect the 2D images analyzed to create 3D models. The NDE data acquisition types vary in levels of difficulty. They are selected based on tasks that a UAV may perform to collect information regarding the health of a structure and whether or not the UAV can process data.

Sankarasrinivasan *et al.* (2015) presented an integrated data acquisition and image processing platform mounted on UAV for crack detection in infrastructure, as shown in Figure 7.12. It was proposed to be used for the inspection of infrastructures for real-time SHM. In this framework, real-time image and data will be sent to the GCS, processed there using an image processing unit to facilitate the diagnosis and inspection process. They combined hue-saturation-value thresholding and Hat Transform technique and used for cracks detection on the concrete surfaces. The method combines the output obtained through two filters, resulting in enhanced cracks to achieve good accuracy.

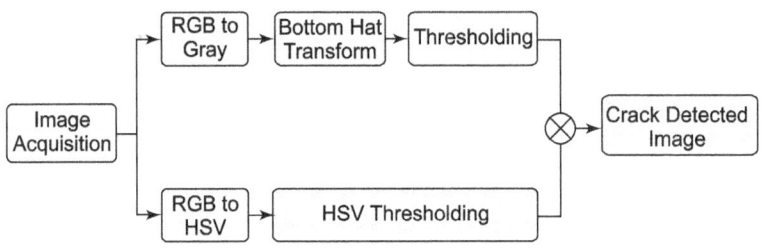

FIGURE 7.12 Proposed approach for crack detection in concrete surfaces (Sankarasrinivasan et al., 2015).

In Table 7.2, Greenwood *et al.* (2019) presented examples of infrastructure in which UAV-based inspection and data collection are carried out. For each category, the relative risk, consequences of a UAV malfunction, and the likelihood of undesirable human interaction while investigating the different types of infrastructure have been rated. These are relative ratings meant to allow users to make a general comparison between the examples, but factors, such as current usage (e.g., highway), will increase further risk. For bridge inspection, a camera or LiDAR system mounted on top of the UAV will be useful for inspecting structural components on the underside of the bridge. Close-range (i.e., within a few meters) inspection and flight around or under-complicated 3D structures emphasize the need for collision avoidance systems. Satellite-free navigation is essential for operations at indoor or underground facilities, such as tunnels.

TABLE 7.2 Infrastructure-specific Considerations for Data Collection and Inspection (Greenwood et al., 2019).

Infrastructure type	Relative risk	Consequences of UAV failure	Likelihood of human interaction	Platform types	Non-universal UAV features or functions	Potential sensors
Roads	Moderate	Moderate	High	Multirotor, fixed-wing	Sub-surface sensing	Camera, LiDAR, IR
Bridges	High	High	High	Multirotor	Collision avoidance, precision positioning, vertical camera	Camera, LiDAR, IR
Levees and dams	Moderate	Moderate	Low	Multirotor, fixed-wing	Sub-surface sensing	Camera, LiDAR, IR, SaR

Infrastructure type	Relative risk	Consequences of UAV failure	Likelihood of human interaction	Platform types	Non-universal UAV features or functions	Potential sensors
Landfills	Low	Low	Low	Multirotor, fixed-wing	Sub-surface sensing	Camera, LiDAR, IR, Gas, SaR
Power lines	High	Very high	Low	Multirotor	Collision avoidance, precision positioning	Camera, LiDAR, IR
Pipelines	Low	Low	Low	Multirotor, fixed-wing	Long-range communication	Camera, LiDAR, IR
Tunnels	High	Low	Moderate	Multirotor	Collision avoidance, satellite-free navigation, vertical camera, sub-surface sensing	Camera, LiDAR, IR
Rail/roads	Low	Low	Low	Multirotor, fixed-wing	Long-range communication	Camera, LiDAR, IR, SaR
Irrigation systems	High	High	Moderate	Multirotor, fixed-wing	Sample collection	Camera, LiDAR, IR, Gas

The net market value of the deployment of UAVs in support of inspection of construction and infrastructure is about 45% of the total UAV market. So there is a growing interest in UAV uses in large construction projects monitoring (Liu et al., 2014) and inspection of power lines, gas pipelines, and GSM towers infrastructure (Jones, 2015).

7.3.3.1 Bridges

Bridge inspection is the most widely approached topic for UAV integration in infrastructure monitoring. Brooks et al. (2014) used UAVs for bridge inspections by including thermal images. Gillins et al. (2016) also demonstrated bridge inspection with a low-cost UAV. Eschmann and Wundsam (2017) developed a multisensory UAV bridge inspection platform with three sensors; long-wavelength IR, optical camera, and LiDAR. Each sensor performed different tasks, which are then fused to create a complete 3D visualization. Surfaces and deformation can be recorded using the LiDAR and images from the optical camera to overlay model textures and monitor the cracks. In contrast, IR cameras can detect moisture around the cracks (Culus *et al.*, 2018).

From the above studies, it is evident that bridge inspection has been implemented using UAVs, but there are some general limitations. It is still challenging to acquire high-resolution images of the most obscure or difficult to reach locations, such as fasteners (Sankarasrinivasan *et al.*, 2015). Future work into obstacle avoidance and localization within the spatial challenges of bridges will help alleviate the limitations.

7.3.3.2 Dams

A dam is a multi-hazard vulnerable structure severely affected by earthquakes, flooding, and terrain stability. In recent years, dam safety has acquired increasing attention because of the high number of accidents and failures of dams. Due to the large size and limited accessibility of dams, UAVs are extremely advantageous for visual investigation of dams and retention walls (Hallermann *et al.*, 2015). The use of the UAV images for inspection and re-creation of the geometry of dams can substantially improve monitoring and survey operations (Ridolfi *et al.*, 2017). Data acquired during the UAV survey can be used to construct a 3D model for its vulnerability studies. Hallermann *et al.* (2015) presented a methodology for inspecting the old structures, combining conventional inspection measurements with modern photogrammetric computer vision methods for 3D modeling and automatic post-flight damage detection.

7.3.4 Forestry

Among many UAV fields, forestry has diverse uses ranging from forest cover assessment to species classification and real-time forest fire monitoring. Several studies, carried out in various parts of the world, indicate that UAVs have ample potential to assess forest resources (Tomaštík *et al.*, 2019). UAVs in the forestry sector supplement the existing methods of remote sensing satellite images to fill the gap of information and improve the quality of forest information (Figure 7.13). It provides accurate, flexible, and cost-effective techniques in forestry (Banu *et al.*, 2016). The UAVs with visible and IR sensors can help to monitor forest health and to map forest diseases

FIGURE 7.13 Forest monitoring from UAV.

and pest infestation. Berie and Buruda (2018) found that UAVs can predict canopy height and above-ground biomass at the same level of precision as the LiDAR systems.

The UAVs have the potential for post-fire recovery monitoring and forest protection. They have generated information about the vegetation condition, water stress, and risk indices before a fire happens. UAVs with GPS and ultra-high-resolution digital cameras can provide images with precise location information. Due to frequent forest fire and loss of revenues, there is a perceived need for accurate, demand-based, and cost-effective forest resources and forest fire monitoring tools in developing countries (Berie and Buruda, 2018). The UAVs have been increasingly used as surveillance tools for wildfire detection and monitoring, especially in difficult topographic settings (Cruz *et al.*, 2016; Zhao *et al.*, 2018). During the fire incidence, visible and IR sensors based UAVs are required for fire detection & monitoring and real-time behavior of forest fire, such as the location and shape of the fire front, rate of spread, and fire flame height (Banu *et al.*, 2016).

The UAVs make it possible to carry out non-destructive and accurate measurements of many attributes of trees (forests) (Berie and Buruda, 2018). The UAVs can also support participatory forest management efforts, particularly useful in developing countries. It can serve as a means of evaluating the impacts of intervention and as a planning tool for new forestry projects. These applications of UAVs are expected to enhance community-based forest management (CBFM) schemes in the tropics, as external funding is provided there to the poor living communities (Paneque-Gálvez *et al.*, 2014). The CBFM program will result in less greenhouse gas emission because of reduced deforestation and degradation and ensure sustainable forest resources management.

Paneque-Gálvez *et al.* (2014) assessed the feasibility of using small, low-cost UAVs in CBFM programs. The key advantages according to technical issues (e.g., sensing capabilities, UAV operation and maintenance skills, image analysis, monitoring capabilities, potential to enhance and ease of CBFM), social issues (implications for people in communities), and environmental issues (implications for the local environment) and their relative importance to communities, partner organizations and end-users, are presented in Table 7.3.

TABLE 7.3 Relative Assessment of Advantages and Disadvantages Expected with a UAV-assisted CBFM (Paneque-G·lvez *et al.*, 2014)

(a) Advantages	Community	Partner organization	End-user
Extremely high-spatial-resolution	*High*	*High*	*High*
Potential for high temporal resolution	*High to medium*	*High*	*High*
Insensitivity to cloud cover	*Medium to low*	*High*	*High*
Potential for three-dimensional UAV image generation	*Low*	*High to medium*	*High*
Potential to ease CBFM and making it more attractive	*High*	*High to medium*	*Low*
Shallow learning curve of UAV users	*High*	*High*	*Low*
Relatively low price of UAV imagery	*Medium to low*	*High*	*High*
High cost-effectiveness within the context of CBFM	*Medium to low*	*High*	*Medium to low*
Data acquisition decentralization	*High*	*High to medium*	*Medium to low*
Enhanced monitoring of illegal activities	*High*	*High to medium*	*Medium to low*
Access to otherwise inaccessible areas	*High*	*Medium to low*	*Low*
Potential environmental benefits	*High to medium*	*Medium to low*	*Medium to low*

(a) Advantages	Community	Partner organization	End-user
Potential social and institutional strengthening of communities	*High*	*Medium to low*	*Medium to low*
Control of data acquisition and ownership would lie with the community	*High*	*High to medium*	*Low*
(b) Disadvantages			
Small payload	*Low*	*Medium to low*	*High to medium*
Low spectral resolution	*Low*	*Medium to low*	*High*
Poor geometric and radiometric performance	*Low*	*Medium to low*	*High*
Low software automation	*Low*	*Medium to low*	*High*
Sensitivity to atmospheric conditions	*Medium to low*	*High to medium*	*High to medium*
Short flight endurance	*High to medium*	*High*	*High*
Possibility of collisions	*High*	*High*	*Low*
Potential problems for repairs and maintenance	*High*	*High*	*Low*
Dependence on external assistance and funding	*High*	*High*	*Medium to low*
Ambiguous or cumber some regulatory environments for flying small UAVs	*High*	*High*	*High*
Safety & security issues	*High*	*High*	*High to medium*
Debatable relevance for community conservation and socio-economic development	*Medium to low*	*High*	*Low*
Potential social impacts	*High*	*High*	*Medium to low*
Ethical issues	*High*	*High*	*High to medium*

The UAV-based technology can also be used to enhance the Decision Support System (DSS) for forest management. As it is growing, the technology can be used to study forest dynamics, stand species composition, disturbances, and other attributes of forests (Banu *et al.*, 2016).

7.3.5 Agriculture

Agriculture is another important sector where UAVs are completely changing the way agricultural operations are carried out. The UAVs have become the main-stream of the farming community. This non-destructive method allows for quicker, simpler, and more accurate assessment than traditional surveying methods. The UAVs can be efficiently used (Figure 7.14) for small crop fields at low altitudes with higher precision and low cost than traditional human-crewed aircraft. High-resolution images can be utilized in Precision Agriculture (PA) for crop management and monitoring, weed detection, irrigation scheduling, disease detection, pesticide spraying, and gathering data from ground sensors (moisture, soil properties, etc.) (Huang *et al.*, 2013; Pflanz *et al.*, 2018; Garg *et al.*, 2020). The UAVs can provide precise and real-time data about specific locations required for crop management. The UAVs can be customized for different farming applications to provide a cost-effective monitoring platform without requiring an expert operator (Shamshiri *et al.*, 2018).

FIGURE 7.14 The UAV in crop and weed mapping.

Figure 7.15 presents the steps involved in using the UAVs for various agricultural activities. There are five major steps involving data collection to data modeling, data analysis through image processing, and GIS software. It also involves the creation of 3D maps and NDVI maps for analysis.

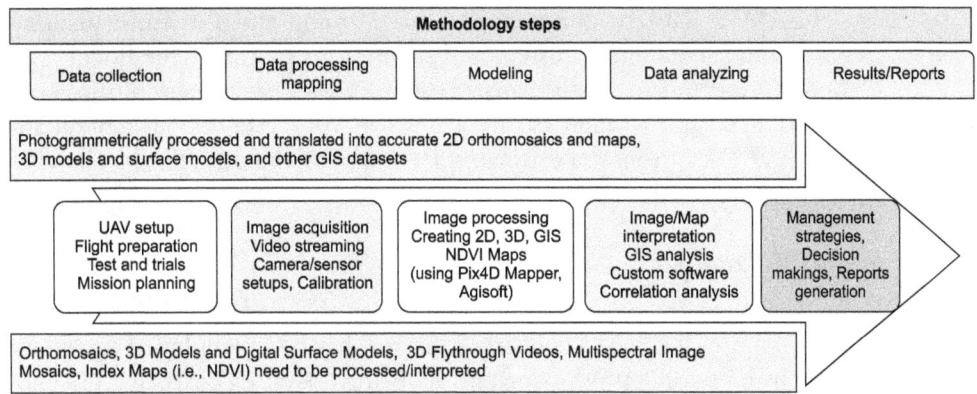

FIGURE 7.15 General procedure involved with a UAV platform for agricultural applications (Shamshiri et al., 2018).

Agriculture professionals can get valuable information for crops with specially designed sensors to (i) monitor nutrients, moisture levels, and overall vigor, (ii) identify and measure crop disease, pest problems, weeds, and water stress, (iii) estimate yields, (iv) characterize soil and vegetative cover, and (v) use data for predictive analysis. This information can help assess and prevent crop damage by routine monitoring and early detection. The UAVs can assist farmers throughout the crop cycle by (Tsouros et al., 2019):

- Producing precise maps for the soil analysis that direct seed planting patterns and inform management of irrigation and nitrogen levels;
- Enabling more accurate, cost-effective, and real-time crop monitoring than using satellite imagery;
- Allowing faster and more accurate assessments of irrigation conditions and measuring crop heat signatures through hyperspectral, multispectral, or thermal sensors; and
- Identifying plants affected by bacteria or fungus and measuring disease, pest problems, weeds, and water stress through the use of visible and IR cameras.

The UAVs can provide multispectral images converted into spectral indices (e.g., NDVI) to predict several physiological variables (Mahajan and Bundel, 2016). Conventional multispectral indices present limitations to assess water; hence, ANN could improve the assessment of the spatial variability of crop water status. Machine learning techniques and ANN models can be applied to multispectral information obtained from multiple types of sensors and platforms for assessing different types of plant stress (Pederi and Cheporniuk, 2015).

The UAVs have revolutionized smart farming and PA, from planting to harvesting, seeding to sensing, and scouting to spaying. The global population is expected to reach over 9 billion by 2050, with agricultural consumption expected to increase by a massive 70%. The global market for agricultural UAV/drones is estimated to reach 3.7 billion USD by the year 2022 (Culus *et al.*, 2018).

7.3.5.1 IoT-based agriculture

Enhancing the rural economy has been one of the world's most challenging issues. In more developed countries, the IoT has been employed to benefit farmers by increasing production and reducing operating costs, and enhancing labor efficiency. However, it is still out of reach for farmers of many developing countries (Sharma and Garg, 2019). Using UAV and WSN, the farmers can quickly obtain real-time data on farmland environments with high accuracy. Information collected from the sensor network is transmitted to the UAV for further processing and analysis, real-time, precise, large-scale, and automated.

In recent years, the IoT has made remarkable progress and is regarded as the most promising technology for propelling agriculture. However, there are no base stations and Wi-Fi stations in most farming areas, which may prevent the IoT application in agriculture. That means the data acquired through the WSN cannot be transmitted using wireless communications. An alternative solution is to employ UAVs to communicate with the WSN in large areas to get real-time data processing and analysis (Sylvester, 2018). The merits and cost-saving of using UAV-WSN in agriculture are presented in Table 7.4. The UAVs working with on ground IoT sensor networks can help agricultural companies monitor land and crops; energy companies survey power lines and operational equipment; and insurance companies monitor properties for claims and policies (DHL, 2014).

TABLE 7.4 Merits and Cost-Saving of Using UAV-WSN in Agriculture (Sylvester, 2018).

Merits	Savings
Through measuring related parameters by the UAV-WSN system, water resources can be saved significantly by accurate irrigation.	Compared to conventional irrigation, the UAV-WSN based system can save water resources by up to 67%.
Savings in human resource expenses, that is, reducing the workforce by one management staff and one worker.	Saving human resource expenses.

Merits	Savings
Compared to the conventional method, the UAV-WSN system can reduce wastage of chemical fertilizer as well as reduce soil pollution dramatically by more accurately measuring the rate of mixing pesticide, chemical fertilizer, and water.	Saving water and chemical fertilizer expenses and increase fertilizer utilization rate by up to 40% annually.
Improves farm live rate and reduces growth period as well as improves product volume significantly.	Farm crops live rate can be improved from 20% to 80%, growth period reduced by up to 30%, and production volume improved 1.5 to 2 times.
Improves labor efficiency significantly and reduces labor intensity. In addition, farmers no longer need to spend significant time on the farm.	This application can get back the UAV-WSN investment and gain profits within one year.

The next generation of UAVs is expected to utilize the new technologies, such as machine learning, in agriculture. The process may involve performing UAV surveys on the agricultural land at critical decision-making points in the growing season. Then, UAV images are uploaded to the cloud before being processed with machine learning techniques. Finally, the mobile app and web-based platform can provide farmers with actionable insights on crop health. The advantages of utilizing UAVs with machine learning technology in agriculture are (i) early detection of crop diseases, (ii) precision weed mapping, (iii) accurate yield forecasting, (iv) nutrient optimization and planting, and (v) plant growth monitoring.

7.3.5.2 Precision agriculture

PA is a farming management concept based on observing, measuring, and responding to the variability of crop fields. It calls for a DSS for farm management that maximizes the outputs and minimizes the inputs. The UAVs have been used in PA to increase crop production and profitability in farming systems and are cost-effective and time-saving (Khanal *et al.*, 2017). The PA requires more sophisticated and more detailed geo-information about nature and environment, health and vigor of crops, yield capacity, quality, physical soil properties, irrigation, fertilizer, and pesticide application rate (Tsouros *et al.*, 2019). The UAVs have been used in different countries to study plant biophysical properties, such as leaf area index and yield, the nutrient content of crops, and estimate the nitrogen status.

The most common features that can be monitored with UAV-based remote sensing for PA are presented in Table 7.5 (Tsouros et al., 2019).

TABLE 7.5 Crop Features that can be Monitored with UAVs.

S.No.	Crop Features	
	Vegetation	Soil
1	Biomass	Moisture content
2	Nitrogen status	Temperature
3	Moisture content	Electrical conductivity
4	Vegetation color	
5	Spectral behavior of chlorophyll	
6	Temperature	
7	Spatial position of an object	
8	Vegetation indices	

Applications of satellite-based observations in PA are often limited due to their coarser spatial resolution (Sharma and Garg, 2019). The PA can be improved by applying UAVs fitted with RGB and multispectral cameras to produce ortho-images of a field and create NDVI maps (Mahajan and Bundel, 2016). With the latest sensors, the U2AV will provide useful geo-information for site-specific development and management of small and medium-scale crop fields. Sankaran et al. (2015) presented several types of sensors that can be used in UAV-based PA, as summarized in Table 7.6. This information would be very useful while selecting a particular sensor or a combination of sensors for PA studies.

TABLE 7.6 The UAV Sensors in PA Applications.

Type of sensor	Operating frequency	Applications	Disadvantages
Digital camera	Visible region	Visible properties, outer defects, greenness, and crop growth	• Limited to visual spectral bands and properties
Multispectral camera	Visible-IR region	Multiple plant responses to nutrient deficiency	• Limited to few spectral bands water stress, diseases among others

Type of sensor	Operating frequency	Applications	Disadvantages
Hyperspectral camera	Visible-IR region	Plant stress, produce quality, and safety control	• Image processing is challenging. • High cost sensors
Thermal camera	TIR region	Stomata conductance and plant responses to water stress and diseases	• Environmental conditions affect the performance • Very small temperature differences are not detectable. • High-resolution cameras are heavier
Spectrometer	Visible-NIR region	Detecting disease, stress, and crop responses	• Background, such as soil, may affect the data quality. • Possibilities of spectral mixing. • More applicable for ground sensor systems
3D camera	IR laser region	Physical attributes, such as plant height and canopy density	• Lower accuracies. • Limited field applications
LiDAR	Laser region	Accurate estimates of plant/tree height and volume	• Sensitive to small variations in path length
SONAR	Sound propagation	Mapping and quantification of canopy volumes, digital control of application rates in sprayers or fertilizer spreader	• Sensitivity limited by acoustic absorption, background noise, etc. • Lower sampling rate than laser-based sensing

There are several challenges in the deployment of UAVs in PA, as mentioned below (Khanal *et al.*, 2017, Tsouros *et al.*, 2019):

- Thermal cameras are expensive and don't have very high-resolution. Depending on the quality and functionality, the majority of thermal cameras have a resolution of 640 pixels by 480 pixels.
- Thermal aerial images can be affected by many factors, such as moisture in the atmosphere, distance from the ground, and other emitted and reflected thermal radiation sources. Therefore, calibration of aerial sensors is critical to extract scientifically reliable surface temperatures of objects.
- Temperature readings through aerial sensors can be affected by crop growth stages. When plants are small and sparse at the beginning of the growing season, temperature measurements can be influenced by reflectance from the soil surface.
- In the event of adverse weather, such as extreme wind, rain, and storms, there is a big challenge of UAV deployment in PA applications. Therefore, small UAVs can't operate and collect data in extreme weather conditions.
- One of the key challenges is the ability of lightweight UAVs to carry a heavy-weight payload, limiting UAVs' ability to carry an integrated system that includes multiple sensors, high-resolution and thermal cameras.
- UAVs have a short battery lifetime, usually less than 1 hour. Therefore, the power limitation of UAVs is one of the biggest challenges of using UAVs in PA.

As recommended by Shamshiri et al. (2018) for PA studies, some salient points of fixed-wing and rotary-wing UAVs are given in Table 7.7. One can make a choice of UAV based on their specifications presented in this table.

TABLE 7.7 Specifications of Some Fixed-wing and Rotary-wing UAVs Recommended for PA (Shamshiri et al., 2018).

Model	Price ($)	Weight (kg) payload	Size (mm)	Camera resolution	Coverage	Flight time (min)	Max altitude (m)	Flight speed (km/h)
Parrot Disco Pro AG Drone	6875	UAV: 0.78 Take-off: 0.94	1150 x 580 x 120 (LWH)	—	—	—	—	—
RF70 UAV	3000	Payload: 3	—	1080 p	600 acres/hour	45–60	—	18
DT-26 Crop mapper	120,000	—	—	1080 p	—	60	—	110

Model	Price ($)	Weight (kg) payload	Size (mm)	Camera resolution	Coverage	Flight time (min)	Max altitude (m)	Flight speed (km/h)
Quad Indigo	25,000	—	—	1080 p	—	45	—	—
AgDrone UAV	10,000	—	—	1080 p	—	60	—	
E384 Mapping Drone	2400	UAV: 2.5 Payload: 1	Wing span: 1900 Length: 1300	—	1000 acres in 100 minutes at 5 cm resolution	90	—	47
Precision Hawk Lancaster 5	—	Payload: 1	—	1 cm/ pixel	300 acres/ flight	45	—	—
Xena observer	—	Take-off: 5	—	—	—	27	5000	—
Xena thermo	—	Take-off: 4.6	—	—	—	32	5000	—
AEE AP10 Drone	299	—	—	1080 p Full HD video at 60 FPS	—	25	500	71
UAV drone crop sprayer	—	UAV: 9 Payload: 10 Take-off: 13	800 x 800 x 70 (LWH)	—	—	16	1000	—
DJI drone sprayer	15,000	—	—	—	10 acres/ hour	—	—	29
Yamaha's helicopters spray & survey	130,000	UAV: 71 Payload: 30	—	—	10 acres	—	—	—
JMR-V1000 6-rotor 5L	665-3799	UAV: 6.5 Take-off: 18	875 x 1100 x 480 (LWH)	—	—	14–18	—	11—22
AG-UAV Sprayers 1	—	UAV: 8 Payload: 6	Height: 650	—	—	8—15	—	—
AG-UAV Sprayers 2	—	UAV: 14.2 Payload: 20	Height: 650	—	—	15-30	—	—
AG-UAV Sprayers 3	—	UAV: 9.5 Payload: 10	Height: 650	—	—	10-20	—	—

(continued)

Model	Price ($)	Weight (kg) payload	Size (mm)	Camera resolution	Coverage	Flight time (min)	Max altitude (m)	Flight speed (km/h)
DJI AGRASMG—1 Sprayer	7999	Payload: 10	—	—	7-10 acres/hour	—	—	—
Hercules Heavy Lift UAV (HL6)	—	UAV: 8 Payload: 6	Height: 660	—	—	30	—	37
Hercules Heavy Lift UAV (HL10)	—	UAV: 9.5 Payload: 10	Height: 660	—	—	30	—	37
Hercules Heavy Lift UAV (HL20)	—	UAV: 14 Payload: 20	Height: 660	—	—	60	—	37
DJI Phantom 3	469	—	—	2.7 K HD videos, 12 MP photo	—	25	—	—
Fixed-Wing UG-II	—	UAV: 11 Take-off: 15	2240 x 1600 x 650 (LWH)	—	—	180	—	65-110
Professional Electric Six Rotor Drone UA-8 Series	—	Payload: 3	860 x 860 x 540	—	—	28	5000	36
Yuneec H520 Hexacopter	2500-4500	—	—	4 K/2 K/ HD video or 20 MP images	—	—	—	—
Ag drone AK-61	6999	Take-off: 22 Payload: 10	—	—	—	1-15	0.5-5	18-36
YM-6160	5000	Take-off: 21.9 Payload: 10	—	—	—	10-15	0.5-5	18-36
Skytech TK110 HW	32-52	—	—	0.3 MP	—	6-7	—	—
JJRC H8D 5.8G FPV RTF RC	169-175	UAV: 0.023	330 x 330 x 115 (LWH)	—	—	8	—	—

Model	Price ($)	Weight (kg) payload	Size (mm)	Camera resolution	Coverage	Flight time (min)	Max altitude (m)	Flight speed (km/h)
X810 Long-Range UAV Sprayer	4000-6500	Payload: 10	2490 x 1645 x 845 (LWH)	—	—	25-40	—	—
Syma X8C	68.99	—	508 x 508 x 165 (LWH)	2 MP HD	—	5-8	—	

7.3.5.3 Soil mapping

Soil properties, such as soil texture, can indicate soil quality, which may influence crop productivity (Garg *et al.*, 2020). The UAV thermal images can quantify soil texture by measuring the differences in land surface temperature under a relatively homogeneous climatic condition (Wang *et al.*, 2015). Crop residue is essential in soil conservation by providing a protective layer on agricultural fields that shields soil from wind and water. An accurate assessment of crop residue is necessary for the proper implementation of conservation tillage practices. The UAV thermal images can provide more accurate variability in crop residue cover than using visible and NIR images. For a detailed study of soil mapping, readers are advised to refer to Garg *et al.* (2020).

7.3.5.4 Irrigation scheduling

Irrigation scheduling can be defined as "the process of determining when to irrigate and how much water to apply, based upon measurements or estimates of soil water or water used by the plant" (Rai *et al.*, 2017). There are four factors that are to be considered to determine the need for irrigation: (i) availability of soil water, (ii) crop water need, which represents the amount of water needed by the various crops to grow optimally, (iii) rainfall amount, and (iv) efficiency of the irrigation system (Rhoads and Yonts, 2000). These factors can be quantified by utilizing UAVs to measure soil moisture, plant-based temperature, and evapotranspiration (Chao *et al.*, 2008). For instance, the spatial distribution of surface soil moisture can be estimated using high-resolution multispectral images captured by a UAV combined with ground sampling. The crop water stress index can also be estimated to determine water-stressed areas by utilizing the thermal UAV images (Poblete *et al.*, 2017).

7.3.5.5 Crop height mapping

The UAV can inspect the crop fields at a much closer range (40-120 m) and provides spatial information at a much higher resolution. Crop height can be estimated in experimental fields, using crop surface models derived from UAV-RGB color images, and related to the extracted biomass (Viljanen *et al.*, 2018). The UAVs have been used in agriculture to provide high-spatial-resolution images to detect individual crops and weeds at a very large-scale (Chang *et al.*, 2017). A LiDAR sensor mounted on a UAV can map the canopy heights to estimate crop biomass in agriculture. For example, Shi *et al.* (2016) used UAV technology to measure the plant height of crops, such as maize and sorghum. The UAVs can also be used practically to monitor crops and determine the harvesting time, particularly when the entire area cannot be harvested in a short time.

7.3.5.6 Crop disease detection

Traditional methods of locating pests in a large area of crops are not effective. The UAVs can collect information to assess the health of the plantations (Uddin *et al.*, 2018). The UAVs-based thermal remote sensing can monitor the spatial and temporal patterns of crop diseases (e.g., early stage development of soil-borne fungus) during various disease development phases, and hence can help farmers reduce crop losses (Sabrol and Satish, 2016). In the USA alone, it is estimated that crop losses caused by diseases result in a loss of about US$ 33 billion every year (Patil and Kumar, 2011).

By aerial imaging the crops using onboard UAV sensors, such as visible RGB camera, NIR, hyperspectral, and multispectral sensors, it is possible to link temporal and spatial reflectance variations with crop health. The NDVI images can calculate the index describing the relative density and health of the crops. At the same time, the thermal camera can show the heat signature of different spots in the plantations. A conceptual demonstration of a UAV remote sensing platform equipped with visible and NIR sensors for crop health assessment is shown in Figure 7.16. The proposed approach in this figure has been successfully used by Shamshiri *et al.* (2018) to assess oil palm plantations' health assessment to spot fungal infection and bacterial disease.

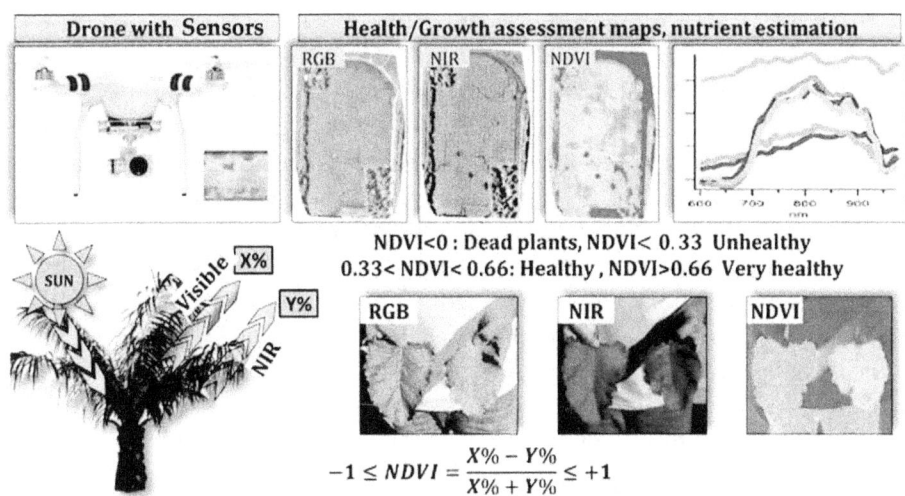

FIGURE 7.16 Conceptual demonstration of a UAV remote sensing platform for a crop health assessment with NDVI images (Shamshiri *et al.*, 2018) [*See Plate 7 for Color*]

7.3.5.7 Spraying operations

Spraying of pesticides and fertilizers is common (Figure 7.17) for the healthy growth of crops. The NDVI maps generated by UAV images can provide information about the crop evolution and the requirements for spraying of pesticides and fertilizers. The UAVs can spot accurate locations of such areas that require the application of pesticides and fertilizers (Kale *et al.*, 2015). Based on the field requirements, the use of pesticides and fertilizers can be optimized.

FIGURE 7.17 Pesticide spray in crop

For pest monitoring, UAV imagery platform equipped with a thermal camera and high-resolution RGB vision sensors may be required to accurately identify the spots detected with specific insects and pests. The spraying UAV can also be equipped with distance-measuring and light detection sensors, such as lasers, ultrasonic echoing, or LiDAR methods to scan the ground and adjust the flight altitude with the varying topography of the plantation (Desale *et al.*, 2019). In pests-affected areas, spraying UAVs can be used for dropping a targeted load of pesticide. This would help to apply the right amount of spraying liquids for even coverage and avoid collision of UAV with the terrain/plants. This approach will increase efficiency and reduce the amount of penetrating spray chemicals in the soil and groundwater. It is estimated that UAV spraying is five times faster than conventional tractor and machinery equipment (Shamshiri *et al.*, 2018).

7.3.5.8 Crop yield mapping

Farmers require accurate, early estimation of crop yield for a number of reasons, including crop insurance, planning of harvest and storage requirements, and cash flow budgeting. The UAV images can be utilized to estimate the yield and total biomass of crops as well as predict the yields (Duan *et al.*, 2019). By implementing UAV technology, farms and agriculture businesses can improve crop yields, save time, increase overall profitability and make land management decisions. Strong correlations have been observed between NDVI data measured at certain crop stages and crop yield (Huang *et al.*, 2013). Hence tracking the crop growth at key stages will help provide an accurate estimate of the crop yield.

Sola-Guirado *et al.* (2017) related the actual yield (kg/tree) with geometrical parameters of the olive tree canopy in southern Spain. They established the regression equation to forecast the actual yield based on manual canopy volume data acquired from different orchard categories and cultivars during different harvesting seasons. Images were acquired with UAV, and orthoimages were created to calculate individual crowns so as to relate them to canopy volume and actual yield. Manual canopy volume was related to the individual crown area of trees calculated by orthoimages acquired with UAV imagery. The complete procedure used is shown in Figure 7.18.

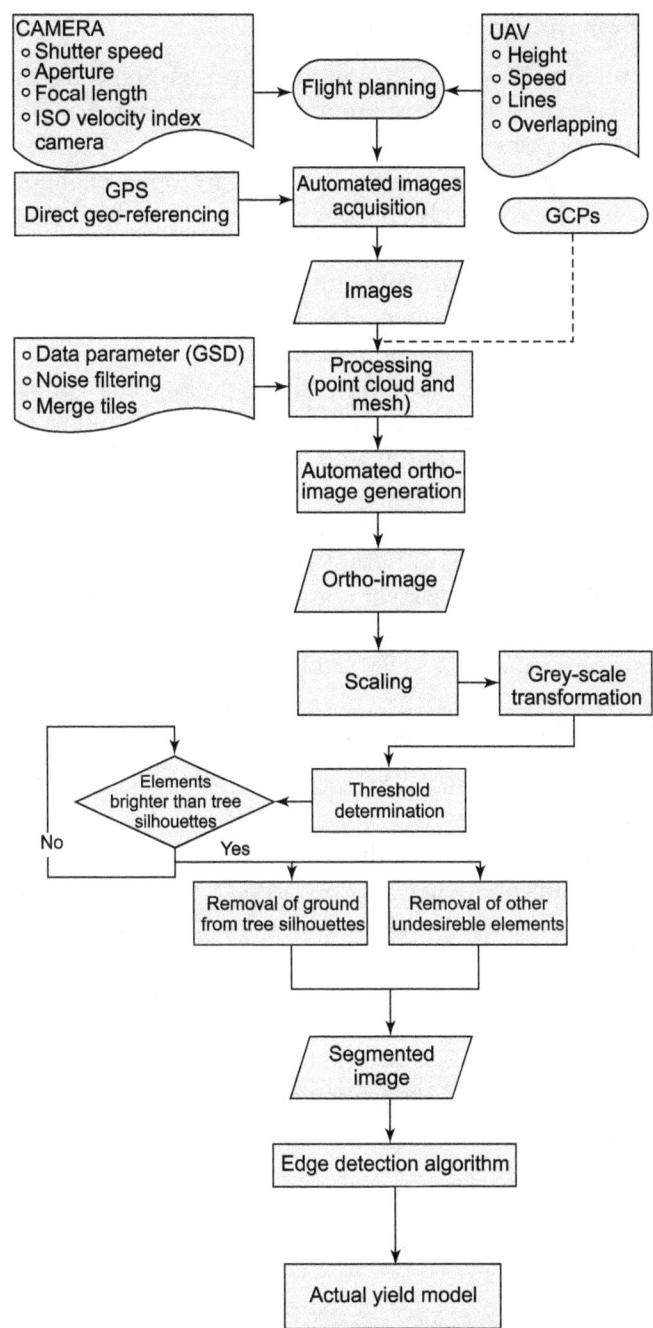

FIGURE 7.18. Orthoimage analysis procedure in an olive orchard using UAV (Sola-Guirado *et al.*, 2017).

7.3.6 Hydromorphological Studies

Hydromorphology is a term used in river basin management to describe the hydrological (water flow, energy, etc.) and geomorphological (surface features) processes and attributes of rivers, lakes, estuaries, and coastal waters. In hydrology, UAVs are used to measure open channel surface velocities. The use of UAVs is also becoming common in flood control, natural hazard management, identify flood damages, and serve humanitarian response services during disaster events in flood-affected areas (Popescu et al., 2017). The emerging technology of UAV can also be applied for coastal wetland mapping and environmental management.

Methodologies based on high-resolution photographs from UAVs have been developed by Casado et al. (2017) for hydromorphological characterization based on the automated recognition of hydromorphological features through ANN. The framework successfully identifies the vegetation, deep water, shallow water, riffles, sidebars, and shadows for most of the reaches and produces 81% accuracy in feature classification. The UAV imagery collected at each reach of river is assessed based on quality and spatial coverage for inclusion in the photogrammetric process and generated standard products (i.e., orthoimage, DTM, and point cloud) using Photoscan Pro software.

7.3.7 3D Modeling-Based Applications

A digital camera on a UAV produces 2D images; however, as the requirements have grown, 3D vision is needed in many applications, such as 3D terrain visualization. Multispectral, LiDAR, photogrammetry, and thermal vision sensors are being used on UAVs to create 3D models of buildings and landscape. The DEMs of land areas are created to provide precision elevation data of the terrain (Qu et al., 2018). There is also a need to reconstruct 3D structures from the UAV collected 2D images, and for this purpose, photogrammetric and computer vision methods are used (Zollhöfer et al., 2018). The 3D models can be generated from several overlapped images using software, such as Agisoft PhotoScan or Pix4D software. Figure 7.19 shows a 3D terrain model.

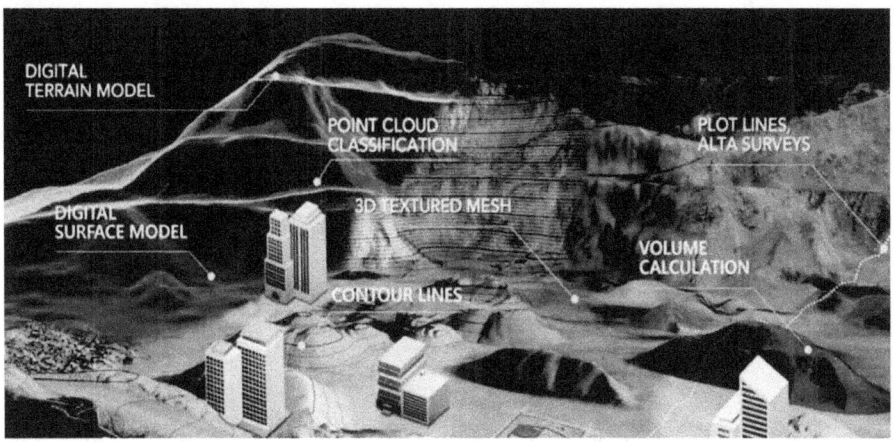

FIGURE 7.19 Uses of a 3D terrain model [*See Plate 8 for Color*].

Several other methods have also been proposed to generate 3D models, but the most important include the SLAM, SfM, and MVS algorithms, which have been implemented in many practical applications. As the number of images and their resolution increases, the algorithms' computational time will also increase significantly, limiting them in some high-speed reconstruction applications. Finally, the software lays texture taken from the original photographs over the 3D mesh. The final outcome is a detailed 3D textured model that can be used for a variety of specialized analyses, including 3D city mapping, archaeological research, creation of flooding models, and disaster damage assessment. The reconstruction of 3D city maps has key applications in a wide range of both non-communications (facility and utility management, city planning, navigation, etc.) as well as communications-related areas (e.g., radio engineering, base station deployment planning, channel quality, coverage prediction, etc.) (Mohammed *et al.*, 2014).

7.3.8 Wireless Network-Based Applications

The UAVs are being used to deliver Internet access to developing countries and provide airborne global Internet connectivity. Qualcomm and AT&T are planning to deploy UAVs for enabling wide-scale wireless communications in the fifth generation (5G) wireless networks (QUALCOMM, 2018). The Amazon Prime Air and Google's Project Wing (Stewart, 2014) initiatives are prominent examples of use cases for cellular-connected UAVs. The 5G is specially designed to provide quality of service (QoS) to users, which means that it can provide the maximum bandwidth for enhanced cellular coverage, reduced delay, error rate, uptime, and increased data rate. Additionally, 5G has reduced delay (Qiao, *et al.*, 2016). Such a linked system can connect smart cities; smart homes by offering high-speed broadband Internet connectivity and can also support e-payments, e-transactions, and other fast electronic transactions (Bor-Yaliniz *et al.*, 2016).

The concept of 5G coverage using UAVs in urban settlements, presented by Al-Turjman *et al.* (2019), is shown in Figure 7.20. Wireless users, in general, expect to have unlimited and affordable Internet access all the time. Increasing the number of BS in a given area is a potential way for satisfying the users and providing extended 5G coverage. They have used a linear optimization problem, using Simulated Annealing and Genetic Algorithm due to their ability in providing fast and efficient solutions to service providers. It is not an easy task as a few of these BSs can have less or no load at a particular time, while other BSs might experience heavy data traffic and unnecessary overhead. The unpredictability characteristic of the users makes it hard to know exactly where and when a BS should be located. This problem can be resolved by designing drone-based stations, as depicted in Figure 7.20. The drone-based station is flexible and can be placed at any particular time it is needed most. Hence, it efficiently provides 5G coverage for the users at all times. There is a growing number of research works being done on drone-based stations in cellular networks. However, one critical challenge that has not been given much attention is finding the optimum number and locations of drone-based stations in a given 3D space to get maximum 5G coverage with guaranteed QoS.

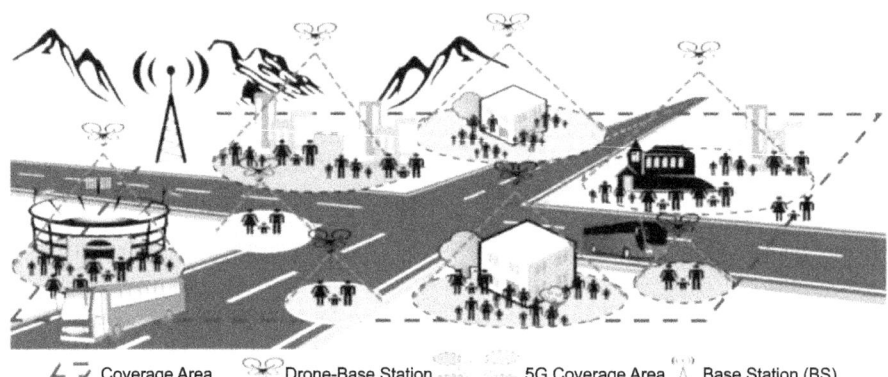

FIGURE 7.20 The UAV-based 5G coverage in the urban area (Al-Turjman *et al.*, 2019).

7.3.9 Smart Cities

The world population is increasing, and it is foreseeable to be doubled by 2050. It, therefore, creates new challenges and opportunities as well as pressure for cities on associated infrastructure. One of the emerging areas of UAV applications and functions is in smart cities. These applications include monitoring traffic flow to measuring and detecting floods and natural disasters by using wireless sensors installed in UAVs. Integrating UAVs with smart cities will create a sustainable business environment and a safe place for living. The UAV systems and smart cities can significantly impact and benefit the population when used effectively and efficiently (Mozaffari *et al.*, 2018). The model for UAV-based application in a smart city is presented in Figure 7.21, where many smart services will be made available to the smart citizens. Here, the role of IoT-based devices connected to UAVs cannot be under-estimated.

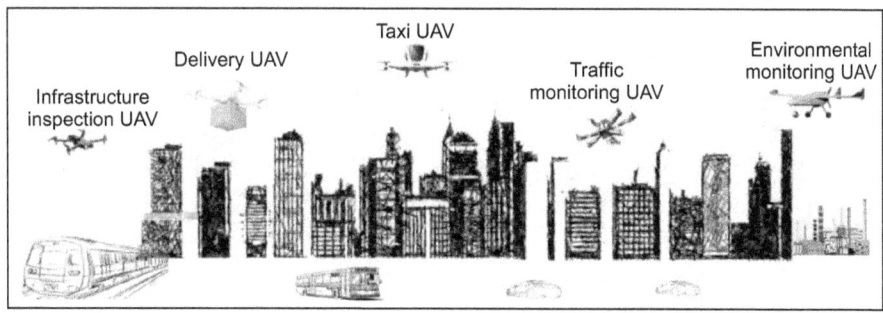

FIGURE 7.21 The UAV-based applications in smart cities.

The global vision of smart and connected communities is a great technological challenge. Smart cities will effectively require integrating UAV-based technologies and services, including an IoT environment, a reliable wireless cellular network, resilience to calamities, and a large amount of data (Menouar *et al.*, 2017). The UAVs can provide several wireless applications in smart cities and gather vast data within a city and deliver them to a central cloud unit for big data analytics purposes. The UAV base stations can also simply enhance the cellular network coverage in a city or respond to specific emergent situations. The UAVs can also be used to sense the radio environment maps (Kosmer and Vilhar, 2014) across a city to assist network operators in their network and frequency planning efforts.

Another key application of UAVs in smart cities is their ability to act as mobile cloud computing systems (Yang *et al.*, 2017). A UAV-mounted cloudlet can provide fog computing and off-loading opportunities for devices that are unable to perform computationally heavy tasks. Within smart cities, UAVs may need to be temporarily positioned themselves on top of buildings for recharge of the battery. So, there will be a need for on-demand site renting management to accommodate the operation of UAVs. In the future, developments are expected to continue in cloud computing, wireless sensors, networked unmanned systems, big data, open data, and IoT to provide substantial opportunities for using UAVs in smart cities (Sharma and Garg, 2019). Several issues need to be considered with the development and use of UAV systems for smart cities, such as proper use of wireless sensors, data communications, application management, resource training and allocation, and power management. Overall, in the future, the UAVs will be an integral part of smart cities from both wireless and operational perspectives.

7.3.10 Traffic Monitoring and Management

The transportation system cannot be automated through vehicles only (Menouar *et al.*, 2017). In fact, other components of the transportation system, such as field support teams, traffic police, and rescue teams, also need to be automated. Smart and reliable UAV can help automate these components, as it is considered a novel traffic monitoring technology to collect information about traffic conditions on roads. Traffic surveillance and monitoring have been explored as one of the first applications for UAVs in civil engineering as they provide aerial perspective for security teams (Shakhatreh *et al.*, 2018). The major applications of UAVs in transportation include security surveillance, traffic monitoring, an inspection of road construction projects, and survey of traffic, rivers, coastlines, pipelines, etc. (Puri, 2015).

Compared to the traditional monitoring devices, such as loop detectors, surveillance video cameras, and microwave sensors, UAVs are cost-effective and can monitor large continuous road segments or focus on a specific road segment. Moreover, as computing, communications infrastructure, or power systems fail during disasters, it may not be possible to control and collect data about the transportation network (Mozaffari *et al.*, 2018). Data collection on the performance of transportation systems within the civil infrastructure is often sparse and can ignore specific details. For example, conventional traffic counting may ignore vehicle types and speed, whereas vision-based data collected by UAVs can provide greater detail when monitoring traffic patterns (Ro *et al.*, 2007). Figure 7.22 shows the setup for a smart UAV-based traffic monitoring system.

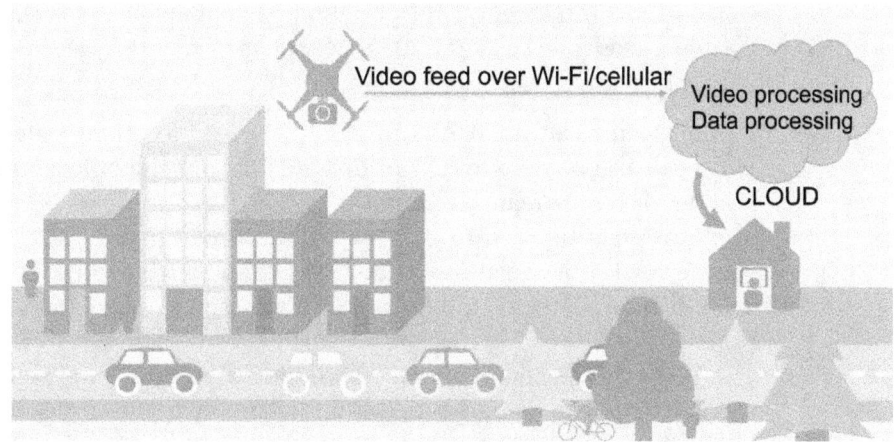

FIGURE 7.22 The UAV-based traffic monitoring system.

Efforts have been made to use UAVs for streamlining roadway condition assessment. For example, Zhang and Elaksher (2012) used UAV collected imagery to produce 3D models of distressed unpaved roads. A 3D model with an absolute resolution of less than 1 cm was used to detect potholes and ruts within the roads. Brooks *et al.* (2014) explored the possible applications for UAVs in transportation engineering and found UAVs to be cost-effective tools for monitoring traffic and inspecting road assets.

Recognition of moving vehicles using UAVs is still a challenging problem. Moving vehicle detection methods depend on the accuracy of image registration methods since the background in the UAV surveillance platform changes frequently. Accurate image registration methods require extensive computing power, which affects the real-time capability of these methods. Qu *et al.* (2016) studied the problem of moving vehicle detection using UAV cameras. They used CNN to identify vehicles more accurately in real-time, using three steps: (i) adjacent frames are matched, (ii) frame pixels are classified as background or candidate targets, and (iii) a deep convolutional neural network is trained over candidate targets to classify them into vehicles or background. They achieved detection accuracy of around 90% when evaluating their method using the UAV dataset.

The battery life and flight time of UAVs may limit the samples and length of traffic data. For example, Ro *et al.* (2007) utilized an autonomous UAV to collect 6 h of continuous data and to deliver high-quality video imagery and sensor data in real-time with a telemetry range of 1 mile and payload capacity of 4 lbs. The UAV technology to collect 24 h of continuous data of the traffic can be highly uneconomical.

Some of the Intelligent Transportation System applications that UAVs can enable are as follows (Puri, 2015; Menouar *et al.*, 2017; Sutheerakul *et al.*, 2017; Shakhatreh *et al.*, 2018):

- Rescue teams can use UAVs to reach accident locations quickly. The UAVs can also be used to deliver first aid kits to accident locations while waiting for rescue teams to arrive.
- The UAVs can be used to fly over different road segments to stop the vehicle for traffic violations. The UAVs can change the traffic light in front of the vehicle to stop it or relay a message to a specific vehicle to stop.
- The UAVs can be complemented with dedicated short-range communication to enable a flying Road Side Unit. The flying Road Side Unit can fly to a specific position to execute a specific application. For example, in an accident on the highway at a specific segment that is not equipped with any Road Side Unit, the traffic management center can activate a UAV to fly to the accident location and land at the proper location to broadcast the information and warn all approaching vehicles about a specific incident.
- The UAVs can recognize suspicious or abnormal behavior of ground vehicles moving along with the road traffic.
- The UAVs can be used as an alternative data collection technique to monitor pedestrian traffic and evaluate demand & supply characteristics. In fact, the collected data can be categorized into four areas; the measurement of pedestrian demand, pedestrian characteristics, traffic flow characteristics, and walking facilities and environment.
- The UAVs may also be employed for a wide range of transportation operations and planning applications, such as incident response, coordination among a network of traffic signals, traveler information, monitoring parking lot utilization, and estimating origin-destination flows.

7.3.11 Swarm-based Applications

There have been many applications and development of UAV swarm, particularly military applications date back to the early 1990s (Campion *et al.*, 2018). The AI applied to the swarm UAVs has diverse applications in the military field. It allows the UAV swarm to identify and attack the enemy defenses in an autonomous way. In addition, it helps the army to search for enemy submarines in the ocean by incorporating IR sensors (Pires *et al.*, 2016). The UAV swarms could disperse over large areas to identify and eliminate hostile surface-to-air missiles and other air defenses. They could potentially serve as novel missile defenses, blocking incoming hypersonic missiles. For homeland security, swarms equipped with chemical, biological, radiological, and nuclear explosives detectors, facial recognition, anti-drone weapons, and other devices offer defenses against a range of threats (Shakhatreh *et al.*, 2016).

Although swarms are largely used for military purposes, their civilian applications have received increased attention recently. Swarm UAVs can also carry out "search and rescue" missions effectively over huge territories. In case of fire, they can assist rescue teams by obtaining real-time information about fire advances. The UAV swarms with coordinated control and communication capabilities would be efficient for many interesting applications (Tahir *et al.*, 2019). Such swarms, surveying entire farmland with little to no operator intervention, would greatly increase efficiency and potentially revolutionize PA (Campion *et al.*, 2018). The most notable example of a UAV swarm includes a swarm of 300 drones developed by Intel which were deployed as a coordinated light show for Super Bowl 51 and the 2018 Winter Olympics (Molina, 2017).

Swarm UAVs have ample potential for applications related to delivery services. Amazon and United Postal Service have indicated their interest in using UAVs for package delivery (Business Insider, 2017; McFarland, 2017). Low-cost drones and their swarms provide a promising platform for innovative projects and future commercial applications to help people in their work and everyday lives.

7.3.12 Surveillance Operations

The UAVs can be used for security, surveillance, and monitoring to improve public safety in areas, such as anti-terror operations, criminal investigations, traffic surveillance, searching for missing persons, emergency communication networks, border area surveillance, coastal area surveillance, anti-piracy operations, and controlling hostile demonstrations and rioting. The detailed information can also be acquired using fixed cameras, but a UAV can be mobilized to many locations where constant or long-term surveillance is required.

Surveillance applications do not require very accurate coordinates, while many surveying applications need accurate coordinates. The usage of UAVs for surveillance has difficulties in exact target identification. Employing UAVs for surveillance introduces several important issues. Table 7.8 summarizes the advantages, disadvantages, and important concerns for the UAV surveillance applications.

TABLE 7.8 Advantages, Disadvantages, and Important Concerns for Surveillance Applications of UAVs.

Advantages	Disadvantages	Important concerns
– Surveillance coverage and range improvement – Better safety for human operators – Robustness and efficiency in surveillance	– High accidental rate – High operating costs – Difficulties in surveillance of risk societies	– Co-operation of multi-UAVs – Post-processing algorithm improvements – Privacy concerns – Law enforcement

An autonomous UAV can perform a mission for border surveillance very effectively. The use of autonomous UAVs for such tasks is advantageous due to their long endurance and vast area coverage (Eaton et al., 2016). The nominal mission requires one or more UAVs to patrol a designated border region and detect intruders.

Figure 7.23 shows the steps which are followed under the mission for border surveillance, where the numbers indicate the sequence followed in the mission. (Stenger et al., 2012). When more UAVs are required to take part in the mission, one UAV in standby receives the mission plan. The aircraft status then changes to available, and the UAV starts following the assigned tasks. The primary task of that UAV is to fly and scan the border region by using an appropriate search strategy which depends on the sensor coverage and on the probability that intruders will cross the border in a specific region (due to landscape or previously encountered violation). As soon as a UAV detects an intruder or a group of intruders, it transmits its current position, as well as live video images to the GCS. The GCS then decides whether the UAV shall further track the intruder or search within surrounding areas for additional violations due to a false alarm as a detected object is not an intruder. A nominal mission is unlikely to occur in reality. Consequently, the robustness of the UAV is tested when confronted with typical stressors (situations), such as system failures or other aerial traffic.

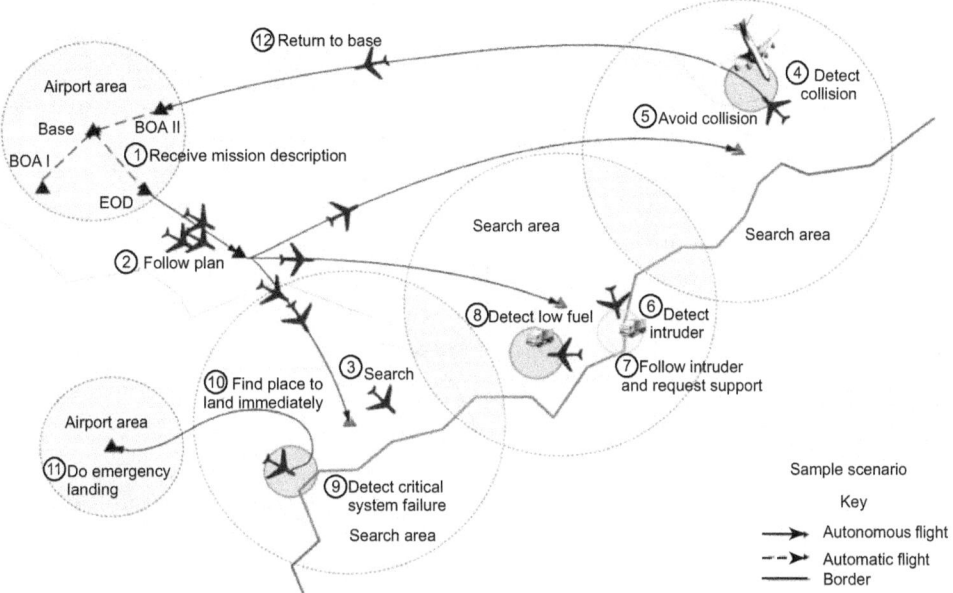

FIGURE 7.23 Stressed mission (Stenger et al., 2012).

The future research works on UAV surveillance can be classified into the following six categories, based on their focus and contents (LeCun, 2015):

- *Multi-UAV cooperation*: the research focuses on developing new orchestration algorithms or architectures in the presence of multiple UAVs.
- *Homeland/military security concerns*: the research provides analysis and suggestions for the authorities regarding homeland/military security considerations.
- *Algorithm improvement*: the research develops new algorithms for more efficient post-processing of the data (e.g., videos or photos captured by the UAVs).
- *New use case*: the research integrates the UAV system into new application scenarios for potential extra benefits.
- *Product introduction*: the research describes mature techniques leveraged by a real UAV surveillance product. Therefore, advanced sensor technologies, such as nano-sensors (small size), ultra-high-resolution image sensors (accurate data), and energy-efficient sensors (low energy consumption), are to be considered to capture more accurate data more efficiently and secretly.

- *New algorithms*: the research develops more efficient and accurate multi-UAV cooperation algorithms or data post-processing algorithms, such as advanced machine learning algorithms (e.g., deep learning), which could be utilized to achieve better performance and faster response. Specifically, multi-UAV cooperation brings more benefits than single UAV surveillance, such as wider surveillance scope, higher error tolerance, and faster task completion time. However, multi-UAV surveillance (swarms) requires more advanced data collection, sharing, and processing algorithms. Applying advanced machine learning algorithms (e.g., deep learning algorithms) could help the UAV system analyze huge amount data and draw more concise and reliable results for better conclusions quickly.

7.3.13 Search and Rescue (SaR) Operations

The SaR operations using traditional aerial systems (e.g., aircraft and helicopters) are typically very expensive. The use of UAVs in SaR operations reduces the costs, resources, and human risks (Figure 7.24). In case of natural or man-made disasters, such as floods, tsunamis, or terrorist attacks, critical infrastructure, including water and power utilities, transportation, and telecommunications systems, can be partially or fully damaged (Adams and Friedland, 2011). It necessitates providing rapid communications coverage to support the SaR operations (Valavanis and Vachtsevanos, 2014). The UAVs can provide timely disaster warnings and assist in speeding up SaR operations when the public communications networks are disrupted. Moreover, UAVs can quickly provide coverage of a large area without ever risking the security or safety of the personnel involved.

FIGURE 7.24 The UAV in SaR operation.

The UAVs offer an immense advantage in support of disaster management. In highly disastrous situations, such as wildfire, leakage of poisonous gases, avalanches, or nuclear fallout, it is a challenging task to rescue as in such types of emergent situations, the rescuers have to put their own lives at risk (Adams and Friedland, 2011), but UAVs can be used to speed up the rescue operations, as they are designed to fly smartly in any environment and withstand extreme temperatures and radiations. Moreover, healthcare workers can use UAVs to deliver medicines, vaccines, and care packages to the people in the affected areas that may be practiced otherwise inaccessible.

There are two types of SaR systems: single UAV systems and multi-UAV systems (i.e., swarms) (Shakhatreh *et al.*, 2016). In a single UAV system, the rescue team defines the search region. Then the search operation is started by scanning the target area using a single UAV equipped with vision or thermal cameras. The real-time aerial videos/images from the targeted areas are sent to the GCS. The rescue team analyzes these videos and images to take appropriate SaR operations.

In multi-UAV systems, UAVs with onboard imaging sensors are used to locate the position of missing persons (Burnap *et al.*, 2015). First, the rescue team carries out the path planning to compute the optimal trajectory of the SaR mission. Each UAV receives its assigned path from the GCS. Second, the search process starts, where all UAVs (swarms) follow their assigned trajectories to scan the targeted region. This process utilizes object detection, video/image transmission, and collision avoidance methods. Thirdly, the detection process starts, where a UAV that detects an object hovers over it, while the other UAVs act as relay nodes to facilitate the coordination between all the UAVs and communications with the GCS. Thereafter, UAVs switch to data dissemination mode and set up a multi-hop communications link with the GCS. Finally, the location of the targeted object and related videos and images are transmitted to the GCS (Bernard *et al.*, 2011).

In recent years, cell phones have become more prolific in SaR operations in the built environment (Grogan *et al.*, 2018). Devices that can passively detect active cell phones, much the same way a laptop detects Wi-Fi, are attached to a UAV. This new tool can remotely sense and approximate the location of individuals who may be trapped in a structure after a disaster. In addition, the use of wireless sensors to detect cell phones can be explored, but its use in urban SaR operations has been underexplored. Installing the wireless sniffer on a drone, or several drones could give responders a means to quickly scan a disaster area and identify potential victims.

The use of UAVs in SaR operations attracted considerable attention and became a topic of interest in the recent past. The UAV can be utilized in SaR missions, as follows:

1. Collect high-resolution images and videos using onboard cameras on UAV to survey a given area. These images are used for post-disaster aerial assessment or damage evaluation. It helps to evaluate the magnitude of the damage in the infrastructure caused by the disaster. The rescue teams then identify the targeted search area and commence SaR operations accordingly.

2. An autonomous UAVs can provide accurate data and without additional risks. The UAVs can also employ thermal infrared imaging devices to find missing persons.

3. The UAVs can be used to deliver food, water, and medicines to the injured people. However, the UAVs still suffer from load-carrying capacity due to the limitations of their payloads. A high-power propellant system can allow the UAV to carry heavy cargo between 10 and 15 kg, including medicines, food, and water.

4. The UAVs can act as effective aerial base stations for rapid service recovery when the communications infrastructure is completely damaged in disaster-affected areas. This helps in SaR operations for outdoor and indoor environments, respectively.

5. Weather conditions pose a challenge to UAVs as they might result in deviations in their pre-determined paths. In cases of some natural disasters, the weather may become a challenging issue. In such scenarios, UAVs may fail in their missions as a result of uncontrolled weather conditions.

6. The SaR operation in low visibility may pose another problem. There is a need to develop low-vision cameras and algorithms to process such images.

7.3.14 Logistics Operations

UAVs have great potential applications in logistics networks (González-Jorge et al., 2017). The potential of UAV technology for logistics is more evident in rural locations with poor infrastructure or isolated populations (e.g., mountains or islands). The UAVs can be used to transport food, packages, emergency medicines, and other goods quickly (Marsh, 2015), as shown in Figure 7.25. With the rapid, massive growth of e-commerce, postal companies have been forced to find new methods to expand beyond their traditional mail delivery business models. Different postal companies have undertaken various UAV trials to test the feasibility and profitability of UAV delivery services (Heutger and Kückelhaus, 2014).

FIGURE 7.25 The UAV technology used for logistics.

The delivery in the urban area is probably the most tangible and spectacular in the logistics industry. In addition, its application has the largest barriers because of privacy and safety concerns arising out in the densely populated urban environment. It is also the most challenging in regulatory framework conditions and infrastructure, especially integration into existing urban infrastructures. Figure 7.26 shows the use of UAVs for the delivery of goods in the dense urban network. The use of UAVs for logistics is still in its early stages (Bamburry, 2015). So far, the payloads are limited, but a network of UAVs could nevertheless support logistics networks in large areas. The challenges associated with the UAV-based delivery system of goods, food, and medicines, are legislations, insurance, weather, air traffic control, and theft.

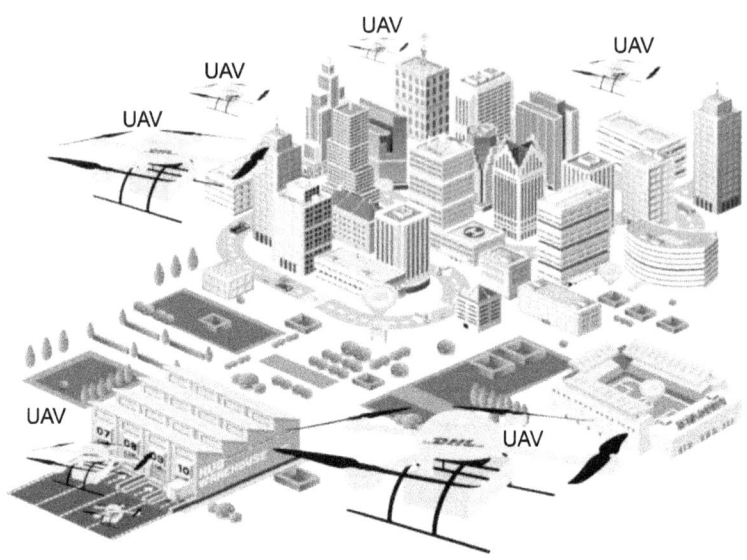

FIGURE 7.26 Logistics in dense urban setup (Heutger and K¸ckelhaus, 2014).

In UAV-based goods delivery system, a UAV can travel between a pick up location and a delivery location. The UAV is equipped with a control processor and GPS module. It receives a transaction packet for the delivery operation that contains the GPS coordinates and the identifier of a package docking device associated with the order. Upon the arrival of a UAV at the delivery location, the control processor checks if the identifier of a package docking device matches the device identifier in the transaction packet. It performs the package transfer operation and sends confirmation of completion of the operation to an originator of the order (Marsh, 2015). Suppose the identifier of a package docking device at the delivery point does not match the device identifier in the transaction packet. In that case, the UAV communication components transmit a request over a short-range network, such as Bluetooth or Wi-Fi. The request may contain the device identifier or network address of the package docking device. Under the assumption that the package docking device has not moved outside of the range of UAV communication, the package docking device having the network address transmits a signal containing the address of a new location. The package docking device may then transmit updated GPS coordinates to the UAV. The UAV is re-routed to the new address based on the updated GPS location (Shakhatreh *et al.*, 2018).

In a goods delivery system, based on the demand of the customer, a UAV leaves the hub and makes delivery direct to the customer location or, in case of returns, picks it up right from the customer. For instance, aerial delivery company Flirtey plans to introduce the worlds-first commercial UAVs for delivery. Student textbook rental service Zookal will use Flirtey to deliver parcels directly to a customer. Customers will receive a smartphone notification that will enable them to track the parcel via GPS and receive the parcel directly at an outdoor location. Once the UAV arrives at the outdoor delivery destination, it hovers and carefully lowers the parcel through a delivery mechanism that is attached to a retractable cord. This aims to significantly reduce waiting times from two to three days to as little as two to three minutes (Heutger and Kückelhaus, 2014; *http://www.atp-innovations.com.au/flirtey/*).

The UAVs used for the transportation of goods can safely take-off and landing in the proximity of buildings. These allow rapid delivery of goods in developing countries without having a good road network and to finally unleash the full potential of their e-commerce telecommunication infrastructure. Such UAVs will also help developed countries to improve the QoS in congested or remote areas. They will enable rescue organizations to quickly deliver medical supplies in the field and on-demand (Park *et al.*, 2017). Large corporates, such as Amazon and DHL, are experimenting with UAV technology to determine its future viability in the transportation of smaller parcels (DHL, 2014; Kimchi, 2015). In the near future, UAVs could automatically pick up the assigned shipment, take-off, fly to the delivery point, land, and provide the product to the customer.

7.3.15 Crowd-Sourcing

Crowd-sourcing involves seeking knowledge, goods, or services from a large body of people. All people in the crowd participate in creating, discussing, refining, and rank meaningful ideas or tasks or contributions *via* the Internet, social media, and smartphone apps. Millions of people, connected by the Internet, contribute ideas and information to projects, big and small. It can cover unpredictable events, produce novel discoveries and raise public awareness. Crowd-sourcing has several benefits for businesses (Hargrave, 2019), such as: putting up many minds together to work in solving tough problems, a greater diversity of thinking, a reduced management burden, faster problem solving, and a rich source of data. The advantages of crowd-sourcing include cost savings, speed, and the ability to work with people who have skills that an in-house team may not have. Research in crowd-sourcing has shown results to be as accurate as traditional research methods (Alwateer *et al.*, 2019).

UAVs are excellent tools to be used in crowd-computing scenarios due to their versatility as a type of mobile device. Crowd-powered devices with a single UAV approach can collect and process data in an efficient manner. Crowd-sourced spatial data collection is also facilitated by devices, such as handheld GPS, digital notebooks, and mobiles equipped with open source software, photos, video, and voice recorders (Parshotam, 2013). These data get synchronized (Figure 7.27) with a central database with/without moderation and are accessible/sharable as web-based databases, services, and maps. Uber taxi service, which pairs available drivers with people who need rides, is an example of crowd-sourced transportation. Facebook and Flickr are other potential examples of crowd-sourcing.

FIGURE 7.27 Representation of crowd-sourcing.

Crowd-sourcing can effectively be used to help out the people affected from the disaster zones. In emergency response situations, open and real-time spatial data can be collected involving crowds along with data fusion for integrating information from a wide variety of sources. It has helped to map the infrastructure and devastation caused by natural disasters (Ghosh *et al.*, 2012). During the disaster-hit rescue operation, UAV photographs are combined with human volunteer spotters (Alwateer *et al.*, 2019). A team of UAVs first explores an area and takes images which are then geotagged, and online virtual volunteers scour the pictures for signs of people. When several virtual team spotters tag an image, a rescue team is sent to the location. By crowd-sourcing their rescue efforts, the UAV technology could help find individuals lost in vast wildernesses that would otherwise be too difficult for the individual team to search (Fu *et al.*, 2018).

Recently, more directed forms of crowd-sourcing for disaster response have been undertaken. Crowd members tagged satellite imagery of destroyed buildings and road accessibility for use by on the ground disaster response teams (Burnap *et al.*, 2015). Efforts along this direction are supported by a multitude of Internet-based communities, including the Humanitarian OpenStreetMap mapping infrastructure.

7.3.16 Post-disaster Reconnaissance

In recent years, the worldwide community has witnessed many natural disasters (e.g., earthquakes, landslides, tsunamis, and tropical storms), causing significant damage to infrastructure systems and loss of human life (Burnap *et al.*, 2015). During natural hazard events, normally, teams of civil engineers and earth science specialists perform reconnaissance studies to document the structural or geotechnical damage and collect other field data. Findings are then used as a critical input to decision-making for rehabilitation and improve the design methods to achieve greater system resiliency. Data collection in such difficult environments presents many problems, including personnel safety, acquiring physical measurements, and inaccessibility to many sites (Greenwood *et al.*, 2019).

The UAVs have recently been incorporated for immediate post-disaster reconnaissance surveys after high-wind, flood, and seismic events (Xu *et al.*, 2014). Small UAVs equipped with conventional optical cameras can be flown to take images and develop 3D point clouds of targets. These 3D models, coupled with terrestrial LiDAR surveys, are used to acquire relative displacement measurements in post-disaster scenarios. The fundamental aspects of UAVs, mainly mobility, make them ideal tools to be leveraged for post-disaster efforts, including rapid risk assessment, hazard identification, and data collection (Bernard *et al.*, 2011). While UAVs can collect and transmit different types of data in real-time, the communication networks soon after the post-disaster events for disseminating the information to stakeholders and decision-makers still require further development.

7.3.17 Security and Emergency Response

The UAV provides real-time insight into security and emergency situations for better control, precise intelligence gathering, comprehensive situational awareness, and more informed decision-making. The UAV can provide a platform that provides security officers and emergency responders with a professional tool for collecting aerial data and providing real-time visibility into security threats and emergency situations (Kosmer and Vilhar, 2014). Security operations are simplified with pre-defined missions, and emergency response is enhanced through on-demand availability. The fully automatic platform deploys and lands the UAV and collects aerial data to be processed and analyzed (Edwin et al., 2019). Real-time aerial video and photos are delivered directly to personnel on the ground, enabling more informed decision-making in emergent situations. The UAV platform eliminates the logistics involved in operations while providing reliable, scheduled, and on-demand aerial data collection, premium processing, and analytics capabilities.

7.3.18 Insurance Sector

The UAVs can play a part at all stages of the insurance lifecycle. They can gather data before a risk is insured, help in preventative maintenance, and assess damage after an event. Creating claims for insurance investigations can be a lengthy process, especially if the damaged areas are hard-to-reach. The UAVs offer insurance companies a growing number of potential benefits, such as help reduce costs, save time, and improved efficiency to the tune of 40–50% (Brown, 2018). The UAVs can be used to capture images to cover wide areas for crop insurance claims or to create a 3D model of major infrastructure damage caused by hurricanes and earthquakes with 24 × 7 surveillance ability of UAVs. With the growing number and severity of these catastrophes, insurance companies are entrusted with quickly assessing the damage and paying out claims to affected ones (Ayranei, 2017). Overall, the real benefit of UAVs for insurance companies comes from the potential gains in safety and efficiency.

An inspection may be particularly dangerous if a building has suffered fire damage. A UAV can photograph the entire roof, including parts of the structure that aren't accessible to humans. A farmer can get faster claims of the insured crops, as the insurance companies use a UAV to inspect the farmer's crops to assess the real damage (Culus et al., 2018). Personalized premium amounts can also be quickly calculated based on far more precise inputs into pre-defined pricing algorithms. The UAVs could also be used to deter insurance fraud. Insurance fraud is a common problem, and mitigating that fraud can help save companies millions of dollars. After an extreme event happens, some policy owners try to claim damage done prior to the disaster. Using UAVs to capture images of insured properties before an extreme event and after an event can help insurance companies protect against such fraud.

The crop insurance scheme of the Government of India, Pradhan Mantri Fasal Bima Yojna (PMFBY), requires accurate estimation of crop yield for administrators to decide the insurance claims. Satellite image-based classification supported with the ground data in the form of Crop Cutting Experiment has been the main criterion for assessing crop yield and crop damage (PMFBY, 2018). This approach will improve further with the improvement in more accurate crop assessment from high-resolution satellite images and the tremendous growth in various other related technologies (e.g., UAV-based imaging, big data analytics, cloud computing, crowd-sourcing, sensor networks, IoT, and AI).

7.3.19 Health Sector

In developing nations and areas with mountains, deserts, or forests, roads are impassable and take a long time to travel. In such areas, lack of access to roads is more critical, particularly for medical supplies, such as vaccines and drugs. Air transport such as a helicopter is the only alternative so far, but it is expensive and not affordable to the patients or the health system. The UAVs can be successfully used in the field of public health as medical couriers (as shown in Figure 7.28) (Howell *et al.*, 2016). The UAVs can fly over vast distances and challenging terrain, enabling just-in-time delivery of life-saving medical supplies to those in hard-to-reach communities.

FIGURE 7.28 The UAV for supply of emergency medicines.

Innovation is only meaningful if it leads to an impact. This is more so true with the potential of UAVs in global health, particularly to impact in the humanitarian space. The benefits of UAVs within global health are extensive, and therefore, the demand for these UAVs to transport health-related materials is increasing. The technology could also improve access to non-emergency commodities for villages where health worker visits and commodity deliveries are not frequent. It could greatly reduce response times in an emergency. Thus, UAVs could bring meaningful benefits to global health through improved supply chain performance and health outcomes.

In the healthcare field, ambulance UAVs can deliver medicines, immunizations, and blood samples, into and out of unreachable places. They can rapidly transport either medical instruments or the patients in the crucial few minutes after cardiac arrests. In addition, they can also provide live video streaming services allowing paramedics to remotely observe and instruct on-scene individuals on the use of medical instruments (Pulver and Wei, 20018). The UAVs can provide a lifeguard for a castaway, a defibrillator for a person who suffers a heart attack, or supply a thermal blanket for a mountaineer waiting to be rescued.

To test the efficacy of UAVs, the United States Agency for International Development (USAID) has invested in numerous pilot health projects in many different countries (USAID, 2018). The USAID's framework was applied to three examples of UAVs; blood delivery, two-way sample transport, and vector control, and identified five specific challenge areas that could prevent UAVs from reaching full scale in global health. Further details are given in Figure 7.29.

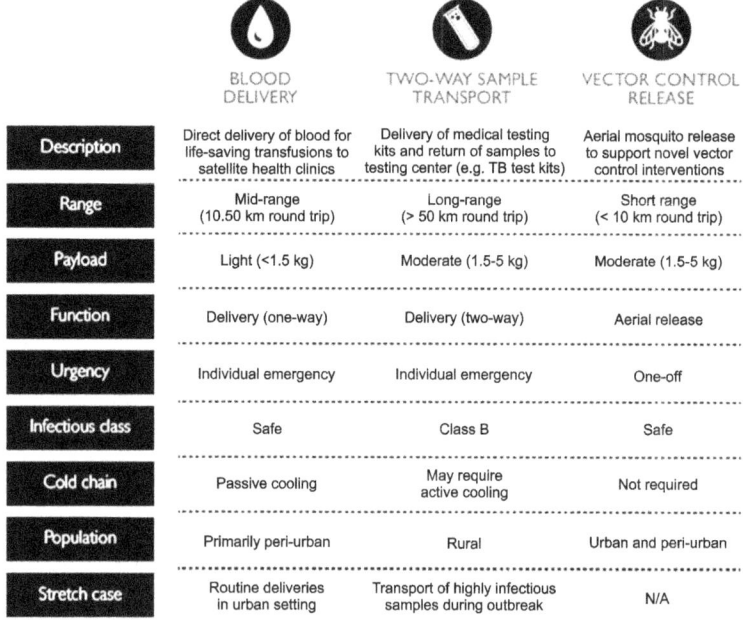

FIGURE 7.29 Use of UAVs in three examples of global health (USAID, 2018).

It is important to acknowledge that the adoption of UAV technology in global health is still at a nascent stage (USAID, 2018). Deeper coordination and collaboration are needed to accelerate the development and appropriate application of this technology within global health. Over time, greater alignment and coordination across innovators, manufacturers, implementers, governments, beneficiaries, funds-providers, and other stakeholders will help UAVs achieve scale. In addition, high upfront costs for UAVs and the related infrastructure mean UAV programs are likely to require startup funding, but donors and other investors may be unwilling to support scale-up without a path to sustainable funding of ongoing costs. To build upon past efforts and unlock UAVs' future potential in global health, a broad community of users, donors, innovators, and implementers are working together to identify the most pressing needs as well as prioritize and coordinate future efforts to ensure that resources are efficiently deployed with maximum benefit.

7.3.20 COVID-19 Pandemic

Public health experts around the globe are scrambling to understand, track and cure a new virus that first appeared in Wuhan, China, at the beginning of December 2019. On 30 January 2020, the World Health Organization declared the outbreak a Public Health Emergency of International Concern, and named the illness caused by the coronavirus as, COVID-19- where "co" stands for corona, "vi" for virus, "d" for disease, and "19" for the year when the disease emerged. It is a pandemic, which has emerged from China, but soon gripped the whole world. Several countries, including India, Spain, Italy, United Kingdom, and the United States, are the worst affected.

The analysis found that the genetic sequence of the COVID-19 virus is 96% identical to one coronavirus found in bats. Sneezing or coughing, cold, chest infection, fever, and breath problems are the major symptoms in affected people. Early evidence suggested that, such as other coronaviruses, the virus jumps between people who are in very close contact with each other. It probably spreads when an infected person sneezes or coughs. If the droplets of mucus and saliva make entry into another person's eyes, mouth or nose, the person can catch the virus and get sick. The viruses in those little droplets can also fall onto surfaces, such as tables or doorknobs and handles, and can remain active there for a long time (Wetsman, 2020). If people touch such surfaces and then touch their eyes, mouth, or nose, they can also get sick, which is why it's important to clean these items as well as wash the hands regularly.

Figure 7.30 presents information about both how severe an illness is and how easily it can spread to determine how "bad" it can be (Wetsman, 2020). Epidemiologists often use this tool to assess new strains of the flu, but it is almost true for coronavirus as well. If an illness is not very severe (and kills only a small percentage of people), but it is highly transmissible, it can still be devastating. An easily transmitted illness that kills a small percentage of the infected people can still cause many deaths, precisely because so many people get sick.

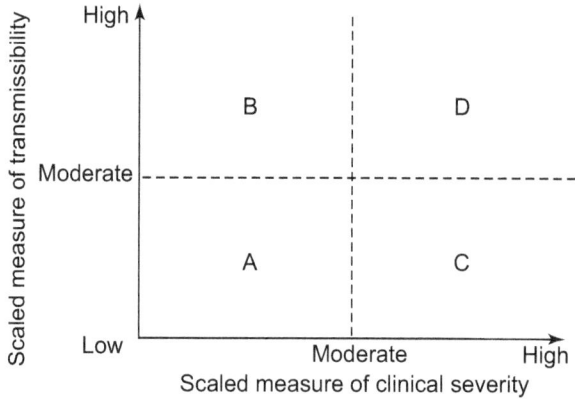

FIGURE 7.30 Correlation between transmissibility and severity of viral disease (Wetsman, 2020).

Technology has found its way in controlling the spread of COVID-19, as drones are used for spraying disinfectant across large areas, robots are dispensing hand sanitizer, mobile apps are analyzing coughs to see if they are "dry" or "wet," and AI techniques are normally used for detecting the tuberculosis being tweaked so that virus can be detected. During the peak of the coronavirus pandemic, China started using drones either for distributing the medicines, disinfecting the streets, or warning people who do not wear masks or break the rules of lockdown. A month later (March/April 2020), several European countries declared public health emergencies and decided to complete their country's lockdown for varying periods. They started using drones in different ways, for example, urge people to maintain a safe distance outside their houses, and discourage travel to impose the lockdown to limit the spread of the virus (Skorup and Haaland, 2020). Spain began to employ manned loudspeaker drones on March 15, 2020, followed by France and other European countries and many more.

In addition, the use of drones for medicines, parcel, and grocery deliveries is proving enormously beneficial during the lockdown period. Drone deliveries help to increase the social distancing, thus reduce the spread of COVID-19 and save lives. Those who are at a high risk to COVID-19 are being urged to avoid highly trafficked areas. The residents rely on public transportation for essential services in any city, such as grocery, medical, etc. In the United States, cities such as New York City and Washington, DC, due to the closure of metro and subway, drones effectively expand the number of options for grocery stores and other services directly to homes, which would help with supply issues and maintaining social distancing. Drones would also greatly assist in rural areas for the supply of essential items, as these places may not be well covered with the transportation mode. With a maximum speed of 60 to 100 mph and no traffic to weave through, drones make an otherwise lengthy supply chain quite short (Skorup and Haaland, 2020).

Much of China's success in containing the virus's spread outside Hubei province was the heavy use of surveillance technology to prevent a resurgence of the virus. It enabled them to adopt a more tailored approach, allowing most people to resume their normal lives while monitoring those who might be infected. For example, the drivers will have to scan the QR code held up by a drone (Figure 7.31) and download an app, "Hangzhou Health Code," in their smartphones while moving around Shenzhen city and entering the city. If the QR code turns red color, it indicates that the person is supposed to be undergoing 14 days of self-quarantine (The Economist, 2020). The yellow color means a lower risk and had to be isolated for seven days. The green color will allow entry. People will have to must produce their phones at checkpoints and show that they have a green QR code.

FIGURE 7.31 Use of drone carrying QR code (The Economist, 2020).

Drones are also being used to bring COVID-19 testing samples to laboratories, which help the quick diagnosis and quarantining of infected citizens. More importantly, drones are used to safely transport medical supplies into hospitals where COVID-19 patients are treated. These areas would be dangerous to the courier, who might contract the disease and then go on to infect even more people, but delivery drones make this task safe. In fact, in medical supply, drones cover a distance of 12 km in 8 minutes, which is 80 times faster than the traditional delivery (Bora, 2020). Medical drone delivery networks are already popping up in places, such as China, Rwanda, and Switzerland. Drones in China reportedly decreased the delivery time of medicines by over 50% compared with road transportation, and it took humans out of the process, which helped decrease the rate of spreading COVID-19.

In addition, drones are being used for the disinfection of areas by spraying chemicals. It is claimed that drones disinfect 50 times more area than the traditional methods in a given time, keeping human operators out of harm. The automated drone is equipped to adjust itself to terrain height and avoid obstacles. The roads and areas can be selected directly from Google maps to perform the task. The drones can carry up to 10 L of disinfectant and cover about 20 km a day, whereas a human being can cover about 4–5 km a day (Bora, 2020).

Other services of drones include detecting the human body temperature in crowded places. While government personnel has been deployed to check the temperatures of individuals in sensitive areas, conducting personnel checks is almost at an equal risk of contracting the virus. The drone camera with a calibrated thermal camera can be used to measure body temperatures while the checking officer remains at a safe distance away. However, the solution is not designed to be used as standard medical procedures and, data needs to be re-verified with manual methods before taking some concrete decision.

With the technological development, a "pandemic drone" that will be equipped with specialized sensors and computer vision system is being developed by the North American company, "Draganfly," based on technology from the University of South Australia and Australia's Department of Defence (Sakharkar, 2020). This drone can monitor the temperature, heart breathing rate, coughing, and even blood pressure of people in crowds, offices, airports, and other places where people are working. California Police force has attached speakers and night vision cameras on DJI drone units to equip them for public safety. The speakers announced the messages, such as "Travel is prohibited unless there are exceptional circumstances" or "Please respect the safety distances" (Gaulkin, 2020).

The contactless services by UAVs in the medical sector seem to be no less than a boon during emergency times. It is evident that the UAVs have great opportunities in public health (Shakhatreh et al., 2018) and can be used to transport blood, specimens, and biologicals, such as vaccines to remote places reduce the travel time. They can be employed in disaster relief to save lives. In the future, UAV delivery systems and UAV delivery ports could be constructed near health care systems (Laksham, 2019). Assessment of public safety and privacy has to be done before scaling up of UAVs in public health. There is also a need for health education to alleviate the apprehension about UAVs in people's mind (Gupta et al., 2018).

7.3.21 Entertainment and Media

Currently, the UAVs are widely preferred by professionals from the media and entertainment industry (Figure 7.32) for high-definition video recording and still photography (DRONEITECH, 2018). They offer new story-telling formats by enabling dramatically different angles, such as shooting video over water or while flying through trees. The UAVs are used in television, advertising, live sports, news reports, and more. Their use by corporations is also increasing for producing corporate video footage and photography for marketing purposes (Swaminathan, 2015).

FIGURE 7.32 Use of UAV in journalism.

The UAVs have already started to be used for information gathering in crime scenes, natural disasters, and other hard-to-film events (Heutger and Kückelhaus, 2014). In journalism, UAVs could be used for accident/incident monitoring to capture the big picture over time, address gathering, interview common people on the street, and do rapid interviews with sports players. In the field of photography, the UAVs have already begun to show their involvement in real estate, hotels and resorts photography, and professional photography of beautiful scenes. Aerial film and photography service providers are probably the heaviest commercial users of UAVs today (Cohn *et al.*, 2017).

The technical requirement in UAV photography is comparatively low in many cases, and off-the-shelf cameras are attached to the UAV with ready-made mountings. Chips are now available, allowing the UAV camera to film in 4K HD video, including electronic image stabilization, removing the need for camera gimbals. Recently, the FAA gave its permission for UAVs to be used on film sets for motion capture and high-definition video recording. Now many movies with UAV-filmed scenes, such as "Skyfall," "The Hunger Games," and "The Dark Knight Rises" (*http://www.washingtonpost.com/blogs/the-switch/wp/2013/08/15/its-a-bird-its-a-plane- its-a-drone-that-makes-movies/*) are available.

The UAVs are the latest technological advancement in the storied history of aerial footage in sports events (Ayranci, 2017). One of the interesting ways UAVs are impacting everyday life, is in journalism, especially amateur investigative journalism, bringing an eye to the sky to document what's going on in the world around us (Baker, 2018). The UAVs offer broadcasters and journalists many benefits that traditional news-gathering techniques cannot, by allowing the journalists to get much closer to the subject than regular cameramen. They can be used to capture photos of extreme and outdoor sports events by making the events stand out from the competition with aerial footage that would amaze the spectator. Many sports benefit from the UAV technology, such as extreme sports; skiing and snowboarding; horse riding and horse racing; water sports, such as sailing, yachting, boat racing, and surfing; motorsports, such as Formula-1 motor racing and motorcycling; and outdoor sports, such as cycling, mountain biking and golfing.

7.3.22 Military

The UAVs that carry cameras, sensors, communications equipment or other payloads have been employed by military units since the late 1950s (Pike, 2003b). Until the last 20 years or so, however, their use was limited because the analog data collected by these UAVs were not accessible until they returned from their missions. Digital technology has changed this paradigm. Since the early 1990s, the Department of Defense (DoD) of the

United States has employed UAVs to carry out surveillance requirements in close-range, short-range, and endurance categories. Initially, close-range was defined to be within 50 km; short-range within 200 km; and endurance range was set as anything beyond 200 km. By the late 1990s, the close- and short-range categories were combined (Blom, 2011). The current classes of these UAVs are the tactical UAVs and the endurance category. Numerous digital multispectral, hyperspectral, and radar sensor platforms are now used onboard both tactical and endurance UAVs for military applications. As the volume of data and its communication has become faster, the army is now able to receive current details of battlefields. UAV technology and innovation have really leaped forward in the past few years.

Figure 7.33 illustrates how unmanned systems are already employed in a significant number of roles. The systems are divided by the military department to illustrate areas with current and potential future collaboration. Reconnaissance, strike, force protection, and signals collection are already being conducted by fielded systems, and acquisition programs are developing systems to support the warfighter in even broader roles (DoD, 2017). The use of aerial photographs and digital images has seen a modest increase in military applications over the past few years. The ability to provide real-time or near-real-time data about the terrain and the enemy has always been a goal of the military intelligence community (Mahnken, 1995).

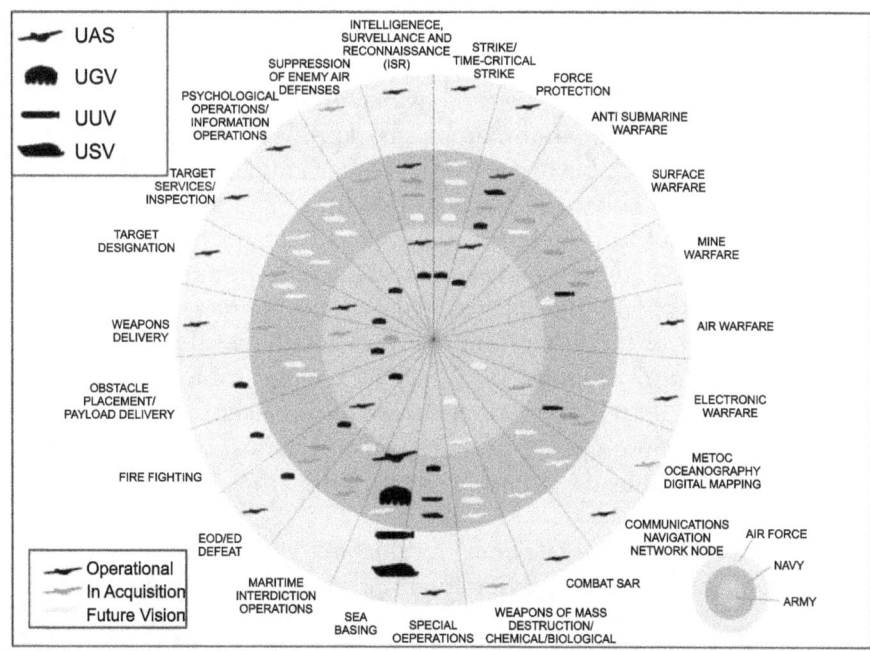

FIGURE 7.33 Military applications of unmanned aerial systems (DoD, 2007).

In UAVs' age, there are obvious advantages that modern military technology can adopt in fighting wars. In particular, UAVs' ability to minimize bloodshed, especially civilian casualties, as compared with previous military technology, such as strategic bombing, can be well appreciated. Under military usage, the UAV technology is also employed for missile launching and bomb-dropping (Kaag and Kreps, 2014). In addressing the ethical challenges that drone warfare presents, the use of UAV technology may create a moral hazard and create the leisure for soldiers and their commanders to reflect on the intricacies of military ethics in unprecedentedly careful ways (Mahnken, 1995). The UAVs demonstrate how technology can radically alter the collective understanding of warfare and its justification. Many factors lead to better decision-making in the use of UAVs, such as time pressure, complexity, and responsibility that need to be understood with reference to their moral and legal implications.

Peeping drones could be spying. With a few modifications, the drone becomes a low cost and simple electronic surveillance tool to gather information. Drone operators could drop a sophisticated microphone into a restricted area for eavesdropping if there are no power, weight, and range issues. Putting a Wi-Fi access point on top of a building, or inside its perimeter, could allow hackers to listen in to data traffic. Drones can also be used to carry traditional spying devices used to eavesdrop on a conversation (e.g., a laser microphone) and perform keylogging (e.g., using a microphone).

Basically, the spy drones will have a drone, a camera, and a monitor to see what the drone sees (and a recorder to record it). Instead of the monitor, FPV goggles can be used, which are more interesting and provide the sensation of being in the drone's cockpit. With a camera in the air vehicle, soldiers will be able to see further and around obstacles that they previously wouldn't be able to see in near real-time. The FPV channel also provides excellent infrastructure for a malicious operator to spy on people without being detected because (Nassi *et al.*, 2019): (i) it eliminates the need for a malicious operator to be close to the drone or target by allowing the operator to maneuver the drone from far away to a target that is also far away from the operator's location, (ii) it can be secured using encryption, and (iii) it supports HD resolutions that enable the attacker to obtain high-quality pictures and close-ups (by using the video camera's zooming capabilities) that are captured by the drone, even when the drone is far from the target point of interest.

The UAVs are becoming a valuable tool for military, army, and ground commanders in the preparation and execution of the missions. With the increasing use of UAVs in the battlespace, coupled with robust communications systems, these data may soon be available to every soldier. The US Army plans to give almost all of its ground combat units the tiny drones which can spy on other forces from the sky (Eriksson, 2018). These UAVs can be further developed to be used as Flying Camouflage drones (that will visually mask everything nearby), Communication disruptors (that will create zero-communication zones over the targeted areas), and battlefield medical supply drones (to provide a source of supplies and equipment for battlefield injuries, instantly).

Military and defense are expected to emerge as the largest end-use segment due to an increase in R&D activities by the defense. The market for anti-drone systems in military and defense applications is expected to cross US$ 900 million by 2024. Significant investments have been made in the development of safer and effective anti-drone technologies by major venture capital firms. The global anti-drone market size is expected to reach US$ 4.5 billion by 2026, expanding at a CAGR of 29.9% from 2019 to 2026. Asia-Pacific is anticipated to witness a CAGR of close to 30.0% from 2019-2026 due to increase in government expenditure for development of aerospace infrastructure across emerging economies. Prominent drone/UAV developers, such as DJI, have begun incorporating "No-fly zone" restrictions onto onboard firmware of drones (Reportlinker, 2018).

Considering the pace of development, the number of UAVs involved, and apparently, the unlimited possibilities offered by UAV technology, there is no doubt that UAV will fundamentally change the applications scenario. Figure 7.34 shows the projected usage of UAV till 2030 in the government/defense applications and civil/commercial applications in the radio communication sector (ITU-R M.2171, 2009).

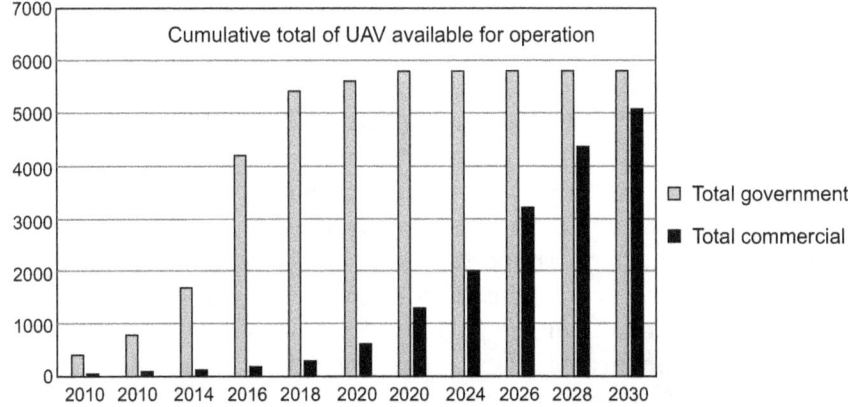

FIGURE 7.34 Government and civil usage of UAV (ITU-R M.2171, 2009).

7.4 SUMMARY

The emergence of UAV technology not only removed the limitations of its use in the military but expanded its wings in commercial applications related to agriculture, recreation, surveillance, delivering goods, transportation, urban, logistics, journalism, and many more. From construction to insurance to real estate to PA, the ability to survey the land and hard-to-reach locations with UAVs is valuable. Agriculture and infrastructure perhaps cover the maximum share of UAVs applications. The utilization of autonomous UAVs in agri-business is reaching out at an energetic pace in PA, crop health monitoring, soil nutrients, crop yield, etc. They help farmers to make smart farming decisions for enhanced productivity on time. By 2020, the market for drone jobs in the United States alone will be US$ 1.3 billion for construction and US$ 1.4 billion for agriculture, according to Culus et al. (2018).

The UAVs can be used as efficient tools to accurately survey and map the landscape to collect necessary information before infrastructure construction to make good plans. The UAVs are efficient in assessing the cracks and thermal leaks and evaluating the conditions of roads and other structures. The UAV data from real-time traffic monitoring are useful for the design of new roads and bridges. Such real-time traffic information will further guide the building of smart cities with respect to reversible lane systems. They are expected to have a worth of US$ 91 billion over the next decade and US$ 45 billion market value usage in civil infrastructure (Intelligence, 2018).

In recent years, technological advances in autonomous and semi-autonomous UAVs vehicles have increased their utility and ease of use in disaster management as well as SaR operations. The UAVs provide evidence of the efficiency of safety inspections in constructions or industry. Their deployment tremendously reduces the safety hazard risks while offering the ability to access hard-to-reach and dangerous areas. The UAVs also have limitations for some applications: limited flying time; limited payload, flying within VLOS, need for multi-drone cooperation (swarm) to address the material delivery, medicine delivery, etc. In addition, they are easily affected by rough weather conditions, such as heavy rain, snow, or strong wind.

Innovative technologies, such as radar and acoustic-based detection, are increasingly implemented to identify and classify small UAVs that are invisible to the eye (Reportlinker, 2018). The use of such technologies helps improve living conditions. The integration of UAVs with other recent technological innovations (e.g., BIM, VR, AR, wearable technologies, photo/videogrammetry, big data, 5G technology, smart connected solutions, AI, and machine learning-based predictive analytics), in addition to the legal and financial aspects of using them, will also lead future research of UAV applications. In an industry that is consistently evolving and integrating new emerging technologies, UAVs have proved to be a safe, yet efficient alternative. However, there are both benefits and challenges associated with the use of UAVs.

REVIEW QUESTIONS

Q.1. Discuss various application areas under each case where (i) photogrammetry UAVs, (ii) LiDAR UAVs, and (iii) autonomous UAVs are most useful.

Q.2. Explain the maturity levels of various applications of UAVs worldwide. What are the emerging technologies identified for various UAV-based applications?

Q.3. Discuss in brief various industrial applications of UAVs. Explain the status of the use of UAVs in the construction industry.

Q.4. Explain how you will deploy UAVs for inspection and monitoring of civil infrastructure as well as in industry.

Q.5. Describe the utility of UAVs in forestry.

Q.6. Explain in brief how the agricultural community has benefited from the use of UAVs in its various activities.

Q.7. What is an NDVI image? Explain its utility in forestry and agriculture.

Q.8. Discuss various practical applications of IR and TIR images.

Q.9. What is a smart city? How UAV and IoT can be deployed to provide wireless services to smart citizens?

Q.10. Write short notes on using UAVs in (i) Traffic monitoring and management, (ii) Logistics, (iii) Insurance sector, and (iv) Entertainment and media.

Q.11. Explain the utility of UAVs and swarm UAVs in (i) Surveillance operations, (ii) Search and rescue operations, (iii) Security/emergency, and (iv) Post-disaster events.

Q.12. What is crowd-sourcing? Which UAV-based applications are most benefited from the use of the crowd-sourcing approach?

Q.13. Discuss the utility of UAVs in the health sector. How UAVs have been deployed worldwide to combat the COVID-19 pandemic?

Q.14. Describe the most popular uses of UAVs in various activities of the military.

CHAPTER 8

THE FUTURE OF UAV TECHNOLOGY

8.1 INTRODUCTION

UAVs have almost limitless possibilities of use in the commercial world, offering great promises in various applications, such as military, surveillance and monitoring, telecommunications, delivery of medical supplies, and rescue operations (Chmaj and Selvaraj, 2015). The UAVs have evolved rapidly in recent years due to focused efforts in the field of electronics, optics, computer science, energy storage, etc. The improvement in technologies, such as GNSS, IMU, LiDAR scanners, imaging sensors, and robotics, has contributed to new applications of UAVs (Bor-Yaliniz et al., 2016). In recent years, data collection and processing research efforts have extended beyond imaging methods to include other sensors, such as gases, biological, temperature, moisture, etc. While hardware and software development are on the rise, the main challenge remains large aerial data handling and real-time processing.

In addition to hardware (UAV, sensors, nano-sensors, ground station, etc.), software solutions and services are integral parts of a UAV system. The use of, for example, machine learning algorithms, AI, IoT (Chan et al., 2014), wireless communications technologies, etc., has further accelerated research trends and benefits in the area of UAV applications. Owing to their autonomy, flexibility, and a broad range of applications, UAVs have been the subject of concerted research over the past few years (Motlagh et al., 2016).

The use of swarms has expanded greatly in recent years in many applications, such as surveying, homeland security, surveillance, or perform complementary operations, such as sensor placement and interrogation. As UAVs become integrated into spaces with other aircraft and UAVs, methodologies for managing the interaction between the UAVs are critical (Campion et al., 2019). Recent discussions in the USA about integrating UAVs in the national airspace at night have raised some interest in investigating the non-vision sensors for autonomous navigation and obstacle avoidance, such as sonar and LiDAR.

Developing and implementing various components of autonomous UAVs face additional challenges. The component of developing a fully autonomous UAV must understand how to integrate human cognition for decision-making. The studies have generally shown that human-UAV interaction is an important direction for research. In the future, UAVs will become powerful autonomous systems having the ability to develop an action plan, collect data, process data, perform computations, analyze results, and make next-step decisions. The key areas for R&D in UAVs include artificial intelligence, UAV detection and avoidance technology, control and communications, image processing, and battery capacity (Al-Turjman et al., 2019).

8.2 FUTURE INSIGHTS

The use of emerging technologies needs to be a core part of every company's corporate strategy to make clear-headed decisions that will sustain revenue growth and enhance business operations. To help companies focus their efforts, Culus et al. (2018) analyzed more than 150 emerging technologies to identify what is known as the "Essential Eight," as shown in Figure 8.1.

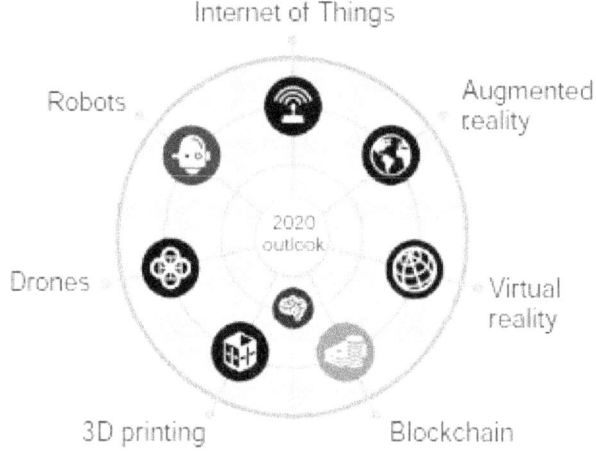

FIGURE 8.1 Essential eight technologies for the future (Culus et al., 2018).

These eight technologies include IoT, AR, VR, blockchain, AI, 3D printing, drones, and robots, which will have the most significant global impact across various industries worldwide in the near future.

There are significant numbers of current and future uses for UAVs throughout the military and civilian world (Kaag and Kreps, 2014). The military has been using and continues to anticipate increased usage of these systems in numerous areas, including intelligence, surveillance, and reconnaissance (Khuwaja, 2018), data and communication interfaces (Bekmezci, 2013), electronic warfare (Safi, 2019), and limited attack roles (Chmaj and Selvaraj, 2015). While UAVs are tactically effective, their strategic effectiveness is limited. The UAVs may kill terrorists with ruthless efficiency, but they leave the fundamental drivers behind terrorism unaddressed, and they may even exacerbate those motivations. Certain technologies, such as robotics, when combined with UAV, may be used to protect the military troops in combat and support them with the best intelligence. However, there are also risks associated with human welfare if UVAs are misused (Boyle, 2015). Legal and ethical restraints are required to be maintained to make full use of the potentials of UAVs.

In civilian applications, UAVs are already being used for the large number of applications, such as post-disaster response, structural damage assessment, infrastructure inspection, urban planning, mining, and construction monitoring (Greenwood *et al.*, 2019). The cargo and transport, search and rescue operations, traffic management, and construction management have been envisioned as future research areas (Reportlinker, 2018). The UAVs play an important role in complex humanitarian emergency services as they can be used for surveillance in difficult or inaccessible areas.

With the expansion of UAV usage, it is important to investigate the trend of future UAV research. However, there are few studies in the literature that have introduced research trends related to UAVs. Suzuki (2018) has proposed the research on UAVs in the robotics field into five categories: (i) research on the structures and mechanisms of UAVs, (ii) research on control technologies of UAVs, (iii) research on novel navigation systems of UAVs, (iv) research on guidance systems of UAVs, and (v) research on payloads and applications of UAVs, as shown in Figure 8.2.

The most interesting UAV research development would involve UAVs interacting with humans and interfacing with the data streams. The physical actuation of infrastructure components is an important research topic. This physical interaction could include deployment of wireless sensor nodes or use of IoT, or stimulation of an infrastructure component, such as model analysis of a structure. Very limited work has been carried out involving the physical interaction of UAVs with civil infrastructure as well as UAV-based sensing of below-surface features (Mckinsey, 2017). Subsurface sensing is necessary for geo-infrastructure applications, but the major limitation is the cost of mobilizing equipment and accessing the remote sites.

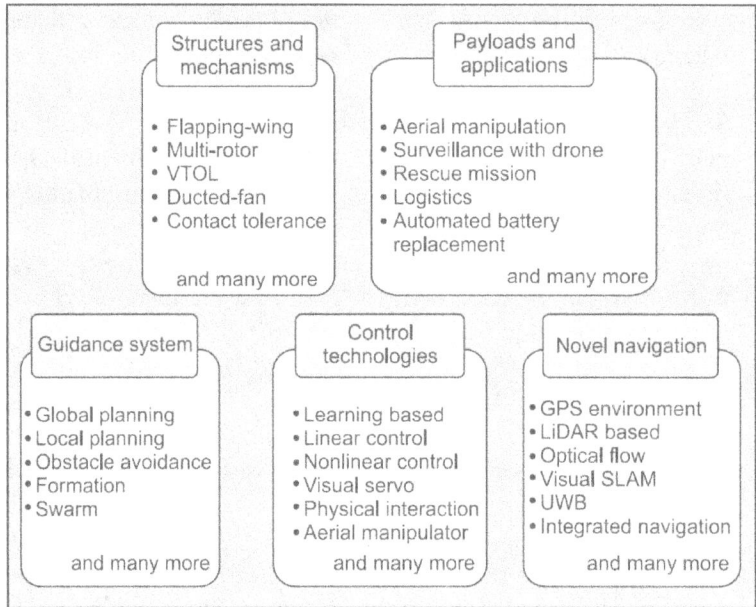

FIGURE 8.2 Classification of UAV research areas (Suzuki, 2018).

8.3 KEY CHALLENGES

Any new technology always has lots of challenges associated with it, UAVs are no different. Scaling a UAV program can be challenging. In case the services are hired, data security remains one of the important issues (White Paper, 2018). Data quality is also an important aspect. The regulatory framework in work area/region/country may be another challenge to allow using UAVs. The gigabytes of data are available by using UAVs, and therefore the degree of automation in data processing has become so important and will continue to gain significance. The reliability, accuracy, training, regulatory hurdles, and process integration are required to be known to the users (Vergouw *et al.*, 2016). It clearly highlights the importance of proper integration into the processes to make UAV operations efficient, safe, and scalable.

The five challenging areas that would likely to come for the next stages of investment in UAV development are given in Table 8.1. These challenge areas are identified which are likely to arise, to varying degrees, as UAVs proceed along the path to scale. Many of these challenges are unique to UAVs and may also be experienced when introducing other new technologies.

TABLE 8.1 Anticipated Challenge Areas.

S.No.	Challenging Area	Requirements
1.	Technology and manufacturability	Solve technical problems for UAV implementation; find manufacturers who can produce to technical specifications at targeted cost
2.	Infrastructure	Build the necessary infrastructure to support UAV (e.g., runways or catapults, cellular and satellite coverage, maintenance, air-traffic management, etc.)
3.	Regulations and policy	Navigate regulatory and policy pathways at a global, national and local level
4.	Community engagement	Engage the community to build trust in the technology and ensure the safety and security of UAVs
5.	Business model and partnerships	Identify business models that enable sustainable delivery of UAV solutions

The possible challenges to exploit the full potentials of UAV technology are described below. The UAV regulations of every country should address all these challenges (Anderson, 2012; Berie *et al.*, 2018).

8.3.1 Topography and Environmental Aspects

Topography, high winds, and weather conditions on the high mountains or near coastal areas might disturb the operation of the UAV system or create noise in the products. The UAV platforms may collide with tall buildings and trees and cause problems unless governed by specific regulations. However, care is required to be taken when flying UAVs in rough topographic settings as the systems may have a limited range of visibility as well as flight altitude.

8.3.2 Legal Aspects

There should be strict rules and regulations to use UAVs. Literature indicates that some countries have well-developed regulations that provide a workable environment for UAV operations, while other countries do not have proper rules and regulations for UAVs to operate in their territories (Jeanneret and Rambaldi, 2016). The absence of well-defined regulations will enhance the scope for misuse of the UAV technology. Some countries have banned the use of UAVs so as to stop the misuse of the technology. Well-designed regulations encompassing the extreme views of different countries together will guide the technology to provide services for the benefit of the community. These regulations should take into account the international laws and enforce them accordingly. The UAV regulations should also be set explicitly so that national security is not endangered (ARC, 2017).

8.3.3 Privacy Aspects

The use of UAVs might be unpleasant when they fly over unwanted areas. Due to flying at a low altitude close to the Earth's surface, they are liable to disturb privacy (Altawy and Youssef, 2016). Each country has to set well-thought-out regulations to avoid human-induced problems and reduce the effect of such challenges. The users can adopt the good things, but the regulations should limit the associated challenges. Hence, the country-specific functional rules need to address these issues so that the technology can provide law-bound services without being influenced by the suspected challenges.

8.3.4 Air Safety and Security Aspects

The safe operation of a UAV system ensures the protection of the surrounding environment, including people, property, and other aircraft, from the unexpected consequences of the operation of the UAV. From a high-level perspective, safe operation constitutes hazard avoidance and safe take-off and landing. The UAVs may be misused by others, such as terrorist attacks (Figure 8.3). Therefore, the government in each country can take the issue seriously and develop regulations that inhibit the misuse of technology. Free flying approach otherwise may cause a disaster.

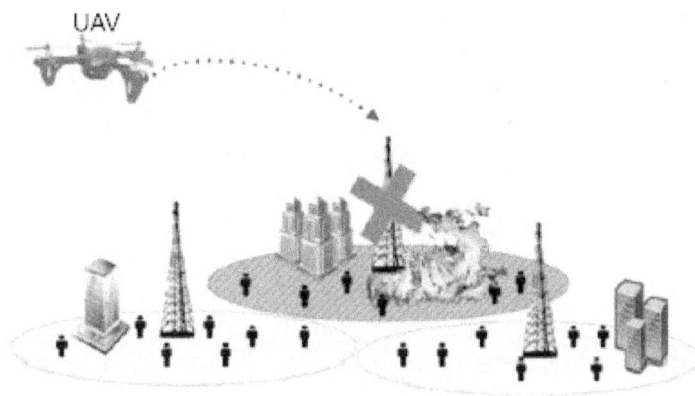

FIGURE 8.3 Unauthorized attack by UAV.

Threats from UAVs can produce severe disruption as they could be highly diverse in nature. The UAV operation strongly depends on its location as well as its distance from critical areas. The frequency of such UAV sighting is also on the increase; for example, various airports in the USA, UK, Ireland, and UAE have experienced major disturbance of operation due to the visibility of unauthorized UAVs in the vicinity of airports (McFarland, 2019). As a result, the regulating authorities worldwide are framing the regulations to reduce the risk by the UAV operation. Regulations may discourage careless or unskilled UAV operations, but probably cannot prevent criminal or terrorist attacks. Rules and supportive technologies are good for those who follow them but not for careless or malicious users. The increase in the number of UAVs in the sky may further challenge this task.

The UAVs may collide with other flying objects or infrastructure. Standardization rules and regulations for UAVs are highly needed worldwide to regulate their operations, reduce the likelihood of collision among UAVs, and guarantee safe hovering (Luo *et al.*, 2013). The UAVs could be integrated with vision sensors, laser range finder sensors, IR, and/or ultrasound sensors to avoid collisions in all directions (Chee and Zhong, 2013).

8.3.5 Cyber-security Aspects

The cyber-security attack may have confidentiality challenges to protect information from being accessed by unauthorized parties. The attackers could compromise the confidentiality of UAV systems by various approaches (e.g., malware, hijacking, social engineering, etc.) utilizing different attack vectors. Attackers then could modify or fabricate the information of UAV systems (e.g., data collected, commands issued, etc.) through communications links, GCSs or compromised UAVs, or GPS signal spoofing attacks (Shakhatreh et al., 2018). Attackers could perform Deny of Service attacks on UAV systems by, for example, flooding the communications links, overloading the processing units, or depleting the batteries.

Figure 8.4 summarizes the identified security threats against UAVs; the corresponding violated security properties, and possible mitigation techniques (Altawy and Youssef, 2016). It categorizes the threats to a UAV system into two types; cyber threats and physical threats. Figure 8.4 describes the types of threat under each category, their security requirement, and mitigation measures that could be taken. A UAV-distributed security architecture presented in Figure 8.4 is a structured collaboration between sets of security services implemented on various components of the UAV system to provide the overall security functionality. There is a need to work in the future to develop the tools and mitigation technologies to safeguard the UAVs and associated systems from cyber threats and physical threats.

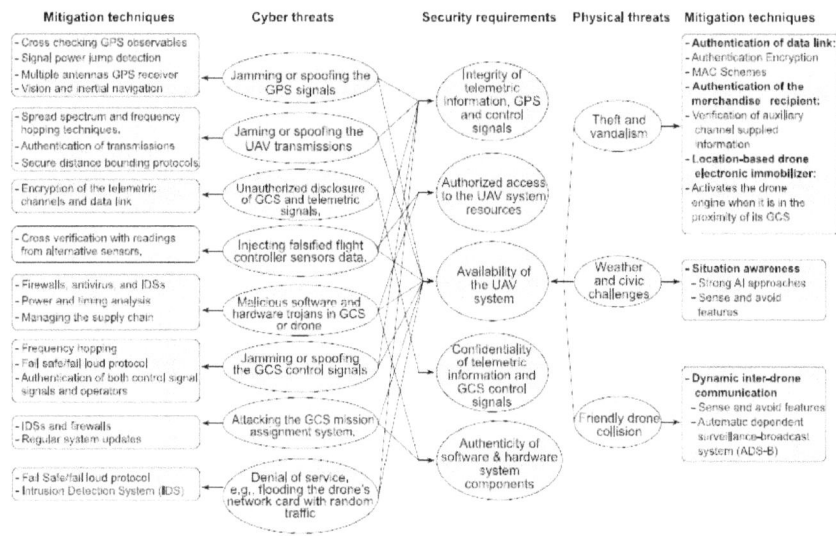

FIGURE 8.4 Cyber and physical security threats against UAV systems (Altawy and Youssef, 2016).

8.3.6 Reliable Communication System

A reliable communication link is necessary for the safe and efficient operation of UAVs. During large-scale natural disasters and unexpected sudden events (e.g., forest fires), the existing terrestrial communication networks may be damaged or even completely destroyed (Vattapparamban et al., 2016). In particular, cellular base stations and ground communications infrastructure can often be damaged during natural disasters. Hence, there is a vital need for efficient communication for public safety as well as between the government and victims for providing relief. Consequently, a robust, fast, and capable emergency communication system is required to enable effective communications during such operations, which will not only contribute to improving connectivity but also will save properties and lives of affected people.

On the other hand, UAVs may be exploited for sensing purposes by leveraging their advantages, such as on-demand deployment, larger service coverage compared with the conventional fixed sensor nodes, and flexible spatial network architecture (Song et al., 2020). This category of UAV applications is known as cellular-assisted UAV sensing. Despite the many benefits of their mobility, UAVs still suffers from some practical constraints, such as limited battery power, restrictions due to no-fly zones, and safety concerns. Therefore, it is essential to develop new communication, signal processing, and optimization frameworks in support of high data-rate UAV communication systems assisting terrestrial cellular communications to enable ultra-reliable and real-time sensing applications in future wireless cellular networks.

Mozaffari et al. (2019) have presented a comprehensive study on UAVs' potential benefits and applications in wireless communications. They have investigated the key UAV challenges, such as three-dimensional deployment, performance analysis, channel modeling, and energy efficiency, which were also identified by Liu et al. (2015). To address the fundamental challenges in UAV communication systems and efficiently use the UAVs for wireless networking applications, various mathematical tools can be used. Table 8.2 summarizes the key challenges, open problems, and analytical tools to analyze, optimize, and design UAV-enabled wireless networks. The analytical framework would be needed to optimize the use of UAVs for wireless networking purposes. As seen from Table 8.2, this is a highly interdisciplinary research area that will require drawing on tools from conventional fields, such as communication theory, optimization theory, network design, and emerging fields, such as stochastic geometry, machine learning, and machine learning and game theory.

TABLE 8.2 Challenges, Open Problems, and Mathematical Tools for Designing UAV-enabled Wireless Networks (Mozaffari et al., 2018).

Research Direction	Challenges and Open Problems	Mathematical Tools and Techniques
Channel modeling	Air-to-ground path loss. Air-to-air channel modeling. Small scale fading.	Ray-tracing techniques. Machine learning. Extensive measurements.
Deployment	Deployment in the presence of terrestrial networks. Energy-aware deployment. Joint 3D deployment and resource allocation.	Centralized optimization theory. Facility location theory.
Performance analysis	Analyzing heterogeneous aerial-terrestrial networks Performance analysis under mobility considerations. Capturing spatial and temporal correlations.	Probability theory. Stochastic geometry. Information theory.
Cellular network planning with UAVs	Backhaul-aware cell planning. Optimizing the number of UAVs. Traffic-based cell association. Analysis of signaling and overhead.	Centralized optimization theory. Facility location theory. Optimal transport theory.
Resource management and energy efficiency	Bandwidth and flight time optimization Joint trajectory and transmit power optimization. Spectrum sharing with cellular networks. Multi-dimensional resource management.	Centralized optimization theory. Optimal transport theory. Game theory and machine learning.
Trajectory optimization	Energy-efficient trajectory optimization. Joint trajectory and delay optimization. Reliable communication with path planning.	Centralized optimization theory. Machine learning.
Cellular connected UAVs	Effective connectivity with down tilted ground base stations. Interference management. Handover management. Ground-to-air channel modeling. Ultra-reliable, low latency communication, and control.	Centralized optimization theory. Machine learning. Optimal transport theory. Game theory. Stochastic geometry.

8.3.7 Swarm-based Operations

Building a large swarm primarily requires the ability to handle a massive amount of information. In general, the more UAVs in a swarm, the more capable the swarm is. More UAVs also mean more inputs that could affect the swarm's behavior and decisions. In addition, more UAVs mean a greater risk of a UAV crashing into another. As the number of UAVs increases, it increases the complexity of the task and the possibility of collision between them. Currently, this technology is in the development stage (Campion et al., 2019).

The UAVs are a cheap way to boost the sheer size of an army force. The UAV swarm technology could significantly impact every area of military competition, from enhancing supply chains to delivering nuclear bombs. In the future, UAVs may one day develop the capacity to carry out targeted killings in swarms. Such UAV swarms could have the capacity to assess targets, divide up tasks and execute them with limited human interaction. For deploying the technology, armed forces are investing millions of dollars in its development (Safi, 2019). The UAV swarms using chemical and biological weapons can improve both defense and offense but appear to favor offense by addressing the key challenges to delivery strongly. In the future, this could weaken the norms against these weapons and encourage proliferation (Kallenborn and Bleek, 2019).

Terrorist organizations are also likely to be interested in UAV-based chemical and biological weapons attacks. At the same time, UAV swarms may also help prevent and respond to chemical and biological weapon attacks. They could be used for coordinate searches along national borders to identify potential smuggling activity, including chemical, biological, radiological, nuclear, and explosive material smuggling, or searches through cities to search for gaseous plumes. Swarms could serve as mobile platforms for chemical or biological detectors with different sensors to mitigate false positives. If an attack is successful, UAVs could even have sprayers to help clean up after an attack without risking harm to humans. As swarms become more sophisticated, they will also be more vulnerable to cyber-attack. Swarm may be hijacked by, for example, feeding it false information, hacking, or generating manipulative environmental signals. Although numerous countersystems are in development, the present system faces scalability challenges, from deployment allocation to training and use (Safi, 2019).

The UAV swarm creates significant vulnerabilities to electronic warfare, and protecting against this vulnerability is critical. It functions inherently depending on the ability of the UAVs to communicate with one another. If the UAVs cannot share information due to jamming, the UAV swarm cannot function as a coherent whole. Advances in technology may also harden the swarm against electronic warfare vulnerabilities (Shakhatreh *et al.*, 2016). Novel forms of communication may weaken or entirely eliminate those vulnerabilities. For example, UAV swarms could communicate based on stigmergy. Stigmergy is an indirect form of communication used by ants and other swarming insects. If an ant identifies a food source, it leaves pheromones for future ants to find. If the next ant also finds food there, it leaves its pheromones, creating a stronger concentration to draw even more. Applied to a UAV swarm, an approach like this that uses environmental cues could mitigate vulnerabilities to jamming.

A future UAV swarm need not consist of UAVs' same type and size but incorporates both large and small UAVs equipped with different payloads. One such swarm may involve a flying UAV collaborating with a walking UAV. Thus, a diverse set of UAVs would be more capable than the individual parts. A single UAV swarm could even operate across the domain, with undersea and surface UAVs or ground and aerial UAVs coordinating their actions. Attack UAVs carry out strikes against targets, while sensors on UAVs collect information about the environment to inform other UAVs, and communication UAVs ensure the integrity of inter-swarm communication (Kallenborn, 2018). Small sensor UAVs can provide reconnaissance for larger attack UAVs, gathering information on adversary targets and relaying it to attack UAVs to carry out strikes. Even the UAVs specifically tasked with conducting attacks can be diverse. A swarm intended to suppress enemy air defenses could include UAVs equipped with anti-radiation missiles for defeating ground-based defenses, while other UAVs might be armed with air-to-air missiles for countering adversary aircraft.

Customizable UAV swarms offer flexibility to commanders, enabling them to add or remove UAVs as needed. This requires common standards for inter-UAV communication so that new UAVs can easily be added to the swarm. Similarly, the swarm must be able to adapt to the removal of UAV, either intentionally or through hostile action. Customization of the UAV swarm would allow commanders to adapt the swarm to the needs of a situation. For missions demanding a smaller profile, a commander may remove extra UAVs. A commander may also vary the swarm capabilities, adding UAVs equipped with different types of sensors, weapons, or other payloads. Research on UAV swarm customization shows that the concept is possible, but its development is still in the early stages. In the future, commanders may be given a collection of UAVs that can be combined in different ways as the mission demands. This enables rapid responsiveness to changes in the military environment (Safi, 2019).

The cost benefits are also important while using UAV swarms. Ultimately, the cost and relevance depend partly on what role the swarm will play and what alternatives are available. Even multimillion-dollar UAV swarms can be cost-effective on the balance if they meaningfully increase the survivability of more expensive or particularly crucial platforms, such as aircraft carriers or nuclear deterrent forces.

8.3.8 IoT-based Operations

The IoT has several important applications when integrated with UAVs (Al-Fuqaha, 2015; Mozaffari *et al.*, 2018). A growing number of devices (e.g., smart phones, bracelets, wearables, different sensors, etc.) need to be almost continuously connected with the IoT paradigm. For example, the UAVs with sensors system can be used effectively for surveillance scenarios, a key application for the IoT. Due to various applications of UAVs with IoT devices, they might be deployed in environments with no terrestrial wireless infrastructure, such as mountains, rural areas, and desert areas.

The UAVs are being planned to be used to deliver Internet access to many developing countries and provide airborne global Internet connectivity. For example, in foreign countries, Qualcomm and AT&T deploy UAVs to enable wide-scale wireless communications in the 5G wireless networks (QUALCOMM, 2018). By 2020, more than one-fifth of the world's countries will have launched 5G services. The 5G focuses on Voice over Internet Protocol devices, which in turn would lead to high levels of data transmissions and call volumes. There is a need for a unified and comprehensive overview on how UAVs can be used as flying wireless base stations in emerging wireless, broadband, and beyond 5G scenarios.

In an IoT-based environment, energy efficiency, ultra-low latency, reliability, and high-speed uplink communications become major challenges as compared to conventional cellular network cases (Lien *et al.*, 2011). In particular, IoT devices are highly battery-driven and cannot typically transmit over a long distance due to their energy constraints. In areas that experience intermittent or poor coverage by terrestrial wireless networks, battery-supported IoT devices may not be able to transmit their data to distant base stations due to their power constraints. In this regard, mobile UAVs may be a promising solution despite a number of challenges associated with IoT networks (Uerkwitz *et al.*, 2016).

UAVs can be deployed as flying base stations in IoT-based scenarios to provide reliable and energy-efficient uplink IoT communications. In fact, due to high altitude, the UAVs can be effectively deployed to reduce the shadowing and blockage effects as the major cause of signal attenuation in wireless links. As a result, the communication channel between IoT devices and UAVs can be significantly improved. Thus, battery-driven IoT devices will need a significantly lower power to transmit their data to UAVs. Hence, IoT networks' connectivity and energy efficiency can be significantly improved by exploiting the unique features of UAVs.

8.3.9 Big Data and Cloud Computing

The UAVs are all about the data that they gather. A number of sensors on UAVs offer completely new possibilities for collecting and analyzing a large amount of valuable data. All of this will be useful only if it can be processed to derive useful results. Computers will help us to process this information. The UAVs collect the big data and transfer it to a computer or cloud. So connectivity, cloud, and devices become part of an IoT ecosystem, and UAVs will be a big part of it in the future. With good wireless communication, such as Wi-Fi or LTE or even 5G, volumes of data will get transferred from one point to another and analyzed to make information useful (Sharma and Garg, 2019).

The challenges of UAV data include analysis, search, sharing, storage, transfer, visualization, and information privacy. Therefore, it is necessary to understand the concept of collecting and organizing all kinds of big data to extract intelligence from it. Google is a great example, as they collected a vast amount of users' data, categorized it, stored it, and developed a way to sift through it for better search results and targeted advertising. The convergence of UAVs and big data in specific application areas creates new opportunities and challenges for IoT. The demand for UAV data services has created a US$100 billion market opportunity between 2106 and 2020, according to Goldman Sachs's (2016) research report.

The new algorithms, such as AI, are more powerful to analyze big data and develop new applications (Sharma and Garg, 2019). Figure 8.5 shows the UAV cloud-based architecture and its linkage within the big data. In the areas of technology based on computers, data processing using UAV has now begun with the expansion into the areas of cloud computing platforms (Edwin et al., 2019). Monitoring the progress of real-time data has been an important aspect in the areas of research. The businesses utilize UAVs as a supplementary tool for collecting data that extends the existing methods.

Therefore, to minimize the cost of acquisition, they often prefer to outsource the work to companies specializing in operating UAVs and analyzing the data.

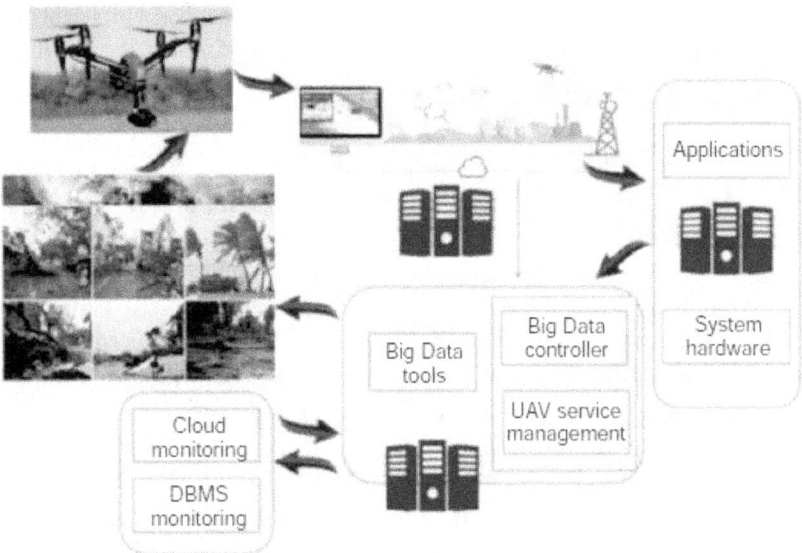

FIGURE 8.5 The UAV cloud service architecture (Edwin *et al.*, 2019).

A cloud for UAVs contains data storage and high-performance computing, combined with big data analysis tools (Bor-Yaliniz and Yanikomeroglu, 2016). It can provide an economic and efficient use of centralized resources for decision-making and network-wide monitoring (Zhou, 2014). The speed at which data is collected has to be matched with data conversion speed into meaningful information and results. The available data analysis techniques are not very efficient in making use of all the contents in UAV collected data. The big heterogeneous data, often unstructured, requires a smart interplay between skilled data scientists and domain experts (Wolfert *et al.*, 2017).

If companies utilize the UAVs data, they can utilize the cloud to gather information from mobile virtual network operators and service providers. These companies have to decide whether to run a UAV service in-house or to outsource. Several factors influence the decision: upfront investment, data security, innovation capabilities, existing know-how, or planned usage of UAVs (utilization). In an outsourcing scenario, the security, privacy, and latency of the data are challenging issues. Operating UAVs is, however, not only about investing in hardware or adhering to government regulations, but also having strong internal analytics capabilities (understanding of big data or machine learning concepts), and a robust IoT architecture. To better exploit the benefits of the cloud, a programmable network can be used to allow dynamic updates based on big data processing, for which network functions. Virtualization and software-defined networking can be good research areas (Bor-Yaliniz and Yanikomeroglu, 2016).

8.3.10 Energy Sources

The energy requirement is one of the most important challenges being faced by UAVs. Batteries constitute the pillar of battery-powered UAVs. They supply power to the propulsion system. Usually, UAVs are battery-powered (such as LiPo, Li-Ion) used for hovering, wireless communications, data processing, and image analysis. The most commonly used battery is the LiPo battery. The next one is Li-Ion, as these batteries are frequently used in Radio Control model vehicles due to their ability to refill. Other batteries, such as lithium-thionyl-chloride batteries (Li-SOCl$_2$), promise two times higher energy density/kg, and lithium-air batteries (Li-Air) provide almost seven times higher energy density/kg as compared to LiPo batteries but are not widely available due to higher costs. Lithium-sulfur batteries (Li-S) may succeed Li-Ion batteries because of their higher energy density (factor 1.8) and reduced costs from the use of sulfur.

Other sources of energy used for UAVs include renewable energy sources based on solar cells, hydrogen-enriched polymer electrolyte membrane (PEM) fuel cells, and laser technologies. The main advantages of fuel cells are their low emissions, high efficiency, modularity, reversibility, fuel flexibility, range of applications, low noise, and low infrared signature (González-Espasandín et al., 2015). However, their disadvantages include cost, the sensitivity of the electrode catalyst to poisons, their experimental state, and the lack of availability of H$_2$. In systems powered by reciprocating and jet engines, fuel only contributes to propulsion in 18–25% of its energy, while in fuel cell-powered aircraft, this efficiency is about 44%.

As battery-driven UAVs cannot fly long distances, solar panels installed on UAVs can harvest solar energy and convert it into electrical energy for long-endurance flights (Sun et al., 2018). Laser power-beaming is also considered as a future technology to provide supplemental energy at night when solar energy is not available or under circumstances of low solar energy at high latitudes during winter. It would enable such UAVs to fly day and night for weeks or possibly months to collect large data or dynamic data of the region.

Table 8.3 lists the advantages and disadvantages of these energy sources.

TABLE 8.3 Advantages and Disadvantages of Batteries, Fuel Cells, and Solar Powered Energy Sources.

Advantages	Disadvantages
LiPo Batteries	
High energy density	Requires special chargers
Lighter than other batteries	Cell loss may occur
Produced in the desired shape and size	More expensive than other batteries (NiCd, NiMH)
Large power capacities	Have issues, such as heating, burning, explosion, etc.
High discharge rates	Care required to use, store and recharge
Fuel Cells	
High efficiency	High initial cost
No moving part	Requires fuel storage for large volume or weight
Easy to install with silent, modular, and compact design	High fuel prices due to the use of hydrogen
Fast energy conversion	Limited life span
Solar Powered	
Free power source	High investment cost
No harmful effects to the environment	Large surface areas needed for panels
Low cost	Energy to be stored continuously
Low maintenance cost	Sun rays must be at right angle
Simple to install with modular structure	Solar panels must be clean
Unlimited source of energy	Cannot use in places having low sun energy

Due to the large number of battery-powered UAVs used for various applications, there is a need to have effective energy management for UAVs. In applications, such as surveillance and monitoring, search and rescue operations, disasters, etc., UAVs need to be operated for extended periods over the regions. Reliable, continuous, and smart management can help UAVs achieve their missions and prevent loss and damage of information and UAVs. Due to the power limitations of UAVs, a decision is required to be taken whether UAVs should perform data and image analysis onboard in real-time or only store the data which can be analyzed at a later stage to reduce the consumption of power (Gupta *et al.*, 2016). The UAVs can provide wireless coverage to remote areas, such as Facebook's Aquila UAVs

(Zuckerberg, 2016). For such applications, UAVs, due to their limited battery capacity, need to return periodically to a charging station for recharging.

The UAV's battery capacity is a key factor for enabling persistent missions. But as the battery capacity increases, its weight increases, which causes the UAV to consume more energy for a given mission. The energy density of Li-Ion batteries is improving by 5–8% per year, and their lifetime is expected to double by the year 2025. These improvements will make commercial UAVs able to hover for more than an hour without recharging, enabling UAVs to deliver more data (Mckinsey, 2017).

To mitigate the limitations of batteries in UAV's, the future research directions may include: (i) UAV battery management, (ii) wireless charging for UAVs, and (iii) solar-powered UAVs (Shakhatreh *et al.*, 2018). These are briefly discussed below.

8.3.10.1 Battery Management

Very little work has been done by the research community on addressing technical hurdles for battery management (Shakhatreh *et al.*, 2018). Research in battery management in UAVs includes planning, scheduling, and replacing the battery to accomplish successful flight missions. Saha *et al.* (2011) found that the depletion of LiPo battery is not only related to the initial start of the charge but also load profile and battery health conditions, which are crucial factors. The use of machine learning techniques can also result in smarter energy management. These techniques are used for path planning and optimization applications, considering energy limitations (Zhang *et al.*, 2017). Similarly, AI, especially deep learning, can promote more advances in UAV power management.

The use of UAVs in long-time enduring missions necessarily requires battery swapping solutions to complete the missions (Shakhatreh *et al.*, 2018). UAVs' autonomous battery swapping system consists of a landing platform, battery charger, battery storage compartment, and micro- controller. Figure 8.6 shows a schematic diagram of *hot-swapping* systems (Boukoberine *et al.*, 2019). *Hot-swapping* means providing continuous power for the UAVs even during battery swapping. The UAV is firstly connected to an external power supply during the swapping process. It will prevent data loss during swapping, as it usually happens in cold swapping. The drained battery is removed and stored in a multi-battery compartment and charging station to be recharged. Then, a charged battery is installed in the UAV, and finally, the external power supply is disconnected. The design for the landing platform is an important part of the swapping system because it can compensate the error in the UAV positioning on the landing point. The swapping time for most of these systems is around one minute. This is a short period as compared to the battery charging average time, which is between 45-60 minutes (Lee *et al.*, 2014).

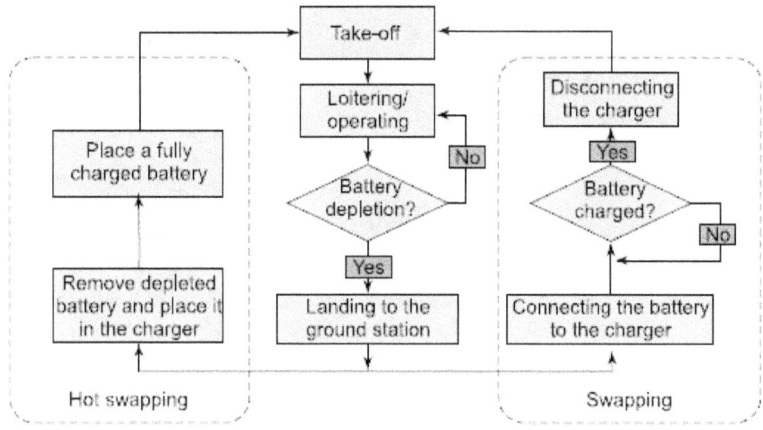

FIGURE 8.6 Illustration of battery hot-swapping system (Boukoberine *et al.*, 2019).

According to Park *et al.* (2017), battery management will be one of the major factors affecting the time-to-delivery of goods. Figure 8.7 shows the battery management of a drone used for the delivery business. The batteries can be attached and detached from the drones to maximize the utilization of both the batteries and the drones. It avoids long wait times for their batteries to be charged after every flight, which will be longer than the flight times themselves. After a delivery is made, batteries are recharged for next use. The state-of-health of batteries degrades due to cycle aging and calendar aging, resulting in reduced battery capacity. If their capacity reduces to a certain threshold, 70%, they are no longer fit for service and are retired to replace the equivalent number of new batteries. The state-of-health degradation of batteries will be a particularly important issue even when compared with other battery-powered applications, for example, electric cars. The drones used to deliver goods are even more weight-sensitive and operate under harsher conditions than electric cars. Most of these drones target to deliver a product within 30 minutes, mandating the batteries to be discharged at a rate of well over 2C, much higher than the typical currents seen in an electric car. This all leads to exploitation of the batteries that accelerates the state-of-health degradation of the batteries. Despite the recent price reduction of Li-Ion batteries, probably the only option for battery-powered drones is to use LiPo batteries as the price of Li-Ion batteries is still quite high.

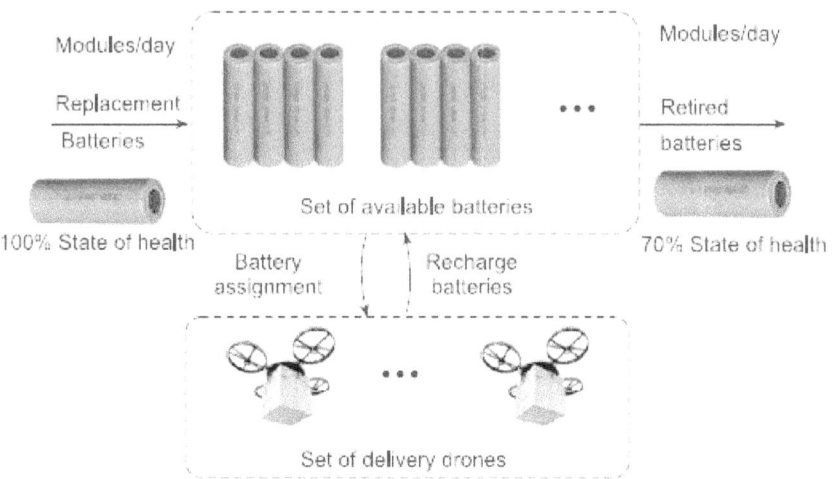

FIGURE 8.7 Battery flow of UAV delivery business (Park *et al.*, 2017).

8.3.10.2 Wireless Charging

Wireless charging is another option that is being explored in UAVs. For example, there is a possibility of energy harvesting from power lines. Simic *et al.* (2015) studied the feasibility of using the circular antenna in UAV to harvest the required energy from power lines while inspecting these power lines. Wang and Ma (2018) proposed an automatic charging system for the UAV, which uses charging stations allocated along the path of the UAV mission. Each charging station consists of a wireless charging pad, solar panel, battery, and power converter, all of which are mounted on a pole. As the power level in the UAV's battery drops to a pre-defined threshold level, the UAV, with the help of GPS and a wireless network module, navigates autonomously to the charging station. In addition, the UAV communicates with the central control room, which will direct the UAV to the closest charging station available at the site.

One of the major challenges that face the deployment of autonomous wireless charging stations is the precise landing of the UAV on the charging pad. The imprecise landing will lead to less efficiency and a long charging time. Different wireless power transfer technologies can be used to overcome this problem and implement automatic charging stations. These technologies include RF Wireless Power Transfer (Dunbar *et al.*, 2015), magnetic resonance coupling technology (Junaid *et al.*, 2016), vision-based target detection (Junaid *et al.*, 2017), and GPS system (Wang and Ma, 2018).

Advance techniques in wireless charging, such as magnetic resonant coupling and inductive coupling techniques, are popular because both of them have acceptable efficiency for small to mid-range distances. The development of battery technologies is expected to positively impact the use of UAVs and allow UAVs to extend flight ranges. While LiPo batteries are still dominant, PEM fuel cells can be more useful in UAVs because of their higher power density (Mostafa *et al.*, 2017).

The majority of the research regarding UAV power management focuses on multirotor UAVs. However, other types of UAVs, such as fixed-wing UAVs, also need attention. Path planning of UAVs is an extremely important factor since it affects the wireless charging efficiency and the average power delivered. Image processing techniques and smart sensors can also play an important role in identifying charging stations and facilitating the precise positioning of UAVs on them. Zhang *et al.* (2017) used deep reinforcement learning to determine the fastest path to a charging station. More research using AI, deep learning, and CNN is required in the path planning and battery scheduling to identify the charging stations, precise landing on a charging station, and precisely predicting the end of the battery charge.

8.3.10.3 Solar-powered UAVs

The first flight of a solar-powered aircraft took place on November 4, 1974, when the remotely controlled Sunrise I, designed by Robert J. Boucher of AstroFlight, Inc., flew after a catapult launch (Figure 8.8). The flight lasted 20 min at an altitude of around 100 m (Noth, 2008).

FIGURE 8.8 First solar-powered plane (De Mattos *et al.*, 2013).

For long-endurance and high-altitude flights, solar-powered UAVs can be an alternate choice. These UAVs use solar power as a primary source for propulsion and battery as a secondary source to be used during night or in the absence of sunlight (Zuckerberg, 2016). Before starting with the design of any solar-powered UAV, it is important to establish how the solar energy will be converted into vehicle motion (propulsion) and how to provide power for the systems. Figure 8.9 provides a scheme of the solar energy conversion to the powerful engine of the UAV. Usually, the wing is an appropriate place to be covered with solar cells, collecting solar radiation and transforming it into electrical energy. Batteries, propulsion, and systems operate under different conditions, and thus a converter is necessary. An appliance called Maximum Power Point Tracking enables the extraction of the maximum power of solar cells and supplies it to the UAV system. There is considerably less energy from the solar cells during the night, and the battery shall fulfill all power needs. Some modern solar cells are able to convert the IR night radiation into electric power, enabling the batteries to be continuously loaded.

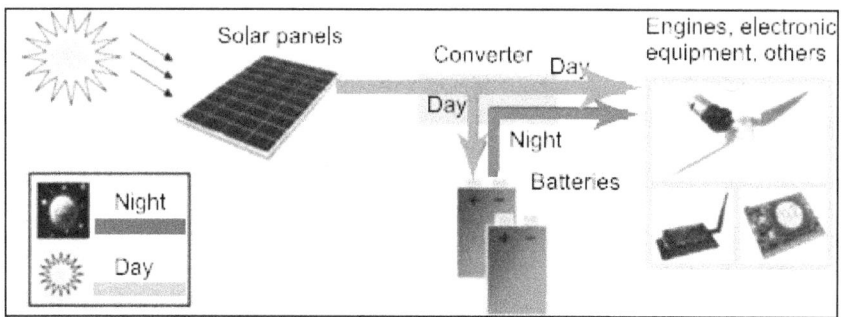

FIGURE 8.9 Solar energy conversion system to power engine of UAV (De Mattos *et al.*, 2013).

Seven small solar UAVs (i.e., So Long, Sky-Sailor, Sun-Sailor, Sun Surfer, AtlantikSolar AS-2, University of Minnesota's UAV, and Cranfield University's Solar UAV) weighing less than 20 kg have been developed till 2018 (Rajendran and Smith, 2018). The latest development from China in solar-powered UAV is that they will soon be capable of taking permanent flight. Such UAVs will use modern high-powered camera-technology and facial recognition tools and can fly around indefinitely until it recognizes a pre-programmed target (Zhi *et al.*, 2020). With a small warhead, that would not require a military-quality drone, and the specific target could be located and automatically eliminated. It may sound like science fiction, but it is possible today.

8.3.11 Drone Taxis

Several companies are competing to bring the flying car/drone in the market, also known as Autonomous Aerial Vehicle (AAV). The concept of the first AAV emerged with prototypes from major players, such as Ehang, whose eco-friendly AAVs aim to serve as autonomous personal transportation devices (Bash, 2018). The Volocopter, a German-based company, is also aiming to help cities resolve growing mobility issues. It features a two-seated drone with 18 rotors, which was chosen to lead Dubai's revolutionary aerial shuttle service, whose testing has already taken place. Even Uber company is joining in the fight against road congestion with its Uber Elevate- a VTOL aircraft, expected to fast-forward the future on-demand urban air transportation. Though the future of drone taxis may not be very appealing to today's professional drivers, these are expected to offer immense benefits. They will be clean, safe, free from roads, and with an infinite number of traffic lanes, making our journeys exponentially faster.

8.4 SWOT ANALYSIS OF UAV TECHNOLOGY

The SWOT is an acronym for Strengths, Weaknesses, Opportunities, and Threats. This analysis is useful to assess the likelihood of uses of a UAV's success or failure (Laksham, 2019).

8.4.1 Strength

UAVs' important advantage is their potential to decrease travel time, as they are a cost-effective alternative to road/rail/water transport in difficult terrains/snow-covered mountains. The UAVs fly close to the surface of the Earth so as to contrast to satellite images and provide very high-resolution images without cloud cover. Companies, such as AeroVironment, Inc. (US), DJI (China), Parrot Drones SAS (France), PrecisionHawk (US), and 3D Robotics, Inc. (US) are the main developers of drones/UAVs for commercial applications. At the same time, Airbus S.E. (France), Boeing (US), General Atomics (US), Lockheed Martin (US), Northrop Grumman Corporation (US), and Thales Group (France) are the main developers of drones/UAVs for military applications (Kenneth Research, 2019).

Currently, there are investments in drones/UAVs by a large number of companies, which would significantly propel market growth. The growing demand for electric drone/UAVs, increasing use of UAVs in the commercial sector, no fuel emissions, and their rising usage for industrial applications are the key factors driving the global drones/UAVs market. Moreover, with the rapid developments in autonomous flight systems, batteries, and electric power, there has been a steep rise in the sale of drones/UAVs.

8.4.2 Weakness

Operating drones/UAVs require trained staff and continuous monitoring from the ground. Lack of infrastructure, such as runway, is a potential problem; however, it can be overcome using VTOL drones/UAVs. Unlike commercial planes and helicopters, UAVs cannot carry heavier payloads or deliver goods for long distances. The payload of a UAV varies between 2 and 4 kg. The disturbance in the signal reception of UAVs due to adverse environmental conditions, such as wind and turbulence, as well as electromagnetic interference, may be of great concern. The safety and efficiency of UAVs are also not well established. The battery life of UAVs is a concern, which can be addressed by using solar-powered UAVs, such as the Aquila by Facebook (Zuckerberg, 2016). Legal permission from the Aviation authorities to fly over the area may pose another problem and restrict its use.

8.4.3 Opportunity

The potential uses of drones/UAVs are enormous and are continuously expanding. In disaster relief areas, drones/UAVs can rescue victims from collapsed buildings or search for fishermen lost in the sea. They can be employed to deliver food, water, and medicines in case of disaster relief and injured in the accidents. A UAV can serve as an ambulance during emergent situations. The global UAV/drone market is estimated to witness a 20.18% CAGR during the forecast period, 2018–2028 (Kenneth Research, 2019). In 2018, the market was led by North America with a 36.02% share, followed by Europe and Asia-Pacific with shares of 22.46% and 20.61%, respectively. Asia-Pacific is the fastest-growing region for the UAVs market. North America would dominate the drones/UAVs market by 2028. It is expected to reach a market size of US$48,092.3 million by 2028. In addition, the UAV industry is expected to generate more than 100,000 jobs (PwC, 2018b).

The UAV/drone market opportunities are the emergence of 3D-printing drones, the introduction of solar-powered UAVs, the increasing need for border monitoring, and usage of UAVs in the energy sector. However, the growth of this market may be hampered due to the possibility of cyber-attacks, regulatory norms, and safety issues. Furthermore, cargo and passenger UAVs/drones have significant implications for transportation companies that are planning next-generation products. As drone/UAV adoption accelerates, the value will quickly shift from UAV/drone makers to the providers of drone-related services. This would significantly boost the global market growth of drones/UAVs.

8.4.4 Threats

In a case of an accident or loss of control, the UAV/drone may fall in a populated area and injure the public. Military drones/UAVs have crashed and caused huge damage to the area. Civilian drones/UAVs may be weaponized and used for terrorist attacks. Hackers can hijack a drone/UAV using GPS jammers.

The SWOT analysis of UAV technology is summarized in Table 8.4. The technological UAV ecosystem has deficiencies (e.g., 4G/LTE bandwidth capacity, processing power for data analysis) that need to be addressed to utilize UAVs efficiently. Regulations are currently limiting the potential, but will eventually be relaxed, enabling quick deployment of UAVs. It is clear that the benefits of using UAV technology outweigh the drawbacks, despite the challenges related to flying regulations. The process of adopting the technology enables UAV operators to capitalize on the value UAVs can create in the present and future (White Paper, 2018). Competing solutions are improving at a high pace while costs of hardware continue to drop. The increasing automation is expected to reduce human error and support results that can be easily reproduced.

TABLE 8.4 The SWOT Analysis of UAVs.

Strength	Weakness	Opportunities	Threats
The high technical maturity of platform and sensors	Lower accuracy compared to ground-based surveying	Facilitate the combination of multiple/multifunctional sensors	Regulatory bodies might limit UAV deployment
Generation of point-cloud, orthophotos, and 3D data	Speed of data processing/analysis	Degree of automation (ease of use)	Increasing quality of competing/emerging solutions

(continued)

Strength	Weakness	Opportunities	Threats
Savings in operational time, costs for better project management	Bandwidth capacity limiting the high expectations of real-time data processing	Early adopters benefit from a competitive advantage	Insufficient data security (encryption)
Increased safety and productivity	UAV technology learning-curve still in progress	Application-specific workflows allow to create of reproducible results, streamline processes, stakeholder management, and pattern recognition (AI)	Lack of trust in UAV technology might hamper the adoption process
Higher flexibility and frequency	Pending regulatory framework and standards limit the true potential of UAVs		

8.5   THE UAVs MARKET POTENTIAL

UAVs have witnessed fast development from 2013 onwards. Many commercial industries adopted UAVs for varied applications due to their low price and ability to take high-resolution images. However, the scale of adoption across these industries differs significantly. As the growth continues, challenges and expectations will continue to rise because users will expect more robust hardware and software. The commercial UAV market as a whole has been able to attract large investments, and these investments in UAV technology over the last four years add up to US$2 billion.

The future growth is expected to occur across three main sectors; (i) consumer UAV shipments which are projected to reach 29 million in 2021, (ii) enterprise UAV shipments which are projected to reach 8.5 million in 2021, and (iii) government UAVs for combat and surveillance activities. About 7.9 million consumer drone shipments and US$3.3 billion revenue are expected by the end of 2020 (Figure 8.10), as compared to 450,000 shipments and US$700 million revenue in 2014.

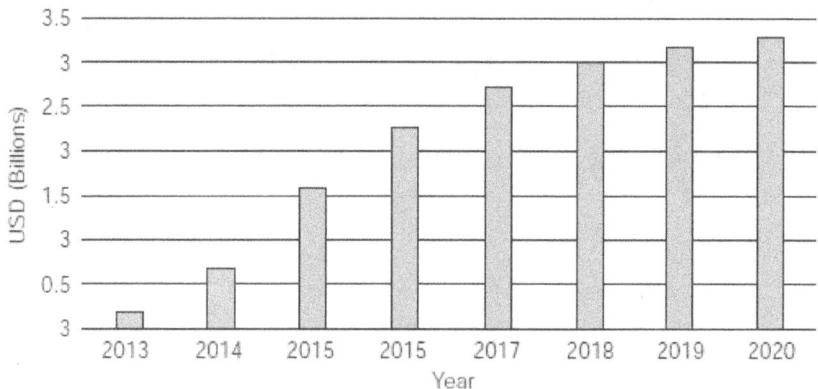

FIGURE 8.10 Revenues from drone shipments (Intelligence, 2018).

Association for Unmanned Vehicle Systems International in 2013 estimated that integration of UAVs into the National Airspace System would create more than 34,000 manufacturing jobs and more than 70,000 new jobs in the first three years of integration, with total job creation to grow to an estimated 104,000 jobs by 2025. This study assessed the economic impact of UAVs in the US. It concluded that the economic impact of the integration of UAVs into the National Airspace System would total more than US$13.6 billion in the first three years of integration and would grow sustainably to more than US$82.1 billion between 2015 and 2025 (Kelly, 2018). To operate a UAV and meet the job requirements, a person should have proper training at universities or specialized institutes or industries.

Over the years, consumer demand will continue to grow. The UAV market has been segmented based on UAV types, applications, classes, and systems, modes of operation, ranges, maximum take-off weight (MTOW), and regions. Based on ranges, the UAV market has been classified into VLOS, EVLOS, and BVLOS. The VLOS segment of the UAV market is projected to grow at the highest CAGR from 2017 to 2023 (6W research, 2016). This growth can be attributed to the increasing use of VLOS in commercial and consumer applications. The increasing use of UAVs in various commercial applications, such as monitoring, surveying and mapping, precision agriculture, aerial remote sensing, and product delivery, also contributes to the growth of the UAV market. The UAVs have ample opportunities in the civil and commercial market, as shown in Table 8.5.

TABLE 8.5 Opportunities in Civil and Commercial UAV Market (Eaton *et al.*, 2016).

Opportunities	Nature of Requirements	Future Requirements
Forest fires Wildlife protection Riots Illegal cross border immigration Asset and property protections Natural disasters Traffic management Road accidents Agricultural activities Communication/connectivity	Small scale Low intensity Unidentified perpetrators Uncertainty Sporadic activity Much wider area Hidden activities Among civilians	Real-time intelligence Constant uninterrupted surveillance Wider area for reconnaissance Immediate and precise action

8.5.1 Based on Regions

Global UAV payload market value is expected to reach US$3 billion by 2027, dominated by North America, followed by Asia-Pacific and Europe (Intelligence, 2018). The payload covers all equipment which is carried by UAVs, such as cameras, sensors, radars, LiDAR, communications equipment, weaponry, and others. Radars and communications equipment segment is expected to dominate the global UAV payload market with a market share of close to 80%, followed by the cameras and sensors segment with around over 11% share, and the weaponry segment with almost 9% share (Kenneth Research, 2019).

Based on application, the drone services market has been segmented into aerial photography and remote sensing, data acquisition and analytics, mapping and surveying, 3D modeling, and others. The aerial photography and remote sensing application segment are expected to grow at the highest CAGR during 2017–2023 (6Wresearch, 2016). There has been an increase in the demand for aerial photography from the infrastructure, architecture, civil engineering, and oil and gas industries worldwide. Commercial drones also have a big market for media and entertainment. Drones offer numerous advantages over other methods of capturing images; some of the benefits include lower costs and improved quality of films and photos. This is the main reason for the growth of the UAVs market for media and entertainment.

The rapid technological advancements in drones and the increase in demand for drone-generated data in commercial applications are the key factors for the growth of the market. Commercial drones are expected to hold the largest market share by 2023 and grow at the highest CAGR of 18.32% during 2017–2023 (Kenneth Research, 2019). This growth is attributed to the rising demand for drones and their data. Also, in some countries, the rules for operating drones in the airspace have been relaxed, thereby boosting the adoption rate of commercial drones.

The drone services market is estimated at US$4.4 billion in 2019 and is projected to reach US$63.6 billion by 2025, at a CAGR of 55.9% from 2019 to 2025 (Grand View Research, 2018). It clearly shows the economic importance of UAVs and their applications in the near future for equipment manufacturers, investors, and business service providers. Smart UAVs will provide a unique opportunity for UAV manufacturers to utilize new technological trends to overcome the current challenges of UAV applications. To spread the UAV services globally, a complete legal framework and institutions regulating the commercial use of UAVs are needed (PwC, 2018).

The small drone market has been segmented into North America, Europe, Asia-Pacific, the Middle East, Latin America, and Africa. North America is estimated to be the largest market in 2024 for small drones (Figure 8.11). The increasing use of small drones for border and maritime surveillance activities in countries, such as the US and Canada, is driving the market growth in North America. Major players in the drone services market include Aerobo (US), Airware (US), Cyberhawk (UK), Deveron UAS (Canada), DroneDeploy (US), Identified Technologies (US), Measure (US), Phoenix Drone Services (US), Prioria Robotics (US), SenseFly (Switzerland), Sharper Shape (US), Sky-Futures (UK), Terra Drone (Japan), The Sky Guys (Canada), and Unmanned Experts (US), among others.

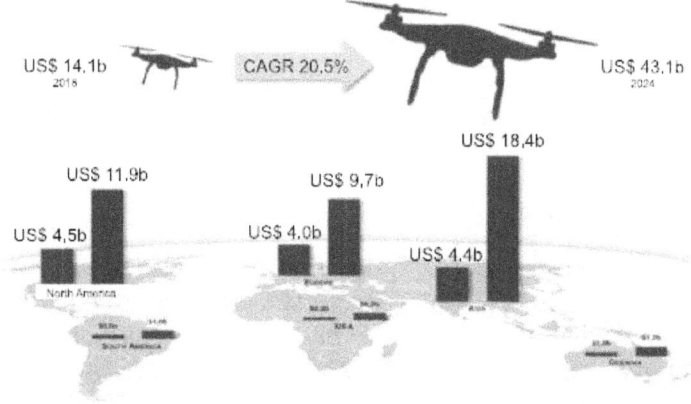

FIGURE 8.11 Drone market size forecast 2018-2024 (White Paper, 2018).

An increasing number of countries are allowing the commercial use of UAVs. Countries, such as the USA, Canada, the UK, Australia, and New Zealand, have come up with varied approval procedures governing the use of UAVs. When combining the total funding of all private drone companies within a country, China is far and away the most well-funded with nearly US$262 million in collective disclosed equity funding since 2015 (CBINSIGHTS, 2017). Canada falls second with just over US$90 million, Israel third with close to US$37 million, and the UK took fourth with a mere US$37.56 million across the same time period. UAVs have been used in diverse environments and high-risk roles in Canada, such as atmospheric research, including weather and atmospheric gas sampling and oceanographic research. The market in Asia-Pacific is expected to grow at the highest CAGR between 2017 and 2023 (6Wresearch, 2016).

The immense potential of UAVs/drones has led to their increasing adoption in India also. Though both the industry and the market in India are at a very nascent stage at the moment, there is the immense growth potential for both. India has been using UAVs for various applications, such as for locating victims of natural disasters. Indian Railways is using UAVs for inspection and tracking of progress of its megaproject, and the largest state-owned natural gas processing and distribution company has implemented UAV for surveillance of its network of gas transmission pipelines. The government of India is making use of drones for the management of a network of National Highways and 3D digital mapping for road widening (Padmanabhan, 2017).

A major thrust is also given by the Indian government to use drones for a variety of purposes, including crop mapping and surveillance of infrastructure projects, pushing the projected value of the domestic industry to approximately US$421 million by 2021 (Padmanabhan, 2017). According to 6Wresearch (2017), the Indian UAV market is projected to grow at a CAGR of 18% during 2017–2023. The adoption of UAVs is increasing in India, and it is projected that the value of industry and market would be around US$885.7 million, while the global market size will touch US$21.47 billion by 2021 (FICCI, 2018).

8.5.2 Based on Applications

The analytics segment is anticipated to witness high growth in the drone software market during 2017–2023. The data captured by drones can be used for analytics and provided to businesses to help them in decision-making. Drone software is developed according to the diverse needs of different applications. Disruptive technologies used in the drone software industry in the region are transforming the way companies do business and helping them gain a competitive advantage. The drone software market in the Asia-Pacific region is expected to witness high growth during 2017–2023, owing to innovations in technology and increasing focus on the manufacturing of commercial drones at low cost (6Wresearch, 2016). As forecasted by Goldman Sachs (2016), the fastest-growing drone/UAV market is the commercial drone market for businesses and governments (Table 8.6). Overall, there are ten largest opportunities for drones in construction, agriculture, insurance claim, offshore, police, fire, coast guard, journalism, customs, and border protection, and real estate. It is estimated that this market will grow to US$13 billion between 2016 and 2020.

TABLE 8.6 The UAV/Drone Market for Various Applications.

Application	US$ (Million)	Application	US$ (Million)
Construction	11,164	Custom and border protection (in USA)	380
Agriculture	5,922	Real estate	265
Insurance claim	1,418	Utilities	93
Offshore, oils/gas and refinery	1,110	Pipelines	41
Police (in USA)	855	Mining	40
Fire (in USA)	881	Clean energy	25
Coast guard (in USA)	511	Cinematography	21
Journalism	480		

Commercial and civil drones represent a significant business opportunity in the future. Between now and 2020, Goldman Sachs (2016) has forecasted a US$100 billion market opportunity for drones, driven by growing demand from the commercial and civil government sectors. They also forecasted that Australia's drone spending for various applications would be an estimated US$3.1 billion between 2017 and 2021.

According to the PwC report "Global Construction 2030," the volume of output from construction is likely to grow by 85% and reach US$15.5 trillion around the world by 2030, with China, the US and India accounting for 57% of this growth. Construction is likely to be one of the most dynamic industrial sectors in the next 1 years, and is utterly crucial to the evolution of prosperous societies around the world. The number is huge, and that translates to the creation of vast numbers of new jobs and significant wealth in certain countries across the globe. As of 2016, the growth in the construction sector has been 2.4% but it is expected to grow around 2.8% during the next five years, as shown in Figure 8.12 (PwC, 2018a). Intensive and extensive activities in this sector should automatically propel the economic engine.

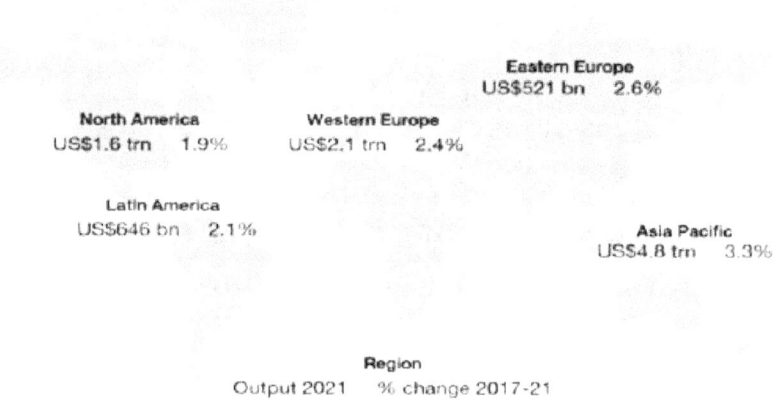

FIGURE 8.12 Global construction output (PwC, 2018a).

8.5.2.1 Civilian Applications

Among all the components of UAVs, the camera is expected to hold the largest share of the market during the 2017 and 2023 period (6W research, 2016). The steep rate of adoption of high-resolution images for military and commercial applications is a key factor driving the growth of the UAV/drones market for cameras. Camera systems are used in UAVs for many applications, such as continuous video monitoring, investigation, remote surveillance, border security, and protection of critical infrastructure, a thermographic inspection of inaccessible buildings, firefighting, and law enforcement. The UAV/drone market for a commercial is estimated to grow to reach over US$1.2 billion by 2020 (Markets and Markets, 2015).

The growth in the civil market is expected to significantly expand as the regulations on civil use of UAVs in the US and around the world are being simplified. The factors expected for the services market's growth include the increasing use of UAV/drone for industry-specific solutions, improved regulatory framework, and increased requirement of qualitative data in various industries. As these markets expand, the need to have systems that can adapt to new missions, sensors, and environments will drive further requirements. The UAVs offer a great market opportunity for equipment manufacturers, investors, and business service providers. According to the PwC report (2018b), the addressable market value of UAV uses is over US$127 billion. Civil infrastructure is expected to dominate the addressable market value of UAV uses, with a market value of US$45 billion. Business Intelligence expected sales of UAVs in the US to reach $ 12 billion in 2021, which is up by a compound annual growth rate of 7.6% from $ 8.5 billion in 2016 (Joshi, 2017). McKinsey reports also estimated that by the year 2026, the commercial UAVs/drones, both corporate and consumer applications, will have an annual impact of US$31 billion to US$46 billion on the GDP of the USA (FICCI, 2018).

8.5.2.2 Military Applications

The UAV market is estimated to be US$20.71 billion in 2018 and is projected to reach US$52.30 billion by 2025, at a CAGR of 14.15% from 2019 to 2025 (Grand View Research, 2018). Although the UAV market is still dominated by military applications, commercial applications are also increasing their market share. The rise in the procurement of military UAVs by defense forces worldwide is one of the most significant factors projected to drive the growth of the UAV market. Goldman Sachs (2016) estimated that drone technologies would reach a total market size of US$100 billion between 2016 and 2020. However, 70% of this figure would be linked to military activities; the commercial business represents the fastest growth opportunity, projected to reach US$13 billion between 2016 and 2020 (FICCI, 2018).

Drones help in capturing images and videos (e.g., footage along the borders) using various payloads, such as camera systems and sensors. For example, Predator drones can silently patrol borders looking for infiltrants, illegal immigrants, human traffickers, and drug peddlers in desolated areas. Of all the major applications of military drones, the market for border security is expected to hold the largest share in 2017. Military drones are being used successfully by defense agencies across the world to guard their borders, enforce the law, and for combat missions. The increasing demand for long-endurance and highly stable drone systems word-wide from the army is a key factor driving the growth of the military drones market for border security applications. The market for spying applications is expected to grow at the highest CAGR during 2017–2023 (6W research, 2016). The military market continues to expand and develop new capabilities and requirements.

In July 2014, the Teal Group predicted that worldwide UAVs expenditure would grow to over US$11 billion per year with a total investment of over US$91 billion by 2024. It is expected that 86% of the market will be in the military and 14% will be in the civil market. The market for military drones with a payload-carrying capacity of up to 150 kg is expected to hold the largest share during 2017–2023 (6W research, 2016) as they have simple operational and maintenance procedures with high tactical capabilities. These UAVs are versatile and can carry out a variety of missions, such as search and rescue, illegal traffic monitoring, support to ground forces, and various intelligence missions. The market for drones with a payload-carrying capacity of up to 600 kg is expected to grow at the highest CAGR during 2017–2023 (6W research, 2016).

8.5.3 Based on Autonomy

Current military and commercial aircraft are becoming increasingly autonomous due to the installation of autonomous components, which have eased the workload of pilots. The autonomous aircraft market is projected to grow from an estimated US$3.6 billion in 2018 to 23.7 billion by the year 2030, at a CAGR of 17.06% from 2018 to 2030 (Markets and Markets, 2015). This growth can be attributed to the increased cost savings, reduction in human error due to increased autonomy, and advancements in AI. The industry's goal is to make these commercial aircraft fully autonomous and introduce new passenger and personal air vehicles along with commercial delivery UAVs/drones that will be fully autonomous. Based on components, the autonomous aircraft market is segmented into flight management computers, air data inertial reference units, sensors, actuation systems, software, intelligent servos, cameras, radars & transponders, and propulsion systems. Based on technology, the autonomous aircraft market has been segmented into increasingly autonomous and fully autonomous (FICCI, 2018). The fully autonomous segment is projected to grow at the highest CAGR due to the increasing autonomy in aircraft and the move toward aircraft being fully autonomous in the future. Thus, the fully autonomous segment is projected to see high growth in the market.

8.5.4 Based on Size

The UAV market has been segmented into small UAVs, tactical UAVs, strategic UAVs, and special purpose UAVs. The small UAVs segment is expected to grow at the highest CAGR during 2017–2023, owing to the demand for these

UAVs from the defense sector for surveillance and reconnaissance applications (6Wresearch, 2016). The demand for small UAVs from the commercial sector is also increasing, as these UAVs are used in monitoring, surveying and mapping, precision agriculture, wildlife survey, search & rescue operations, firefighting, law enforcement, photography, product delivery, and disaster management. The small drones market was valued at US$9.71 billion in 2017 and is projected to reach US$40.31 billion by 2025, at a CAGR of 17.04% (Grand View Research, 2018). The rise in the procurement of small military drones by defense forces worldwide is one of the most significant factors projected to drive the growth of the small drones market.

The small drone market has been segmented on the basis of type, power source, application, MTOW, payload, and region. Based on MTOW, the small drones market has been classified into <5 kg, 5–25 kg, and 25–150 kg. The 25–150 kg of MTOW segment of the market is projected to grow at the highest CAGR from 2017 to 2025. This growth can be attributed to the increasing use of small drones in commercial and consumer applications. Based on application, the small drones market has been segmented into military, civil & commercial, homeland security, and consumer. The civil and commercial segment of the market is projected to grow at the highest CAGR from 2017 to 2023 (6Wresearch, 2016). This growth can be attributed to the increased adoption of small drones for various civil & commercial applications, such as monitoring, surveying, mapping, and remote sensing.

8.5.5 Based on Hardware and Software

Global investment in drone hardware is shown in Figure 8.13. The drone software market is estimated to be US$2.85 billion in 2016 and is projected to reach US$12.33 billion by 2022, at a CAGR of 27.63% from 2016 to 2022 (6Wresearch, 2016). Factors, such as growing investments and the increasing use of drones in commercial and military applications, are expected to drive the market's growth in the near future. Based on architecture, the open-source segment is also expected to dominate the market. Key players profiled in the drone software market report include Pix4D (Switzerland), Precision Hawk (USA), ESRI (USA), Airware, Inc. (USA), 3D Robotics (USA), Dreamhammer Inc. (USA), Drone Volt (France), and DroneDeploy (USA), among others. Acquisitions, collaborations & partnerships are among the key strategies adopted by the leading market players to sustain their position in the market. This strategy has been adopted by companies to increase their reach and enhance their revenue and service offerings.

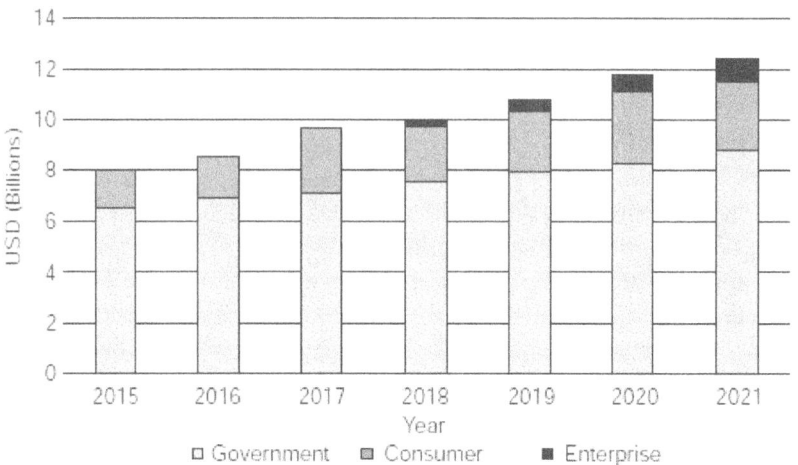

FIGURE 8.13 Global investments estimated in drone hardware (6W research, 2016).

Open-source software is available freely, and many companies are opting for such software. The recent trend in UAV software is toward open-source software, as the end-users can customize this software for operating the UAVs. This segment is projected to witness high growth in the UAV software market during 2017–2023 (6Wresearch, 2016). App-based software is more practical when UAVs are used in remote areas or places where carrying a desktop/laptop is impossible.

8.6 EMERGING RESEARCH AREAS

There are several emerging areas for future research of UAVs, such as automated employee check-in/check-out; automated safety checks; scanning RFID tags on materials in laydown areas for inventory; material delivery; goods delivery; remote job walks; preview views from a building prior to its construction; thermal scanning of utility-scale PV plants and interior missions; industrial inspection; and maintenance and repair of oil and gas platforms, assets, utilities, and infrastructure and energy parks. Other areas may include structural analysis in assessment; archaeology and heritage monument inspection; documentation with aerial imaging; as first responders in accident fire or crisis; UAV/drone-based flight dynamics and control theory research; flight automation and reproduction; computer vision and simultaneous localization and mapping research; and swarming intelligence and networking. Figure 8.14 shows an overview of the research areas in the UAV domain. There are two broad categories of research areas; in the technology domain and in operations, which are further sub-categorized to provide more details.

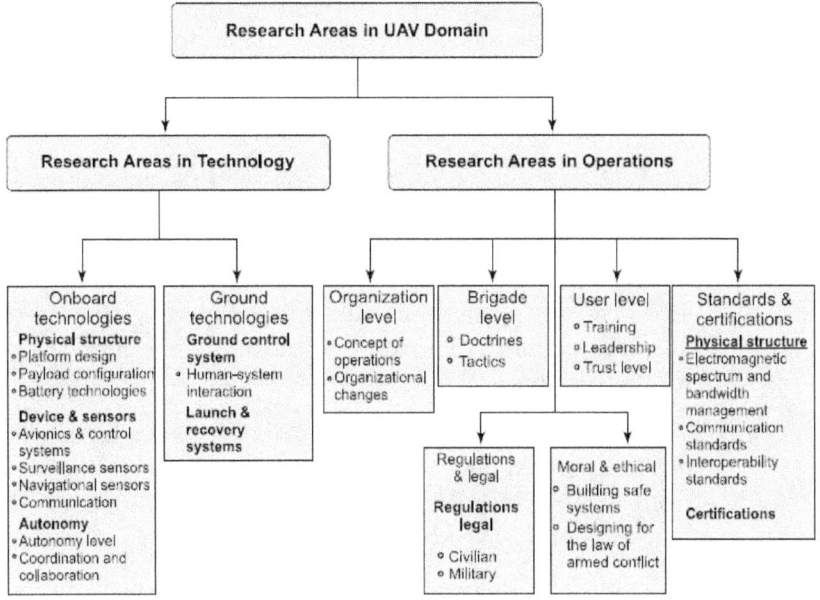

FIGURE 8.14 Research areas in UAVs (Demir et al., 2015).

In ITS applications, there are several challenges that need future research and focus on facilitating the use of UAVs, as given below (Mohammed et al., 2014; Menouar et al., 2017). Swarm plays an important role in some of these applications.

(i) Restrict the privacy of sensitive information (e.g., location) from other vehicles and drones. Since usually there is no encryption on UAVs onboard chips, they can be hijacked and subjected to man-in-middle attacks originating up to 2 km away.

(ii) Develop precise coordination algorithms to enable ITS-UAVs.

(iii) Develop surveillance and live feeds from wireless sensors for controlling traffic systems.

(iv) Develop techniques for data fusion of information from diverse sensors, automating image data compression, and stitching of aerial images.

(v) Develop network-centric infrastructure, enabling teams of operators to control the UAV and retrieve imagery and sensor information in real-time.

(vi) Limited energy, processing capabilities, and signal transmission range are also main issues in UAVs, requiring further R&D.

(vii) The UAVs have slower speeds compared to vehicles driving on highways. Finding the optimal altitude of UAV in support of ITS applications is a research challenge.

(viii) Long operational time of a battery is another challenge. To overcome this limitation, the deployment of UAVs, recharge stations, and ground roadside units jointly is an interesting but complicated optimization problem.

(ix) The detection of multiple vehicles at the same time is another challenge. Computer-vision-based algorithms are required to be developed for this purpose.

(x) Truly autonomous operation of UAV swarms is a big challenge. Therefore, the development of swarm intelligence algorithms that fuse data from diverse sources, including location sensors, weather sensors, accelerometers, gyroscopes, Radars, LiDAR, etc., are needed.

Another form of research could be to perform a qualitative and quantitative comparison of UAV-based measurement techniques to other traditional methodologies, such as manual inspection, embedded sensors, and aircraft or helicopter-based surveying. The pros and cons of each of these methods can be compared by using operations cost, accessibility, quality of data, the total time required, etc., as some of the benchmark attributes.

There is also a need to undertake more concentrated research on the following aspects:

(i) New lighter materials, efficient batteries, and energy harvesting solutions which can contribute to the potential use of UAVs in long-duration missions (Vergouw *et al.*, 2016).

(ii) Algorithms that support UAV autonomy and swarm coordination are needed. These algorithms can include flight route determination, path planning, collision avoidance, and swarm coordination (Safi *et al.*, 2019).

(iii) In multi-UAV systems, many coordination and communications challenges need to be overcome. These challenges include QoS communications between the swarm of UAVs over multi-hop communications links and with the GCS (Scherer *et al.*, 2015).

(iv) More accurate localization an mapping systems and algorithms are required to support precise location-based applications. New algorithms are needed for the fusion of data received from multiple sensors to achieve more precise localization and mapping.

(v) The use of UAVs as aerial base stations is still nascent. More research is needed to study the use of such systems to provide communications coverage when the public communications network is disrupted.

(vi) Systems that help UAVs to avoid collisions and obstacles are expected to emerge with strong solutions (Mckinsey, 2017).

(vii) The UAVs currently travel below the height of commercial aircraft to avoid a collision. The methods that can track UAVs and communicate with air-traffic control systems for typical aircraft are to be developed, making high-altitude missions possible for UAVs (Mckinsey, 2017).

(viii) More research is required to address UAV design that can be used in a wide range of conditions and whose capability rivals that of commercial airliners. It would require a significant effort, ingenuity and contributions from experts in diverse areas (D'Andrea, 2014).

(ix) With the increasing use of UAVs in various commercial applications, there has been a rise in hacking incidents. Thus, strong security mechanisms are required to be developed.

(x) More research is needed for the integration of low-cost sensors and localization systems so that UAVs work in different operating conditions (D'Andrea, 2014).

(xi) More simulation based research is needed to address the UAVs coordination so that a large number of UAVs in the air utilize the same resources, such as charging stations and operating frequency (D'Andrea, 2014).

(xii) Work is needed to enhance the flight time of the UAVs without increasing the cost.

(xiii) Developing robust indoor navigation in scenarios without GNSS coverage and improving control systems under active payloads (i.e., fluid spraying, robotics arms) are required.

8.7 SUMMARY

The UAVs and associated technologies have a wide number of uses providing benefits in terms of time, money, and detailed and latest information for organizations/industries. Over the past few years, UAVs/drones have become central to the functions of various businesses and organizations and have managed to pierce through areas where certain industries were either stagnant or lagging behind. The UAV offers many benefits to industries globally; some of them may include increasing work efficiency and productivity, decreasing workload and production costs, improving accuracy, refining services, and resolving security issues.

The use of UAVs/drones has rapidly changed over the past years, and future opportunities in the field are limitless. The emerging research trends in UAV may include (i) growth in UCAV, (ii) increase in endurance limit, (iii) increase in HALE UAVs, (iv) increase in mission capabilities of UAVs, and (v) increase in the endurance of UAVs for various applications. The next generation of UAVs/drones will be powered by AI, which will allow UAVs/drones to make decisions and operate themselves on behalf of their human controllers. As the demand for autonomous UAVs grows, it is expected that the future research activity in the area of autonomous control of UAVs will be very active, with commercial R&D aimed at enriching the technological capabilities of products to better compete in a growing and demanding market.

The UAV/drone industry is booming. The ongoing investment trend toward software, hardware, and improving the entire process of aerial data acquisition and analysis underlines the importance associated with UAVs. However, there are some issues and challenges which need to be resolved before organizations can leverage the full potential of these technologies. As true with other technological innovations, the early adopters of UAVs will be profiting the longest and increasing their competitive advantages. Today, UAVs/drones are well known as data capture devices, but in the future, we will see them perform the full lifecycle: capturing data, analyzing this information, and then acting on it. So, the future of UAVs/drones not only looks bright but very promising.

REVIEW QUESTIONS

Q.1. Discuss briefly the "essential eight technologies" identified for future development.

Q.2. Write in short various key challenging areas for using UAV technology in the future.

Q.3. What are cyber-security threats to a complete UAV system? How do you propose to minimize them?

Q.4. Write short note on (i) Swarm-based applications in the future, (ii) Big data analytics, (iii) UAV-IoT applications for the future, and (iv) Drone Taxi.

Q.5. Discuss the battery energy sources used in UAVs. What are their limitations? Explain the concept of battery management to collect long-term continuous data from UAVs.

Q.6. Write a short note on (i) Wireless charging of batteries and (ii) Solar-powered UAVs.

Q.7. Carry out a SWOT analysis of UAV technology.

Q.8. Explain the market potential of UAV in the future. Support your answer with the statistics (future projections).

Q.9. Enumerate future research areas of UAV technology and its applications.

REFERENCES

AA (2017), Aviation Authorities: *https://en.wikipedia.org/wiki/List_of_civil_aviation_authorities*.

Adams, S.M.; and Friedland, C.J. (2011), A Survey of Unmanned Aerial Vehicle (UAV) Usage for Imagery Collection in Disaster Research and Management, In 9th International Workshop on Remote Sensing for Disaster Response.

Aicardi, Irene; Chiabrando, F.; Grasso, Nives; Lingua, Andrea M.; Noardo, Francesca; and Spanò, A. (2016), UAV Photogrammetry With Oblique Images: First Analysis On Data Acquisition And Processing, The International Archives of the Photogrammetry, Remote Sensing and Spatial Information Sciences, Volume XLI-B1, 2016 XXIII ISPRS Congress, 12–19 July 2016, Prague, Czech Republic June 2016, DOI: 10.5194/isprsarchives-XLI-B1-835-2016.

Albeaino, Giles; Gheisari, Masoud; and Franz, Bryan W. (2019), A Systematic Review of Unmanned Aerial Vehicle Application Areas and Technologies in the AEC Domain, Journal of Information Technology in Construction – Vol. 24, ISSN 1874–4753.

Alexopoulos, A.; Kandil, A.; Orzechowski, P.; and Badreddin, E. (2013), A Comparative Study of Collision Avoidance Techniques for Unmanned Aerial Vehicles, In International Conference on Systems, Man, and Cybernetics (SMC). IEEE, 2013, pp. 1969–1974.

Al-Fuqaha, A.; Guizani, M.; Mohammadi, M.; Aledhari, M.; and Ayyash, M. (2015), Internet of Things: A Survey on Enabling Technologies, Protocols, and Applications, IEEE Commun. Surveys Tuts., Vol. 17, no. 4, pp. 2347–2376, 4th Quart.

Al-Mousa, Amjed; Sababha, Belal H.; Al-Madi, Nailah; Barghouthi, Amro; and Younisse, Remah (2019), UTSim: A Framework and Simulator for UAV Air Traffic Integration, Control, and Communication, International

Journal of Advanced Robotic Systems, September–October: 1–19, DOI: 10.1177/1729881419870937.

Al-Najjar, Husam A.H.; Kalantar, Bahareh; Pradhan, Biswajeet; Saeidi, Vahideh; Halin, Alfian Abdul; Ueda, Naonori; and Mansor, Shattri (2019), Land Cover Classification from fused DSM and UAV Images Using Convolutional Neural Networks, Remote Sensing, Vol. 11, p. 1461, DOI: 10.3390/rs11121461.

Altawy, Riham; and Youssef Amir Mohamed (2016), Security, Privacy, and Safety Aspects of Civilian Drones: A Survey, ACM Transactions on Cyber-Physical Systems, November, Article No. 7. *https://doi.org/10.1145/3001836*.

Al-Turjman, Fadi; Poncha Lemayian, Joel; Alturjman, Sinem; and Mostarda, Leonardo (2019), Enhanced Deployment Strategy for the 5G Drone-BS Using Artificial Intelligence, IEEE Access, Special Section on Roadmap to 5G: Rising to the Challenge, Vol. 7, DOI: 10.1109/ACCESS.2019.2921729.

Alwateer, Majed; Loke, Seng W. Loke; and Fernando, Niroshinie Fernando (2019), Enabling Drone Services: Drone Crowdsourcing and Drone Scripting, IEEE Access, Open access journal, Volume 7, Digital Object Identifier 10.1109/ACCESS.2019.2933234.

Anderson, C. (2012), Here Come the Drones! August Issue: Wired Magazine, *http://www.wired.co.uk/article/here-come-the-drones*.

Andrews, Lena Simone (2020), Attack of the Drones: Ethical, Legal and Strategic Implications of UAV Use, *https://cis.mit.edu/publications/audits/attack-drones-ethical-legal-and-strategic-implications-uav-use*.

Anwar, N.; Najam, F.; and Izhar, M.A. (2018), Construction Monitoring and Reporting Using Drones and Unmanned Aerial Vehicles (UAVs), The Tenth International Conference on Construction in the 21st Century (CITC-10), July.

ARC (2017), UAS Identification and Tracking Aviation Rulemaking Committee Recommendations, Final Report, September 2017, *https://www.faa.gov/regulations_policies/rulemaking/committees/documents/media/UAS%20ID%20ARC%20Final%20Report%20with%20Appendices.pdf*.

Atoev, Sukhrob; Kwon, Oh-Heum; Lee, Suk-Hwan; and Kwon, Ki-Ryong (2018), An Efficient SC-FDM Modulation Technique for a UAV Communication Link, Electronics, Vol. 7, p. 352, DOI: 10.3390/electronics7120352.

AT&T (2016), Detail Network Testing of Drones in Football Stadiums," *https://www.androidheadlines.com/2016/09/att-detail-network-testing-of-drones-in-football-stadiums.html.*

Austin, Reg (2010), Unmanned Aircraft Systems – UAVs Design, Development and Deployment, John Wiley and Sons, Ltd.

Ayranci, Zehra Betul (2017), Use of Drones in Sports Broadcasting, Entertainment and Sports Lawyer, Spring, Volume 33, No. 3.

Azevedo, Miguel Itallo B.; Coutinho, Carlos; Toda, Eylon Martins; Carvalho, Tassio Costa; and Jailton, Jose (2019), IntechOpen, Chapter on Wireless Communications Challenges to Flying Ad Hoc Networks (FANET).

Bagla, G. (2015), India Leads in Import of Unmanned Aerial Vehicles. *http://www.theindiaexpert.com/tag/sipri.*

Baker, John (2018), 8 Open Source Drone Projects, February, *https://opensource.com/article/18/2/drone-projects.*

Bamburry, D. (2015), Drones: Designed for Product Delivery, Design Management Review, Vol. 26, No. 1, pp. 40–48.

Banu, Tiberiu Paul; Borlea, Gheorghe Florian; and Banu, Constantin (2016), The Use of Drones in Forestry, Journal of Environmental Science and Engineering B5, 557–562, DOI: 10.17265/2162–5263/2016.11.007.

Barik, Soumyarendra (2019), DGCA Certifies 'Small' Category Drone for Complying with 'No-Permission, No-Takeoff' Protocol, MediaNama, September 5, *https://www.medianama.com/2019/09/223-dgca-certifies-small-drone/.*

Barnes, Jonathan (2018), Drones (Photogrammetry) vs Terrestrial LiDAR – What Kind of Accuracy Do You Need?, April, *https://www.commercialuavnews.com/infrastructure/drones-vs-terrestrial-lidar-accuracy-need.*

Bartoš, Karol; Pukanská, Katarina; and Sabová, Janka (2014), Overview of Available Open-Source Photogrammetric Software, Its Use and Analysis, International Journal for Innovation Education and Research, Vol. 2, No. 4.

Bash, Yariv (2018), Drone Trends to Watch in 2018: Big Data, Flying Taxis, and Home Security, VB, *https://venturebeat.com/2018/01/13/drone-trends-to-watch-in-2018-big-data-flying-taxis-and-home-security/.*

Bejiga, M.B.; Zeggada, A.; Noufdj, A.; and Melgani, F. (2017), A Convolutional Neural Network Approach for Assisting Avalanche Search and Rescue Operations With UAV Imagery, Remote Sens., Vol. 9, No. 2, p. 100.

Bekmezci, I.; Sahingoz, O.K.; and Temel, S. (2013), Flying Ad-Hoc Networks (FANETs): A Survey, Ad Hoc Networks, Vol. 11, No. 3, pp. 1254–1270.

Berie, Habitamu Taddese; and Buruda, Ingunn (2018), Application of Unmanned Aerial Vehicles in Earth Resources Monitoring: Focus on Evaluating Potentials for Forest Monitoring in Ethiopia, European Journal of Remote Sensing, Vol. 51, No. 1, 326–335, *https://doi.org/10.10 80/22797254.2018.1432993.*

Bernard, M.; Kondak, K.; Maza, I.; and Ollero, A. (2011), Autonomous Transportation and Deployment with Aerial Robots for Search and Rescue Missions. Journal of Field Robotics, Vol. 28, pp. 914–931, DOI: 10.1002/rob.20401.

BIS Research (2018), Simultaneous Localization and Mapping (SLAM) Technology for Robotics, UAVs, AR, and Cars, August, *https:// marketresearch.com/simultaneous-localization-and-mapping-slam-technology-for-robotics-uavs-ar-and-cars.*

Blom, John David (2011), Unmanned Aerial Systems: A Historical Perspective, Occasional paper-37, Combat Studies Institute Press, US Army Combined Arms Center, Kansas.

Bor-Yaliniz, I.; and Yanikomeroglu, H. (2016), The New Frontier in Ran Heterogeneity: Multi-tier Drone-cells, IEEE Communications Magazine, Vol. 54, No. 11, pp. 48–55.

Bor-Yaliniz, R.I.; El-Keyi, A.; and Yanikomeroglu, H. (2016), Efficient 3-D Placement of an Aerial Base Station in Next Generation Cellular Networks, In Proc. IEEE Int. Conf. Commun. (ICC), Kuala Lumpur, Malaysia, May, pp. 1–5.

Bora, Garima (2020), Covid-19: In the Times of 'Touch-Me-Not' Environment, Drones Are the New Best Friends. *https://Economictimes. Indiatimes. com/small-biz/startups/features/covid-19-in-the-times-of-touch-me-not environment-drones-are-the-new-best-friends/articleshow/74924233. cms?utm_source=contentofinterest &utm_medium=text&utm_campaign= cppst.*

Boskovic, J.D.; Prasanth, R.; and Mehra, R.K. (2002), A Multilayer Control Architecture for Unmanned Aerial Vehicles. In Proceedings of the American Control Conference, Anchorage, AK, USA, 8–10 May, pp. 1825–1830.

Boucher, P. (2015), Domesticating the Drone: The Demilitarisation of Unmanned Aircraft for Civil Markets. Sci Eng Ethics, 21, 1393–1412, *https://doi.org/10.1007/s11948-014-9603-3.*

Boukoberine, Mohamed Nadir; Zhou, Zhibin; and Benbouzid, Mohamed (2019), Power Supply Architectures for Drones – A Review, Conference Paper, October, DOI: 10.1109/IECON.2019.8927702.

Brooks, C.; Dobson, R.J.; Banach, D.M.; Dean, D.; Oommen, T.; Wolf, R.E.; Havens, T.C.; Ahlborn, T.M.; and Hart, B. (2014), Evaluating the Use of Unmanned Aerial Vehicles for Transportation Purposes. Michigan Tech Research Institute Final Rep. No. RC-1616. Houghton, MI: Michigan Tech Research Institute.

Brown, Peter (2018), Drones Changing the Landscape of the Insurance Market, IEEE, *https://www.ieee.org/content/dam/ieee-org/ieee/web/org/members/fap/articles/drone-article-final-igs-with-tag-lm-revised-2018.pdf*.

Buczkowski, Aleks (2018), Drone LiDAR or Photogrammetry? Everything You Need to Know, January 6, *https://geoawesomeness.com/drone-lidar-or-photogrammetry-everything-your-need-to-know/*.

Burnap, Alex; Barto, Charlie; Johnson-Roberson, Matthew; Ren, Max Yi; Gonzalez, Richard; and Papalambros, Panos Y. (2015), Crowdsourcing for Search of Disaster Victims: A Preliminary Study for Search System Design, International Conference on Engineering Design, ICED15, 27–30 July, Politecnico Di Milano, Italy.

Business Insider (2017), Amazon Takes Critical Step Toward Drone Delivery, *http://www.businessinsider.com/amazon-takes-critical-step-toward-drone-delivery-2017-5*.

Cai, G; Chen, B.M.; and Lee, T.H. (2011), Unmanned Rotorcraft Systems. New York: Springer Science & Business Media; pp. 01–267.

Campion, Mitch; Ranganathan, Prakash; and Faruque, Saleh (2019), A Review and Future Directions of UAV Swarm Communication Architectures, IEEE, 978-1-5386-5398-2/18, *https://ieeexplore.ieee.org/stamp/stamp.jsp?tp= &arnumber=8500274*.

Cao, Lin; Liu, Hao; Fu, Xiaoyao; Zhang, Zhengnan; Shen, Xin; and Ruan, Honghua (2019), Comparison of UAV LiDAR and Digital Aerial, Photogrammetry Point Clouds for Estimating Forest Structural Attributes in Subtropical Planted Forests, Forests, Vol. 10, p. 145, DOI: 10.3390/f10020145.

Carrio, A.; Sampedro, C.; Rodriguez-Ramos, A.; and Campoy, P. (2017), A Review of Deep Learning Methods and Applications for Unmanned Aerial Vehicles, J. Sensors, Vol. 2017, Aug., Art. no. 3296874.

CAS (2002), Unmanned Aircraft and Rockets: Model Aircraft. Advisory Circular AC101-3(0). Civil Aviation Safety Authority Australia. Canberra, Australia.

Casado, Mónica Rivas; González, Rocío Ballesteros; Ortega, José Fernando; Leinster, Paul; and Wright, Ros (2017), Towards a Transferable UAV-Based Framework for River Hydromorphological Characterization, Sensors, MDPI, Vol. 17, p. 2210, DOI: 10.3390/s17102210.

Caughey, David A. (2011), Introduction to Aircraft Stability and Control, Sibley School of Mechanical & Aerospace Engineering, Cornell University, Ithaca, New York.

CBINSIGHTS (2017), Drone Planet: The Most Well-Funded Private Drone Companies In One Map, July 26, Research Briefs, *https://www.cbinsights.com/research/most-well-funded-private-drone-companies/*.

Chandrasekharan, S. (2016), Designing and Implementing Future Aerial Communication Networks, IEEE Commun. Mag., Vol. 54, No. 5, pp. 26–34.

Chang, A.; Jung, J.; Maeda, M.M.; and Landivar (2017), J. Crop Height Monitoring With Digital Imagery from Unmanned Aerial System (UAS), Comput. Electron. Agric., Vol. 141, pp. 232–237.

Chao, H; Baumann, M; Jensen, A.; Chen, Y.; Cao, Y.; Ren, W.; and McKee, M. (2008), Band-reconfigurable Multi-UAV Based Cooperative Remote Sensing for Real-time Water Management and Distributed Irrigation Control. IFAC Proceedings Volume, Vol. 41, No. 2, pp. 11744–11749.

Chatzikyriakou, Chara; and Rickerby, Patrick (2019), Comparing Drone LiDAR and Photogrammetry, *https://terra-drone.eu/en/articles-en/comparing-drone-lidar-and-photogrammetry*.

Chee, K.; and Zhong, Z. (2013), Control, Navigation and Collision Avoidance for an Unmanned Aerial Vehicle, Sensors and Actuators A: Physical, Vol. 190, pp. 66–76.

Chmaj, G.; and Selvaraj, H. (2015), Distributed Processing Applications for UAV/Drones: A Survey, Progress in Systems Engineering. Springer, pp. 449–454.

Cohn, Pamela; Green, Alastair; Langstaff, Meredith; and Roller, Melanie (2017), Commercial Drones Are Here: The Future of Unmanned Aerial Systems, December, McKinsey and Company, *https://www.mckinsey.com/industries/capital-projects-and-infrastructure/our-insights/commercial-drones-are-here-the-future-of-unmanned-aerial-systems*.

Colomina, I.; and Molina, P. (2014). Unmanned Aerial Systems for Photogrammetry and Remote Sensing: A Review, ISPRS Journal of Photogrammetry and Remote Sensing 92 (June), pp. 79–97, DOI: 10.1016/j.isprsjprs.2014.02.013.

Corrigan, F. (2016), Introduction to UAV Photogrammetry and LiDAR Mapping Basics, *http://www.dronezon.com/learn-about-drones-quadcopters/introduction-to-uav-photogrammetry-and-lidar-mapping-basics/*.

Corrigan, F. (2019), 12 Top LiDAR Sensors For UAVs, LiDAR Drones And So Many Great Uses, December 31, *https://www.dronezon.com/learn-about-drones-quadcopters/best-lidar-sensors-for-drones-great-uses-for-lidar-sensors/*.

Corrigan, Fintan (2020), How Do Drones Work and What Is Drone Technology, February, DroneZon, *https://www.dronezon.com/learn-about-drones-quadcopters/what-is-drone-technology-or-how-does-drone-technology-work/*.

Cramer, M.; and Stallmann, D. (2002), System Calibration for Direct Georeferencing. Int. Arch. Photogramm. Remote Sens. Spatial Inf. Sci., XXXIV-3/A, 79–84.

Crommelinck, Sophie; Bennett, Rohan; Gerke, Markus; Yang, Michael Ying; and Vosselman, George (2017), Contour Detection for UAV-Based Cadastral Mapping. Remote Sensing, 9, 171, 10.3390/rs9020171.

Cruz, Henry; Eckert, Martina; Meneses, Juan; and Martinez, José-Fernán (2016), Efficient Forest Fire Detection Index for Application in Unmanned Aerial Systems (UASs), Sensors (Basel), June, Vol. 16, No 6, p. 893, DOI: 10.3390/s16060893.

Culus, Joline; Schellekens, Yves; Agoria; and Smeets, Yannick (2018), A Drone's Eye View, May, PwC, AGORIA, Belgium.

D'Andrea, R. (2014), Can Drones Deliver?, IEEE Transactions on Automation Science and Engineering, Vol. 11, No. 3, pp. 647–648.

Dalamagkidis, K. (2014), Classification of UAVs. In Handbook of Unmanned Aerial Vehicles; Springer: Houten, The Netherlands; pp. 83–91.

Dalamagkidis, Konstantinos; Valavanis, K.; and Piegl, Les A. (2012), On Integrating Unmanned Aircraft Systems into the National Airspace System – Issues, Challenges, Operational Restrictions, Certification, and Recommendations, Springer.

Dawson, Philip (2018), Developing a Global Framework for Unmanned Aviation, United Aviation, July 29. *https://www.unitingaviation.com/ strategic-objective/safety/developing-a-global-framework-for-unmanned- aviation/Delair.aero*.

Dellaert, F.; Seitz, S.M.; Thorpe, C.E.; and Thrun, S. (2000), Structure from Motion without Correspondence. Proceedings of the IEEE International Conference on Computer Vision and Pattern Recognition (CVPR), Hilton Head, SC, USA, 13–15 June, pp. 2557–2564.

De Mattos, Bento Silva; Secco, Ney Rafael; and Salles, Eduardo Francisco (2013), Optimal Design of a High-Altitude Solar-Powered Unmanned Airplane, J. Aerosp. Technol. Manag., Vol. 5, No. 3, pp. 349–361, Jul.–Sep., DOI: 10.5028/jatm.v5i3.223.

Demir, Kadir Alpaslan Demir; Cicibas, Halil; and Arica, Nafiz (2015), Unmanned Aerial Vehicle Domain: Areas of Research, Defence Science Journal, Vol. 65, No. 4, July 2015, pp. 319–329, DOI: 10.14429/dsj.65.8631.

Dempsey, Caitlin (2015), Drones and GIS: A Look at the Legal and Ethical Issues, September 10, *https://www.gislounge.com/ drones-and-gis-a-look-at-the-legal-and-ethical-issues/*.

Dering, Gregory M.; Micklethwaite, Steven; Thiele, Samuel T.; Vollgger, Stefan A.; and Cruden, Alexander R. (2019), Review of Drones, Photogrammetry and Emerging Sensor Technology for the Study of Dykes: Best Practices and Future Potential, Journal of Volcanology and Geothermal Research, Vol. 373, PP. 148–166, *https://doi.org/10.1016/j. jvolgeores.2019.01.018*.

Desale, Rahul; Chougule, Ashwin; Choudhary, Mahesh; Borhade, Vikrant; and Teli, S.N. (2019), Unmanned Aerial Vehicle For Pesticides Spraying, IJSART, Vol. 5, No. 4. April.

DGCA (2018). Requirements for Operation of Civil Remotely Piloted Aircraft System (RPAS), Director General of Civil Aviation, *http://www.dgca.nic.in/misc/draft%20cars/CAR%20-%20UAS%20 (Draft_Nov2017).pdf*.

DHL (2014), Unmanned Aerial Vehicle- A DHL Perspective on Implications and Use Cases for the Logistics Industry.

DHL (2017), DHL Drone Logistics. *http://www.dhl.com/content/dam/ downloads/g0/about_us/logistics_insights/DHL_TrendReport_UAV.pdf*.

Dimc, Franc; and Magister, Tone (2006), Mini UAV Communication Link Systems, Conference paper Dec., *https://www.researchgate.net/ publication/244477850*.

Ding, G.; Wu, Q.; Zhang, L.; Lin, Y.; Tsiftsis, T.A.; and Yao, Y.D. (2018), Anamateur Drone Surveillance System Based on the Cognitive Internet of Things, IEEE Communications Magazine, Vol. 56, No. 1, pp. 29–35, Jan.

D'mello, Anasia (2019), Innovative Ways to Use IoT-enabled Drones in the Near Future, August 2, IoTNOW, *https://www.iot-now.com/2019/08/02/97962-innovative-ways-use-iot-enabled-drones-near-future/*.

DoD (2007), Unmanned Systems Roadmap: 2007–2032, Department of Defense, Createspace Independent Pub. *https://www.globalsecurity.org/intell/library/reports/2007/dod-unmanned-systems-roadmap_2007-2032.pdf*.

Dronethusiast (2020), The History of Drones (Drone History Timeline from 1849 to 2019), *https://www.dronethusiast.com/history-of-drones/*.

DroneApps (2019), Price Wars: Counting the Cost of Drones, Planes and Satellites, *https://droneapps.co/price-wars-the-cost-of-drones-planes-and-satellites/*.

DRONEITECH (2018), How Drones are Transforming the Media Industry *https://www.dronitech.com/how-drones-are-transforming-the-media-industry/*.

D'Sa, R.; Henderson, T.; Jenson, D.; Calvert, M.; Heller, T.; Schulz, B.; Kilian, J.; and Papanikolopoulos, N. (2017), Design and Experiments for a Transformable Solar-UAV, Proceedings – IEEE International Conference on Robotics and Automation, pp. 3917–3923.

Duan, Bo; Fang, Shenghui; Zhu, Renshan; Wu, Xianting; Wang, Shanqin; Gong, Yan; and Peng, Yi (2019), Remote Estimation of Rice Yield With Unmanned Aerial Vehicle (UAV) Data and Spectral Mixture Analysis, Front. Plant Sci., 27 February, *https://doi.org/10.3389/fpls.2019.00204*.

Dunbar, S.; Wenzl, F.; Hack, G.; Hafeza, R.; Esfeer, H.; Defay, F.; Prothin, S.; Bajon, D.; and Popovic, Z. (2015), Wireless Far-Field Charging of a Micro-UAV, pp. 7–10.

Dupont, Quentin F.M.; Chua, David K.H.; Tashrif, Ahmad; and Abbott, Ernest L.S. (2016), Potential Applications of UAV Along the Construction's Value Chain, Science Direct, 7th International Conference on Engineering, Project, and Production Management, DOI: 10.1016/j.proeng.2017.03.155.

Duro, D.C.; Franklin, S.E.; and Dubé, M.G. (2012), A Comparison of Pixel-Based and Object-Based Image Analysis With Selected Machine Learning Algorithms for the Classification of Agricultural Landscapes Using Spot-5 HRG imagery. Remote Sens. Environ, 118, 259–272.

EASA (2017), *https://www.easa.europa.eu/*.

Eaton, Christopher M.; Chong, Edwin K.P.; and Maciejewski, Anthony A. (2016), Multiple-Scenario Unmanned Aerial System Control: A Systems Engineering Approach and Review of Existing Control Methods, Aerospace, MDPI, Vol. 3, DOI: 10.3390/aerospace3010001, www.mdpi.com/journal/aerospace.

Economic Times (2018), India Fastest Growing Market for Unmanned Aerial Vehicles. Retrieved from *https://economictimes.indiatimes.com/news/defence/india-fastest-growing-market-for-unmanned-aerial-vehicles/articleshow/63466658.cms (26 March)*.

Edwin, E. Bijolin; Thanka, M. Roshni; and Deula, Shiny (2019), An Internet of Drone (IOD) Based Data Analytics in Cloud for Emergency Services, International Journal of Recent Technology and Engineering (IJRTE), ISSN: 2277-3878, Vol. 7, No. 5S2.

Encyclopedia Britannica (1999), *https://www.britannica.com/technology/*.

European Commission (2014), A New Era for Aviation: Opening the Aviation Market to the Civil Use of Remotely Piloted Aircraft Systems in a Safe and Sustainable Manner, Report COM, 207.

Eriksson, Nikala (2018), Conceptual Study of a Future Drone Detection System, Countering a Threat Posed by a Disruptive Technology, Master's Thesis, Department of Industrial and Materials Science Chalmers, University of Technology, Gothenburg, Sweden.

Eschmann, C.; and Wundsam, T. (2017), Web-based Georeferenced 3D Inspection and Monitoring of Bridges with Unmanned Aircraft Systems, J. Surv. Eng., 143 (3): 04017003, *https://doi.org/10.1061/(ASCE)SU.1943-5428.0000221*.

FAA. (2010), A Brief History of the FAA, 03 Ocak 2015'de, *http://www.faa.gov/about/history/brief_history*.

FAA (2016), Federal Aviation Administration, *https://www.faa.gov/*.

Fahlstrom, Paul Gerin; and Gleason, Thomas James (2012), Introduction to UAV Systems, Fourth Edition, John Wiley & Sons, Ltd.

Faied, M.; Mostafa, A.; and Girard, A. (2010), Vehicle Routing Problem Instances: Application to Multi-UAV Mission Planning, Proceedings of AIAA Guidance, Navigation and Control Conference, August.

Fan, Jin; and Saadeghvaziri, M. Ala (2019), Applications of Drones in Infrastructures: Challenges and Opportunities, Conference Paper, October, DOI: 10.5281/zenodo.3566281.

Faraz, Aiman (2020), Data on wings – A close look at drones in India, PwC India, *https://www.pwc.in/consulting/financial-services/fintech/fintech-insights/data-on-wings-a-close-look-at-drones-in-india.html*.

Federman, A.; Quintero, M. Santana Quintero; Kretz, S.; Gregg, J.; Lengies, M.; Ouimet, C.; and Laliberte, J. (2017), UAV Photgrammetric Workflows: A Best Practice Guideline, The International Archives of the Photogrammetry, Remote Sensing and Spatial Information Sciences, Volume XLII-2/W5, 2017, 26th International CIPA Symposium 2017, 28 August–01 September, Ottawa, Canada.

Feist, Jonathan (2020), Best RTH Drones – The Best GPS Function in the Air, Drone Rush, *https://dronerush.com/rth-drones-return-to-home-10625/*.

FICCI Report (2018), Make in India Unmanned Aircraft Systems, 2018, Ernst & Young LLP, India.

Finn, Rachel; and David Wright (2012), Unmanned Aircraft Systems: Surveillance, Ethics and Privacy in Civil Applications. Computer Law & Security Review, Vol. 28, No. 2, pp. 184–194.

Floreano, Dario; and Wood, Robert J. (2015), Science, Technology and the Future of Small Autonomous Drones, Nature, Vol. 521, pp. 460–466.

Fonstad, M.A.; Dietrich, J.T.; Courville, B.C.; Jensen, J.L.R.; and Carbonneau, P.E. (2013), Topographic Structure from Motion: A New Development in Photogrammetric Measurement. Earth Surf. Process. Landforms, Vol. 38, pp. 421–430.

Freudenrichn Craig (2020), How Blimps Work, *https://science.howstuffworks.com/transport/flight/modern/blimp4.htm*.

Fu, Kui; Li, Jia; Shen, Hongze; and Tian, Yonghong (2018), How Drones Look: Crowdsourced Knowledge Transfer for Aerial Video Saliency Prediction, Computer Vision and Pattern Recognition, DOI: arXiv:1811.05625.

Furukawa, Yasutaka; and Hernández, Carlos (2015), Multi-View Stereo: A Tutorial, NOW, Boston, *http://dx.doi.org/10.1561/0600000052*.

Gajdamowicz, K.; Ohman, D.; and Horemuz, M. (2007), Mapping and 3D Modelling of Urban Environment Based on LiDAR, Gps/Imu and Image Data, Proceeding of 5th International Symposium on Mobile Mapping Technology, Padova, Italy, 28–31 May.

Garcia-Aunon, Pablo; Peñas, Matilde Santo; and García, Jesus Manuel de la Cruz (2017), Parameter Selection Based on Fuzzy Logic to Improve UAV Path-following Algorithms, Journal of Applied Logic, Vol. 24, pp. 62–75. *http://dx.doi.org/10.1016/j.jal.2016.11.025*.

Garg, P.K. (2019), Theory and Principles of Geoinformatics, Khanna Book Publishing House., Delhi.

Garg, Pradeep Kumar; Garg, Rahul Dev; Shukla, Gaurav; and Srivastava, Hari Shanker (2020), Digital Mapping of Soil Landscape Parameters, Geospatial Analyses using Machine Learning and Geomatics, Springer.

Gartner (2018), Hype Cycle for Emerging Technologies, Gartner Identifies Five Emerging Technology Trends That Will Blur the Lines Between Human and Machine, *https://www.gartner.com/en/newsroom/press-releases/2018-08-20-gartner-identifies-five-emerging-technology-trends-that-will-blur-the-lines-between-human-and-machine*.

Gaulkin, Thomas (2020), Drone Pandemic: Will Coronavirus Invite the World to Meet Big Brother?, Bulletin of the American Scientists, April 1, *https://thebulletin.org/2020/04/drone-pandemic-will-coronavirus-invite-the-world-to-meet-big-brother/*.

Ghosh, Sujata; Prsada Raju, PVSP; Saibaba, J.; and Vardhan Geeta (2012), CyberGIS and Crowdsourcing – A New Approach in E-Governance, Geospatial World – 05/14/2012.

Gibril, M.B.A.; Idrees, M.O.; Yao, K.; and Shafri, H.Z.M. (2018), Integrative Image Segmentation Optimization and Machine Learning Approach for High Quality Land-use and Land-cover Mapping Using Multisource Remote Sensing Data, J. Appl. Remote Sens., 12, 016036.

Gillins, M.N., Gillins, D.T.; and Parrish, C. (2016), Cost-Effective Bridge Safety Inspections Using Unmanned Aircraft Systems (UAS), In Proc., Geotechnical and Structural Engineering Congress, 1931–1940. Reston, VA: ASCE.

Giyenko, Andrey; and Cho, Young Im (2016), Intelligent UAV in Smart Cities using IoT, In 16th International Conference on Control, Automation and Systems (ICCAS 2016), October 16–19, HICO, Gyeongju, Korea.

Glover, John M. (2014), Drone University, DroneUniversity, Edition.

Goldman Sachs (2016), UAV Drones: Reporting for Work – Goldman Sachs. www.goldmansachs.com › insights › technology-driving-innovation.

González-Espasandín, Óscar; Leo, Teresa J.; and Navarro-Arévalo, Emilio (2015), Fuel Cells: A Real Option for Unmanned Aerial Vehicles Propulsion, The Scientific World Journal, Vol. 2015, Article ID 419786; *https://doi.org/10.1155/2015/419786*.

González-Jorge, Higinio; Martínez-Sánchez, Joaquin; Bueno, Martín; and Arias Pedor Arias (2017), Unmanned Aerial Systems for Civil Applications:

A Review, Drones, MPDI, Vol. 1, No. 2, DOI: 10.3390/drones1010002; www.mdpi.com/journal/drones.

Grand View Research (2018), UAV Market Research Report, Gran View Research, June, *https://www.grandviewresearch.com/industry-analysis/uav-payload-market*.

Greenwood, William W.; Lynch, Jerome P.; and Zekkos, Dimitrios (2019), Applications of UAVs in Civil Infrastructure, J. Infrastruct. Syst., 25(2): ASCE, 04019002, DOI: 10.1061/(ASCE)IS.1943–555X.0000464.

Grogan, Sean; Gamache, Michel; and Pellerin, Robert (2018), The Use of Unmanned Aerial Vehicles and Drones in Search and Rescue Operations – A Survey, Researchgate, Conference Paper, September, *https://www.researchgate.net/publication/327755534*.

GSMA (2018), Mobile-Enabled Unmanned Aircraft, GSMA White Paper, Available at: *https://www.gsma.com/iot/wp-content/uploads/2018/02/Mobile-Enabled-Unmanned-Aircraft-web.pdf*.

Gu, Y.; Saad, W.; Bennis, M.; Debbah, M.; and Han, Z. (2015), Matching Theory for Future Wireless Networks: Fundamentals and Applications, IEEE Communications Magazine, Vol. 53, No. 5, pp. 52–59.

Guillen-Perez, Antonio; Sanchez-Iborra, Ramon; Cano, Maria-Dolores; Sanchez-Aarnoutse, Juan Carlos; and Garcia-Haro, Joan (2016), WiFi Networks on Drones, Conference Paper, November, DOI: 10.1109/ITU-WT.2016.7805730.

Gupta, L; Jain, R.; and Vaszkun, G. (2016), Survey of Important Issues in UAV Communication Networks, IEEE Communications Surveys & Tutorials, Vol. 18, No. 2, pp. 1123–1152.

Gupta, Medha; Bohra, Devender Singh; Raghavan, Raamesh Gowri; and Khurana, Sukant (2018), A beginners' guide to understanding Drones, *https://medium.com/@sukantkhurana/a-beginners-guide-to-drones-38d215701c4e*.

Guvenc, I.; Koohifar, F.; Singh, S.; Sichitiu, M.L.; and Matolak, D. (2018), Detection, Tracking, and Interdiction for Amateur Drones, IEEE Commun. Mag., Vol. 56, No. 4, pp. 75–81, Apr.

Hallermann, N.; Morgenthal, G.; and Rodehorst, V. (2015), Unmanned Aerial Systems (UAS) – Case Studies of Vision Based Monitoring of Ageing Structures. In Proceedings of the International Symposium Non-Destructive Testing In Civil Engineering, Berlin, Germany, 15–17 September.

Ham, Y.; Han, K.K.; Lin, J.J.; and Golparvar-Fard, M. (2016), Visual Monitoring of Civil Infrastructure Systems Via Camera-Equipped Unmanned Aerial Vehicles (UAVs): A Review of Related Works, Visualization in Engineering, Vol. 4, No. 1, p. 1.

Hargrave, Marshall (2019), Crowdsourcing, Business, July 8, *https://www.investopedia.com/terms/c/crowdsourcing.asp*.

Harwin, S.J. (2015), Multi-view stereopsis (MVS) from an Unmanned Aerial Vehicle (UAV) for Natural Landform Mapping, PhD Thesis, University of Tasmania.

Hassanalian, M.; and Abdelkefi, A. (2017), Classifications, Applications, and Design Challenges of Drones: A Review, Progress in Aerospace Sciences, Vol. 91, May 2017, pp. 99–131.

Heutger, Matthias; and Kückelhaus, Markus (2014), Unmanned Aerial Vehicle in Logistics – A DHL Perspective on Implications and Use Cases for the Logistics Industry, Published by DHL Customer Solutions & Innovation, Germany.

Howell, C.; Jones, F.; Thorson, T.; Grube, R.; Joyce, L.; and Coggin, J. (2016), The First Government Sanctioned Delivery of Medical Supplies by Remotely Controlled Unmanned Aerial System (UAS), New Orleans, LA: United States: NASA Technical Reports Server.

Huang, H.-M.; Pavek, K.; Ragon, M.; Jones, J.; Messina, E.; and Albus, J. (2007), Characterizing Unmanned System Autonomy: Contextual Autonomous Capability and Level of Autonomy Analyses. In Unmanned Systems Technology IX; International Society for Optics and Photonics: San Diego, CA, USA, p. 65611N.

Huang, Y.; Wang, H.; and Yao, P. (2016), Energy-optimal Path Planning for Solar-powered UAV with Tracking Moving Ground Target, Aerospace Science and Technology, Vol. 53, pp. 241–251, *http://dx.doi.org/10.1016/j.ast.2016.03.024*.

Huang, Y.; Thomson, S.J.; Hoffmann, W.C.; Lan, Y.; and Fritz, B.K. (2013), Development and Prospect of Unmanned Aerial Vehicle Technologies for Agricultural Production Management, International Journal of Agricultural and Biological Engineering, Vol. 6, No. 3, pp. 1–10.

Hugenholtz, Chris; Brown, Owen; Walker, Jordan; Barchyn, Thomas; Nesbit, Paul; Kucharczyk, Maja; and Myshak, Steve (2016). Spatial Accuracy of UAV-Derived Orthoimagery and Topography: Comparing Photogrammetric Models Processed with Direct Geo-Referencing and Ground Control Points. Geomatica., Vol. 70, pp. 21–30. 10.5623/cig2016-102.

Hui, N.B.; Mahendar, V.; and Pratihar, D.K. (2006), Time-Optimal, Collision-Free Navigation of a Car-Like Mobile Robot Using Neuro-fuzzy Approaches, Fuzzy Sets and Systems, Vol. 157, No. 16, pp. 2171–2204.

ICAO (2011), Unmanned Aircraft Systems (UAS), Circular 328 AN/190 (2011), *https://www.trafikstyrelsen.dk/~/media/Dokumenter/05%20Luftfart/Forum/UAS%20-%20droner/ICAO%20 Circular%20328%20Unmanned%20Aircraft %20Systems%20UAS.ashx, archived at https://perma.cc/J5EM-TSAY*.

ICAO (2015), International Civil Aviation Organization (ICAO). Manual on Remotely Piloted Aircraft Systems (RPAS); ICAO: Montréal, QC, Canada.

Iglhaut, Jakob; Cabo, Carlos; Pulit, Stefano; Piermattei, Livia; O'Connor, James; and Rosette, Jacqueline (2019), Structure from Motion Photogrammetry in Forestry: A Review, Current Forestry Reports, Vol. 5, pp. 155–168, *https://doi.org/10.1007/s40725-019-00094-3*.

Iizuka, Kotaro; Itoh, Masayuki; Shiodera, Satomi; Matsubara, Takashi; Dohar, Mark; and Watanabe, Kazuo (2018), Advantages of Unmanned Aerial Vehicle (UAV) Photogrammetry for Landscape Analysis Compared with Satellite Data: A Case Study of Postmining Sites in Indonesia, Cogent Geoscience, Vol. 4, No. 1, DOI: 10.1080/23312041.2018.1498180.

Intelligence, S.D. (2018), The Global UAV Payload Market 2017–2027, *https://www.researchandmarkets.com/research/nfpsbm/the global uav*.

ITU-R M.2171 (2009), Characteristics of Unmanned Aircraft Systems and Spectrum Requirements to Support their Safe Operation in non-segregated Airspace, Report ITU-R M.2171 (12/2009).

IWM (2018), A Brief History of Drones, *https://www.iwm.org.uk/history/a-brief-history-of-drones*.

Jaradat, M.A.K.; Garibeh, M.H.; and Feilat, E.A. (2009), Dynamic motion Planning for Autonomous Mobile Robot Using Fuzzy Potential Field, In Proceedings of the 6th International Symposium on Mechatronics and its Applications (ISMA '09), pp. 1–6, March.

Javaid, A.Y.; Sun, W.; Devabhaktuni, V.K.; and Alam, M. (2012), Cyber Security Threat Analysis and Modeling of an Unmanned Aerial Vehicle System, IEEE Conference on Technologies for Homeland Security (HST), IEEE, 2012, pp. 585–590.

Jawhar, I., Mohamed, N.; Al-Jaroodi, J.; and Zhang, S. (2014), A Framework for Using Unmanned Aerial Vehicles for Data collection in Linear Wireless Sensor Networks. J. Intell. Robot. Sys., Vol. 74, Nos. 1–2, pp. 437–453, *https://doi.org/10.1007/s10846-013-9965-9*.

Jawhar, Imad; Mohamed, Nader; Al-Jaroodi, Jameela; Agarwal, Dharma P.; and Zhange, Sheng (2017), Communication and Networking of UAV-based Systems: Classification and Associated Architectures, Journal of Network and Computer Applications, February, DOI: 10.1016/j.jnca.2017.02.008.

Jeanneret, C.; and Rambaldi, G. (Eds.) (2016). Drone Governance: A Scan of Policies, Laws and Regulations Governing the Use of Unmanned Aerial Vehicles (UAV) in 79 Countries. CTA Working Paper.

Jiang, S.; and Jiang, W. (2017), On-board GNSS/IMU Assisted Feature Extraction and Matching for Oblique UAV Images. Remote Sensing, Vol. 9, p. 813.

Jiang, C.; Zhang, H.; Ren, Y.; Han, Z.; Chen, K.C.; and Hanzo, L. (2017), Machine Learning Paradigms for Next-generation Wireless Networks, IEEE Wireless Communications, Vol. 24, No. 2, pp. 98–105.

Jha, A.R. (2016), Theory, Design, and Applications of Unmanned Aerial Vehicles. New York: CRC Press.

Johnson, Mitch (2015), Components for Creating an Unmanned Aerial Vehicle – Application Note, March, *https://www.egr.msu.edu/classes/ece480/capstone/spring15/group14/uploads/4/2/0/3/42036453/johnsonappnote.pdf*.

Jones, D. (2015), Power Line Inspection- A UAV Concept, In The IEE Forum on Autonomous Systems. (Ref. No. 2005/11271). IET, p. 8.

Jones, K. (2019), The Drones are Coming! The Drones are Coming! (to the Construction Site). *http://constructiondatacompany.com/drones-coming-drones-coming-construction-site/*.

Jones, R.M.; Tambe, M.; Laird, J.E.; and Rosenbloom (1993), Intelligent Automated Agents for Flight Training Simulators, In Proceedings of the Third Conference on Computer Generated Forces and Behavioral Representation, P.S., Orlando, FL.

Jordan, Sophie; Moore, Julian; Hovet, Sierra; Box, John; Perry, Jason; Kirsche, Kevin; Lewis, Dexter; and Tse, Zion Tsz Ho (2018), State-of-the-Art Technologies for UAV Inspections, Vol. 12, No 2, February 2018, pp. 151 – 164, DOI: 10.1049/iet-rsn.2017.0251.

Joshi, D. (2017), Commercial Unmanned Aerial Vehicle (UAV) Market Analysis Industry Trends, Companies and What You Should Know, http://www.businessinsider.com/ commercial-uav-market-analysis.

Juan, Víctor San; Santos, Matilde; and Andujar, José Manuel (2018), Intelligent UAV Map Generation and Discrete Path Planning for Search and Rescue Operations, Complexity Volume 2018, Article ID 6879419, https://doi.org/10.1155/2018/6879419.

Junaid, A.B.; Konoiko, A.; Zweiri, Y.; Sahinkaya, M.N.; and Seneviratne, L. (2017), Autonomous Wireless Self-charging for Multi-rotor Unmanned Aerial Vehicles, Energies, Vol. 10, No. 6, pp. 1–14.

Junaid, A.B.; Lee, Y.; and Kim, Y. (2016), Design and Implementation of Autonomous Wireless Charging Station for Rotary-Wing UAVs, Aerospace Science and Technology, Vol. 54, pp. 253–266. http://dx.doi.org/10.1016/j.ast.2016.04.023.

Kaag, John; and Kreps, Sarah (2014), Drone Warfare, Polity Press, USA.

Kaess, M.; Johannsson, H.; Roberts, R.; Ila, V.; Leonard, J.J.; and Dellaert, F. (2012), iSAM2: Incremental Smoothing and Mapping Using the Bayes tree. Int. J. Robot. Res., Vol. 2012, No. 31, pp. 216–235, DOI: 10.1177/0278364911430419.

Kale, S.D.; Khandagale, S.V.; Gaikwad, S.S.; Narve, S.S.; and Gangal, P.V. (2015), International Journal of Advanced Research in Computer Science and Software Engineering Agriculture Drone for Spraying Fertilizer and Pesticides. Int. J. Adv. Res. Comput. Sci. Softw. Eng., Vol. 5, pp. 804–807.

Kallenborn, Zachary (2018), The Era of the Drone Swarm Is Coming, and We Need to Be Ready for It, October, Modernwar Institute, https://mwi.usma.edu/era-drone-swarm-coming-need-ready/.

Kallenborn, Zachary; and Bleek, Philipp C. (2019), Drones of Mass Destruction: Drone Swarms and the Future of Nuclear, Chemical, and Biological Weapons, War on the Rocks, https://warontherocks.com/2019/02/drones-of-mass-destruction-drone-swarms-and-the-future-of-nuclear-chemical-and-biological-weapons/ February.

Kaminsky, Ryan S.; Snavely, Noah; Seitz, Steven M.; and Szeliski, Rick (2009), Alignment of 3D Point Clouds to Overhead Images, Second IEEE Workshop on Internet Vision, June.

Kan, M.; Okamoto, S.; and Lee, J.H. (2018), Development of Drone Capable of Autonomous Fight Using GPS. In: Proceedings of the International Multi Conference of Engineers and Computer Scientists, Vol. 2.

Katukam, Ravi (2015), Industrial Applications of Drones: An Insight, International Journal of Engineering Sciences & Management, April–June.

Kaushal, H.; and Kaddoum, G. (2017), Optical Communication in Space: Challenges and Mitigation Techniques, IEEE Communications Surveys & Tutorials, Vol. 19, No. 1, pp. 57–96.

Keane, John; and Carr, Stephen (2013). A Brief History of Early Unmanned Aircraft. Johns Hopkins Apl Technical Digest, Vol. 32, pp. 558–571.

Kelly, T. (2018), The Booming Demand for Commercial Drone Pilots, *https://www.theatlantic.com/technology/archive/2017/01/drone-pilot-school/515022/*.

Kendoul, F. (2013), Towards a Unified Framework for UAS Autonomy and Technology Readiness Assessment (atra). In Autonomous Control Systems and Vehicles: Intelligent Unmanned Systems; Nonami, K., Kartidjo, M., Yoon, K.-J., Budiyono, A., Eds.; Springer: Tokyo, Japan; pp. 55–71.

Kenneth Research (2019), Global Drones Market Innovative Growth With SWOT Analysis and Forecast from 2018–2028, America News Hour. August 2, *https://www.marketwatch.com/press-release/global-drones-market-innovative-growth-with-swot-analysis-and-forecast-from-2018-2028-2019-08-02*.

Khanal, S.; Fulton, J.; and Shearer, S. (2017), An Overview of Current and Potential Applications of Thermal Remote Sensing in Precision Agriculture, Computers and Electronics in Agriculture, Vol. 139, pp. 22–32.

Kim, S.H.; and Bhattacharya, R. (2007), Multi-layer Approach for Motion Planning in Obstacle Rich Environments, In Proceedings of the AIAA Guidance, Navigation and Control Conference and Exhibit (GNC '07), pp. 2656–2666, South Carolina, SC, USA.

Kimchi, Gur (2015), Amazon's Top Prime Air Executive Outlines Plans for Delivery Drones to Navigate Skies, *https://www.supplychain247.com/article/amazon_outlines_plans_for_delivery_drones_to_navigate_skies/pinc_solutions*.

Klochkov, V.; and Nikitova, A. (2008), A Simplified Approach to Economic Efficiency Analysis of UAV Pipeline Surveillance, Transp. Eng., Vol. 27, pp. 172–180.

Koci, Jack; Jarihani, Ben; Leon, Javier X.; Sidle, Roy C.; Wilkinson, Scott N.; and Bartley, Rebecca (2017), Assessment of UAV and Ground-Based Structure from Motion with Multi-View Stereo Photogrammetry in a Gullied Savanna Catchment, ISPRS Int. J. Geo-Inf., Vol. 6, p. 328, DOI: 10.3390/ijgi6110328.

Kosmer, J.; and Vilhar, A. (2014), Base Stations Placement Optimization in Wireless Networks for Emergency Communications, In Proc. of IEEE International Conference on Communications (ICC), Sydney, Australia, June.

Krishnan, Armin (2009), Killer Robots, Ashgate, Burlington, VT, p. 45.

Lakshmi, Esakkipandian Subbu; and Yarrakula, Kiran (2018), Review and Critical Analysis on Digital Elevation Models, GEOFIZIKA, Vol. 35, No. 2, pp. 129–157.

Laksham, K.B. (2019), Unmanned Aerial Vehicle (drones) in Public Health: A SWOT Analysis, J Family Med Prim Care, Vol. 8, pp. 342–346.

LeCun, Y.; Bengio, Y.; and Hinton, G. (2015), Deep Learning, Nature, Vol. 521, No. 7553, p. 436.

Lee, B.; Kwon, S.; Park, P.; and Kim, K. (2014), Active Power Management System for an Unmanned Aerial Vehicle Powered by Solar Cells, a Fuel Cell, and Batteries, IEEE Transactions on Aerospace and Electronic Systems, Vol. 50, No. 4, pp. 3167–3177.

Lee, Sungjae; and Choi, Yosoon (2016), Reviews of Unmanned Aerial Vehicle (drone) Technology Trends and Its Applications in the Mining Industry, Journal Geosystem Engineering, Vol. 19, No. 4, pp. 197–204, *https://doi.org/10.1080/12269328.2016.1162115*.

Lee, S.; and Choi, Y. (2015), Topographic Survey at Small-scale Open-Pit Mines Using a Popular Rotary-wing Unmanned Aerial Vehicle (Drone), Tunnel & Underground Space, Vol. 25, pp. 462–469.

Leitgeb, E.; Zettl, K.; Muhammad, S.S.; Schmitt, N.; and Rehm, W. (2007), Investigation in Free Space Optical Communication Links Between Unmanned Aerial Vehicles (UAVs), In 9th International Conference on Transparent Optical Networks, Rome, pp. 152–155.

Lien, S.Y.; Chen, K.C.; and Lin, Y. (2011), Toward Ubiquitous Massive Accesses in 3GPP Machine-to-machine Communications, IEEE Commun. Mag., Vol. 49, No. 4, pp. 66–74, April.

Lin, Y.; Hyyppa, J.; and Jaakkola, A. (2011), Mini-UAV-borne LiDAR for Fine-Scale Mapping. IEEE Geosci. Remote Sensing, Vol. 8, pp. 426–430.

Liu, Dan; Geng, Yishuang; Liu, Guanxiong; Zhou, Mingda; and Pahlavan, Kaveh Pahlavan (2015), An Energy Efficient Relay Algorithm for Wireless Capsule Endoscopy, In Vehicular Technology Conference (VTC Fall), 82nd IEEE, pp. 1–5.

Liu, Guanxiong; Shakhatreh, Hazim; Khreishah, Abdallah; Guo, Xiwang; and Ansari, Nirwan (2018), Efficient Deployment of UAVs for Maximum Wireless Coverage Using Genetic Algorithm, IEEE 39th Sarnoff Symposium.

Liu, H.; Wei, Z.; Chen, Y.; Pan, J.; Lin, L.; and Ren, Y. (2017), Drone Detection Based on an Audio-assisted Camera Array, In Proc. IEEE 3rd Int. Conf. Multimedia Big Data (BigMM), April, pp. 402–406.

Liu, M.; Huang, S.; Dissanayake, G.; and Wang, H. (2012), A Convex Optimization Based Approach for Pose SLAM Problems, In Proceedings of the IEEE/RSJ International Conference on Intelligent Robots and Systems; Vilamoura-Algarve, Portugal, 7–12 October; pp. 1898–1903.

Liu, P.; Chen, A.Y.; Huang, Y.N.; Han, J.Y.; Lai, J.S.; Kang, S.C.; Wu, T.; Wen, M.; and Tsai, M. (2014), A Review of Rotorcraft Unmanned Aerial Vehicle (UAV) Developments and Applications in Civil Engineering, Smart Struct. Syst., Vol. 13, No. 6, pp. 1065–1094.

LOC (2019), Regulation of Drones: Comparative Analysis, *https://www.loc.gov/law/help/regulation-of-drones/comparative.php*.

Lort, Marc; Aguasca, Albert; Lopez-Mart1nez, Carlos; and Marin, Tomas Mart1nez (2017), Initial Evaluation of SAR Capabilities in UAV Multicopter Platforms, IEEE Journal of Selected Topics in Applied Earth Observations and Remote Sensing, IEEE, 1939–1404.

Luo, C.; McClean, S.I.; Parr, G.; Teacy, L.; and De Nardi, R. (2013), UAV Position Estimation and Collision Avoidance Using the Extended Kalman Filter, IEEE Transactions on Vehicular Technology, Vol. 62, No. 6, pp. 2749–2762.

Lucintel (2011), Growth Opportunity in Global UAV Market, Lucintel – Creating the Equation for Growth, January, *https://www.lucintel.com/LucintelBrief/UAVMarketOpportunity.pdf*.

Ma, L.; Fu, T.; Blaschke, T.; Li, M.; Tiede, D.; Zhou, Z.; and Chen, D. (2017), Evaluation of Feature Selection Methods for Object-Based Land Cover Mapping of Unmanned Aerial Vehicle Imagery Using Random Forest and Support Vector Machine Classifiers, ISPRS Int. J. Geo Inf., Vol. 6, p. 51.

Mahajan, U.; and Bundel, B.R. (2016), Drones for Normalized Difference Vegetation Index (NDVI) to Estimate Crop Health for Precision Agriculture: A Cheaper Alternative for Spatial Satellite Sensors. In Proceedings of the International Conference on Innovative Research in Agriculture, Food Science, Forestry, Horticulture, Aquaculture, Animal Sciences, Biodiversity, Ecological Sciences and Climate Change, New Delhi, India, 22 October; pp. 38–41.

Mahnken, T.G. (1995), War in the Information Age, Joint Force Quarterly, Vol. 10, pp. 39–43.

Manousakis, J.; Zekkos, D.; Saroglou, F.; and Clark, M. (2016), Comparison of UAV-Enabled Photogrammetry-Based 3D Point Clouds and Interpolated DSMs of Sloping Terrain for Rockfall Hazard Analysis, Int. Arch. Photogramm. Remote Sens. Spatial Inf. Sci., XLII-2/W2, 71–77, https://doi.org/10.5194/isprs-archives-XLII-2-W2-71-2016.

Mansfield, K.; Eveleigh, T.; Holzer, T.H.; and Sarkani, Shahryar (2013), Unmanned Aerial Vehicle Smart Device Ground Control Station Cyber Security Threat Model. In Proceedings of the IEEE International Conference on Technologies for Homeland Security. IEEE, 722–728.

Marchi, Niccolo; Pirotti, Francesco; and Lingua, Emanuele (2018), Airborne and Terrestrial Laser Scanning Data for the Assessment of Standing and Lying Deadwood: Current Situation and New Perspectives, Remote Sensing, Vol. 10, p. 1356, DOI: 10.3390/rs10091356.

Markets and Markets (2015), http://www.marketsandmarkets.com/Market-Reports/commercial-dronesmarket-195137996.html.

Marques, Pascual (2013), Emerging Technologies in UAV Aerodynamics. International Journal of Unmanned Systems Engineering, Vol. 1, Nos. 3–4, DOI: 10.14323/ijuseng.2013.2.

Marsh (2015), Drones – A View Into the Future for the Logistics Sector?, Marsh & Mclennan Companies. UK.

Marsh, Randall (2013), Goodyear Replacing Its Current Blimp Fleet With Zeppelins, Aircraft, https://newatlas.com/goodyear-blimp-replacement-zeppelins/28335/.

Martinez-Caro, J.M.; and Cano, M.D. (2019), IoT System Integrating Unmanned Aerial Vehicles and LoRa Technology: A Performance Evaluation Study, Wireless Communications and Mobile Computing, Vol. 2019, Article ID 4307925, 12 pp., https://doi.org/10.1155/2019/4307925.

Matthew, Fendy Santoso; Garratt, A.; and Anavatti, Sreenatha G. (2018), State-of-the-Art Intelligent Flight Control Systems in Unmanned Aerial Vehicles, IEEE Transactions on Automation Science and Engineering, Vol. 15, No. 2, April.

McFarland, M. (2017), UPS Drivers May Tag Team Deliveries with Drones, CNN Money, 2017. http://money.cnn.com/2017/02/21/technology/ups-drone-delivery/index.html.

McFarland, M. (2019), Airports Scramble to Handle Drone Incidents. https://www.cnn.com/2019/03/05/tech/airports-drones/index.html.

Mckinsey (2017), Commercial drones are here: The Future of Unmanned Aerial Systems, *https://www.mckinsey.com/industries/*.

Menouar, H.; Guvenc, I.; Akkaya, K.; Uluagac, A.S.; Kadri, A.; and Tuncer, A. (2017), UAV-enabled Intelligent Transportation Systems for the Smart City: Applications and Challenges, IEEE Communications Magazine, Vol. 55, No. 3, pp. 22–28.

Mian, O.; Lutes, J.; Lipa, G.; Hutton, J.J.; Gavelle, E.; and Borghini, S. (2015), Direct Georeferencing on Small Unmanned Aerial Platforms for Improved Reliability and Accuracy of Mapping Without the Need for Ground Control Points. Int. Arch. Photogramm. Remote. Sens. Spat. Inf. Sci., XL-1/W4, pp. 397–402.

Micheletti, N., Chandler, J.; and Lane S. (2015), Structure from Motion (SfM) Photogrammetry. British Society for Geomorphology. Geomorphological Techniques, Chap. 2, Sec. 2.2.

Miller, P.C. (2016), Agriculture and Power Grid Inspection Are Among the Fastest Growing UAV Applications from Now to 2020, Virtual-Strategy Magazine. UAS Magazine. *http://www.virtual-strategy.com/2015/09/18/agriculture-and-powergrid-inspection-are-among-fastest-growing-uav-applications-now-2020#axzz3mFEx0FHz*.

Mohammed, Farhan; Idries, Ahmed; Mohamed, Nadar; Al-Jaroodi, Jameela; and Jawhar, Imad (2014), UAVs for Smart Cities: Opportunities and Challenges, 2014 International Conference on Unmanned Aircraft Systems (ICUAS), May 27–30. Orlando, FL, USA.

Molina, B. (2017), Drones from Super Bowl 51 Halftime Show, USA Today.

Molina, Colomina P. (2014), Unmanned Aerial Systems for Photogrammetry and Remote Sensing: A Review, ISPRS Journal of Photogrammetry and Remote Sensing, Vol. 92, June, pp. 79–97, *https://doi.org/10.1016/j.isprsjprs.2014.02.013*.

Mostafa, T.M.; Muharam, A.; and Hattori, R. (2017), Wireless Battery Charging System for Drones via Capacitive Power Transfer, IEEE PELS Workshop on Emerging Technologies: Wireless Power Transfer, WoW.

Motlagh, N.H.; Taleb, T.; and Arouk, O. (2016), Low-altitude Unmanned Aerial Vehicles-based Internet of Things Services: Comprehensive Survey and Future Perspectives, IEEE Internet Things J., Vol. 3, No. 6, pp. 899–922, December.

Mozaffari, Mohammad; Saad, Walid; Bennis, Mehdi; and Debbah, M´erouane (2017), Mobile Unmanned Aerial Vehicles (UAVs) for Energy-Efficient Internet of Things Communication, arXiv:1703.05401v2 [cs.IT] 12 September.

Mozaffari, Mohammad; Saad, Walid; Bennis, Mehdi; Nam, Young-Han; and Debbah, Mérouane (2018), A Tutorial on UAVs for Wireless Networks: Applications, Challenges, and Open Problems, arXiv:1803.00680v1 [cs.IT] 2 March. DOI 10.1109/COMST.2019.2902862.

Mujumdar, A.; and Padhi, R. (2009), Nonlinear Geometric and Differential Geometric Guidance of UAVs for Reactive Collision Avoidance, IISc Bangalore (India), Tech. Report.

Nagai, M.; Shibasaki, R.; Kumagai, H.; and Ahmed, A. (2009), UAV-borne 3-D Mapping System by Multisensor Integration. IEEE T. Geosci. Remote, Vol. 47, pp. 701–708, *https://doi.org/10.1109/TGRS.2008.2010314*.

Nassi, Ben; Shabtai, Asaf; Masuoka, Ryusuke; and Elovici, Yuval (2019), SoK – Security and Privacy in the Age of Drones: Threats, Challenges, Solution Mechanisms, and Scientific Gaps, arXiv:1903.05155v1 [cs.CR] 12 March.

Newcome, L.R. (2004), Unmanned Aviation: A Brief History of Unmanned Aerial Vehicles, AIAA.

Nex, Francesco; and Remondino, Fabio (2014), UAV for 3D Mapping Applications: A Review, Appl Geomat., Vol. 6, Nos. 1–15, DOI 10.1007/s12518-013-0120-x.

NIAS (2018), The Evolution of Commercial Drone Technology, *https://nias-uas.com/evolution-commercial-drone-technology/*.

Nishar, Abdul; Richardsb, Steve; Breena, Dan; Robertsona, John; and Breena, Barbara (2016), Thermal Infrared Imaging of Geothermal Environments and by an Unmanned Aerial Vehicle (UAV): A Case Study of the Wairakei – Tauhara Geothermal Field, Taupo, New Zealand, Renewable Energy, Vol. 86, February, pp. 1256–1264, *https://doi.org/10.1016/j.renene.2015.09.042*.

Noth, A. (2008), Design of Solar Powered Airplanes for Continuous Flight, Ph.D. Thesis, École Polytechnique Fédérale de Lausanne, Swiss.

Ortiz, Nora Raboso (2018), LiDAR Technology, *https://www.cursosteledeteccion.com/donde-descargar-datos-lidar/lidar-technology/*.

Padmanabhan, Ananth (2017), Civilian Drones and India's Regulatory Response, CARNEGIE India, March, *https://carnegieindia.org/2017/03/10/civilian-drones-and-india-s-regulatory-response-pub-68218*.

Paloz-Sanchez, P.R.; Arenas-Marquez, F.J.; and Aguayo-Camacho, M. (2017), Cloud Computing (SaaS) Adoption as a Strategic Technology: Results of an Empirical Study, Mobile Inf. Syst., p. 20.

Paneque-Gálvez, Jaime; McCall, Michael K.; Napoletano, Brian M.; Wich, Serge A.; and Koh, Lian Pin (2014), Small Drones for Community-Based Forest Monitoring: An Assessment of Their Feasibility and Potential in Tropical, Forests, Vol. 5, pp. 1481–1507, DOI: 10.3390/f5061481.

Park, J.W.; Oh, H.D.; and Tahk, M.J. (2008), UAV Collision Avoidance Based on Geometric Approach, in SICE Annual Conference, IEEE, pp. 2122–2126.

Park, S.; Zhang, L.; and Chakraborty, S. (2017), Battery Assignment and Scheduling for Drone Delivery Businesses, Proceedings of the International Symposium on Low Power Electronics and Design. DOI: 10.1109/ISLPED.2017.8009165, Corpus ID: 23424708.

Parshotam, K. (2013), Crowd computing: A Literature Review and Definition, Proc. South Afr. Inst. Comput. Scientists Inf. Technologists Conf. (SAICSIT), New York, NY, USA, pp. 121–130, DOI: 10.1145/2513456.2513470.

Pascarella, Domenico; Venticinque, Salvatore; and Aversa, Rocco (2013), Agent-Based Design for UAV Mission Planning, Eighth International Conference.

Pastor, Enric; Lopez, Juan; and Royo, Pablo (2007), A Hardware/Software Architecture For UAV Payload and Mission Control, IEEE Aerospace and Electronic Systems Magazine, July, DOI: 10.1109/MAES.2007.384074.

Patel, Y.; Gaurav, A.; Srinivas, K.; and Singh, Y. (2017), A Review on Design and Analysis of the Propeller Used in UAV. Int. J. Adv. Prod. Ind. Eng., Vol. 605, pp. 20–23.

Patil, J.K.; and Kumar, R. (2011), Advances in Image Processing for Detection of Plant Diseases, Journal of Advanced Bioinformatics Applications and Research, Vol. 2, No. 2, pp. 135–141.

Patterson, John (2018), An Aerial View of the Future – Drones in Construction, Geospatial World, 09 May.

Pederi, Y.A.; and Cheporniuk, H.S. (2015), Unmanned Aerial Vehicles and New Technological Methods of Monitoring and Crop Protection in Precision Agriculture. Proceedings of the 2015 IEEE 3rd International Conference Actual Problems of Unmanned Aerial Vehicles Developments, APUAVD 2015—Proceedings, Kiev, Ukraine, 13–15 October; pp. 298–301.

Pflanz, M.; Nordmeyer, H.; and Schirrmann, M. (2018), Weed Mapping with UAS Imagery and a Bag of Visual Words Based Image Classifier, Remote Sens., 10, 1530.

Pires, Rayner de Melo; Arnosti, Sergio Zumpano; Pinto, Alex Sandro Roschildt; and Branco, Kalinka R.L.J. Castelo (2016), Experimenting Broadcast Storm Mitigation Techniques in FANETs, 49th Hawaii International Conference on System Sciences, IEEE Computer Society, pp. 1530–1605/16, DOI: 10.1109/HICSS.2016.726.

Petkovic, Milica; and Narandzic, Milan (2019), Overview of UAV Based Free-Space Optical Communication Systems, International Conference on Interactive Collaborative Robotics, ICR 2019: Interactive Collaborative Robotics, pp. 270–277.

Pérez-Rastelli, J.; and Santos, M. (2015), Fuzzy Logic Steering Control of Autonomous Vehicles Inside Roundabouts, Appl. Soft Comput., Vol. 35, pp. 662–669.

PMFBY (2018), Pradhan Mantri Fasal Beema Yogna (PMFBY), Govt of India, *https://pmfby.gov.in/*.

Poblete, Thomas; Ortega-Farías, Samuel; Moreno, Miguel Angel; and Bardeen, Matthew (2017), Artificial Neural Network to Predict Vine Water Status Spatial Variability Using Multispectral Information Obtained from an Unmanned Aerial Vehicle (UAV), Sensors, MDPI, 2017, Vol. 17, p. 2488, DOI: 10.3390/s17112488.

Popescu, Dan; Ichim, Loretta; and Stoican, Florin (2017), Unmanned Aerial Vehicle Systems for Remote Estimation of Flooded Areas Based on Complex Image Processing, Sensors, Vol. 17, No. 3, p. 446; *https://doi.org/10.3390/s17030446*.

Puliti, Stefano; Ørka, Hans; Gobakken, Terje; and Næsset, Erik (2015), Inventory of Small Forest Areas Using an Unmanned Aerial System. Remote Sensing, Vol. 7, pp. 9632–9654. DOI: 10.3390/rs70809632.

Pulver, A.; and Wei, R. (2018), Optimizing the Spatial Location of Medical Drones. Appl Geogr., Vol. 90, pp. 9–16.

Puri, A. (2015), A Survey of Unmanned Aerial Vehicles (UAV) for Traffic Surveillance, Department of Computer Science and Engineering, University of South Florida.

PwC (2018a), Flying High, PwC-Engineering Council of India, November, *https://www.pwc.in/assets/pdfs/publications/2018/flying-high.pdf*.

PwC (2018b), Clarity from Above, PwC Global Report on the Commercial Applications of Drone Technology, *https://www.pwc.pl/pl/pdf/clarity-from-above-pwc.pdf*.

Qiao, J.; He, Y.; and Shen, S. (2016), Proactive Caching for Mobile Video Streaming in Millimeter Wave 5G Networks, IEEE Trans. Wireless Commun., Vol. 15, No. 10, pp. 7187–7198, Oct.

Qu, Y.; Jiang, L.; and Guo, X. (2016), Moving Vehicle Detection With Convolutional Networks in UAV Videos, In 2nd International Conference on Control, Automation and Robotics (ICCAR), April, pp. 225–229.

Qu, Yufu; Huang, Jianyu; and Zhang, Xuan (2018), Rapid 3D Reconstruction for Image Sequence Acquired from UAV Camera, Sensors (Basel), Jan; Vol. 18, No. 1, p. 225. DOI: 10.3390/s18010225.

QUALCOMM (2018), Paving the Path to 5G: Optimizing Commercial LTE Networks for Drone Communication, *https://www.qualcomm.com/news/onq/2016/09/06/paving-path-5goptimizing-commercial-lte-networks-drone-communication.*

Rai, Raveendra Kumar; Singh, Vijay P.; and Upadhyay, Alka (2017), Chapter 12- Irrigation Scheduling, Planning and Evaluation of Irrigation Projects, Methods and Implementation, Academic Press, pp. 385–412.

Rajendran, Parvathy; and Smith, Howard (2018), Development of Design Methodology for a Small Solar-Powered Unmanned Aerial Vehicle, International Journal of Aerospace Engineering, Volume 2018, Article ID 2820717, 10 pp. *https://doi.org/10.1155/2018/2820717.*

Reportlinker (2018), The Global Anti-drone Market Size Is Anticipated to Reach USD 1.85 Billion by 2024, *https://www.reportlinker.com/p05445430.*

Reudenbach, Chris (2018), Unmanned Aerial Vehicle R based Mission Planning, Oct., *https://cran.r-project.org/web/packages/uavRmp/vignettes/uavRmp_1.pdf.*

Rhoads, F.M.; and Yonts, C.D. (2000), Irrigation Scheduling for Corn: Why and How, National Corn Handbook.

Richard, S. Stansbury; Olds, Joshua Lloyd; and Coyle, Eric Joe (2014), Ethical Concerns of Unmanned and Autonomous Systems in Engineering Programs, In 121st ASEE Annual Conference and Exposition, Indianapolis, IN, June 15–18.

Ridolfi, Elena; Buffi, Giulia; Venturi, Saran; and Manciola, Piergiorgio (2017), Accuracy Analysis of a Dam Model from Drone Surveys, Sensors, MDPI.

Ro, K.; Oh, J.S.; and Dong, L. (2007), Lessons Learned: Application of Small UAV for Urban Highway Traffic Monitoring, In 45th AIAA Aerospace Sciences Meeting and Exhibit, pp. 2007–596.

Rodriguez, Ricardo Schacht (2019), UAV mission planning based on Prognosis & Health Management, Automatic, PhD Thesis, Université de Lorraine; Centro Nacional de Investigación y Desarrollo Tecnológico (Cuernavaca, Mor., México).

Ross, T. (2010), Fuzzy Logic With Engineering Applications, John Wiley & Sons, West Sussex, UK.

Rothermel, M.; Wenzel, K.; Fritsch, D.; and Haala, N. (2012), SURE: Photogrammetric Surface Reconstruction from Imagery, In Proceedings LC3D Workshop, Berlin, Germany.

Rubio, Juan Carlos; Vagners, Juris; and Rysdyk, Rolf (2004), Adaptive Path Planning for Autonomous UAV Oceanic Search Missions, In AIAA 1st Intelligent Systems Technical Conference 20–22 September, Chicago, Illinois.

SACAA (2017), South African Civil Aviation Authority (SACAA) Technical Guidance Material for RPAS, Part 101, *http://www.caa.co.za/Pages/RPAS/ Remotely%20Piloted%20Aircraft%20Systems.aspx*.

Sabo, Chelsea; and Cohen, Kelly (2012), Fuzzy Logic Unmanned Air Vehicle Motion Planning, Advances in Fuzzy Systems, Vol. 2012, Hindawi Publishing Corporation, Article ID 989051, 14 pp., DOI: 10.1155/2012/989051.

Sabrol, H.; and Satish, K. (2016), Tomato Plant Disease Classification in Digital Images Using Classification Tree, In Proceedings of the IEEE International Conference on Communication and Signal Processing, ICCSP 2016, Melmaruvathur, India, 6–8 April; pp. 1242–1246.

Safi, Hossein; Dargahi, Akbar; and Cheng, Julian (2019), Spatial Beam Tracking and Data Detection for an FSO Link to a UAV in the Presence of Hovering Fluctuations, arXiv:1904.03774v1 [eess.SP], 7 April.

Safi, Michael (2019), Are Drone Swarms the Future of Aerial Warfare, Dec., Guardian Weekly, *https://www.theguardian.com/news/2019/dec/04/are-drone-swarms-the-future-of-aerial-warfare*.

Saha, B.; Koshimoto, E.; Quach, C.C.; Hogge, E.F.; Strom, T.H.; Hill, B.L.; Vazquez, S.L.; and Goebel, K. (2011), Battery health Management System for Electric UAVs, IEEE Aerospace Conference, pp. 1–9.

Sakharkar, Ashwini (2020), UniSA Develops Pandemic Drone to Detect Coronavirus in the Crowd, *https://www.techexplorist.com/unisa-develops-pandemic-drone-detect-coronavirus-crowd/31152/*.

Salah-Ddine, Krit; Jalal, Laassiri; and Said, El Hajji (2012), Design Methodology of Energy Consumption for Wireless Sensor Networks Using Renewable Energy, Journal of Computing, Vol. 4, 5, May, ISSN 2151-9617.

Samaniego, Franklin; Sanchis, Javier; García-Nieto, Sergio; and Simarro, Raúl (2019), Recursive Rewarding Modified Adaptive Cell Decomposition (RR-MACD): A Dynamic Path Planning Algorithm for UAVs, Electronics, Vol. 8, p. 306, DOI: 10.3390/electronics8030306.

Sankaran, S.; Khot, L.R.; Espinoza, C.Z.; Jarolmasjed, S.; Sathuvalli, V.R.; Vandemark, G.J.; Miklas, P.N.; Carter, A.H.; Pumphrey, M.O.; and Knowles, N.R. (2015), Low-altitude, High-resolution Aerial Imaging Systems for Row and Field Crop Phenotyping: A Review, European Journal of Agronomy, Vol. 70, pp. 112–123.

Sankarasrinivasan, S.; Balasubramanian, E.; Karthik, K.; Chandrasekar, U.; and Gupta, R. (2015), Health Monitoring of Civil Structures with Integrated UAV and Image Processing System, Procedia Computer Science, Vol. 54, pp. 508–515.

Santoso, Frendy; Garratt, Matthew A.; and Anavatti, Sreenatha G. (2018), State-of-the-Art Intelligent Flight Control Systems in Unmanned Aerial Vehicles, IEEE Transactions on Automation Science and Engineering, Vol. 15, No. 2, April, p. 613.

Saunders, J.B. (2009), Obstacle Avoidance, Visual Automatic Target Tracking, and Task Allocation for Small Unmanned Air Vehicles. Brigham Young University.

Saura, Jose Ramon; Reyes-Menendez, Ana; and Palos-Sanchez, Pedro (2019), Mapping Multispectral Digital Images Using a Cloud Computing Software: Applications from UAV Images, Heliyon, Vol. 5, No. 2, February.

Sawalmeh, Ahmad H.; and Othman, Noor Shamsiah (2018), An Overview of Collision Avoidance Approaches and Network Architecture of Unmanned Aerial Vehicles (UAVs), International Journal of Engineering & Technology.

Scherer, J.; Yahyanejad, S.; Hayat, S.; Yanmaz, E.; Andre, T.; Khan, A.; Vukadinovic, V.; Bettstetter, C.; Hellwagner, H.; and Rinner, B. (2015), An Autonomous Multi-UAV System for Search and Rescue, in Proceedings of the First Workshop on Micro Aerial Vehicle Networks, Systems, and Applications for Civilian Use. ACM, pp. 33–38.

SERA (2017), Standardized European Rules of the Air. *https://www.eurocontrol.int/sites/default/files/article/content/documents/single-sky/mandates/20090814-sera-parta-ec-mandate.pdf*.

SES (2017), Single European Sky, *http://www.eurocontrol.int/dossiers/single-european-sky*.

Shakhatreh, H.; Khreishah, A., Chakareski, J.; Salameh, H.B.; and Khalil, I. (2016), On the Continuous Coverage Problem for a Swarm of UAVs, in Sarnoff Symposium, 37th IEEE, pp. 130–135.

Shakhatreh, Hazim; Khreishah, Abdallah; and Ji, Bo (2017), Providing Wireless Coverage to High-rise Buildings Using uavs, arXiv preprint arXiv:1705.09770.

Shakhatreh, Hazim; Sawalmeh, Ahmad; Al-Fuqaha, Ala; Dou, Zuochao Dou; Almaita, Eyad; Khalil, Issa; Othman, Noor Shamsiah; Khreishah, Abdallah; and Guizani, Mohsen (2018), Unmanned Aerial Vehicles: A Survey on Civil Applications and Key Research Challenges, arXiv:1805.00881v1 [cs.RO] 19 Apr 2018.

Shamshiri, Redmond Ramin; Hameed, Ibrahim A.; Balasundram, Siva K.; Ahmad, Desa; Weltzien, Cornelia; and Yamin, Muhammad (2018). Fundamental Research on Unmanned Aerial Vehicles to Support Precision Agriculture in Oil Palm Plantations, Agricultural Robots – Fundamentals and Applications, Jun Zhou and Baohua Zhang (Eds.), IntechOpen, DOI: 10.5772/intechopen.80936.

Sharma, L.; and Garg, P.K. (2019) (Eds), From Visual Surveillance to Internet of Things- Technology and Applications, Springer. CRC Press, London, ISBN: 9780367221942.

Shen, S. (2013), Accurate Multiple view 3D Reconstruction using Patch-based stereo for Large-scale Scenes. IEEE Trans. Image Process; 22:1901–1914, DOI: 10.1109/TIP.2013.2237921.

Sheridan, T.B.; and Verplank, W.L. (1978). Human and Computer Control of Undersea Teleoperators. Technical Report, Man-Machine Systems Laboratory, Department of Mechanical Engineering, Massachusetts Institute of Technology, Cambridge. Mass.

Shi, Yeyin; Thomasson, J. Alex; Murray, Seth C.; Pugh, N. Ace; Rooney, William L.; Shafian, Sannaz; Rajan, Nithya; Rouze, Gregory; Morgan, Cristine L.S.; Neely, Haly L.; Rana, Aman; Bagavathiannan, Muthu V.;

Henrickson, James; and Yang, Chenghai (2016), Unmanned Aerial Vehicles for High-Throughput Phenotyping and Agronomic Research, PLoS One, July 29, *https://doi.org/10.1371/journal.pone.0159781.*

Sholes, E. (2007), Evolution of a UAV Autonomy Classification Taxonomy. In Proceedings of the IEEE Aerospace Conference, Big Sky, MT, USA, 3–10 March.

Siebert, Sebastian; and Teizer, Jochen (2014), Mobile 3D Mapping for Surveying Earthwork Projects Using an Unmanned Aerial Vehicle (UAV) System. Automation in Construction, Vol. 41, pp. 1–14, DOI: 10.1016/j.autcon.2014.01.004.

Simic, M.; Bil, G.; and Vojisavljevic, V. (2015), Investigation in Wireless Power Transmission for UAV Charging, Proceed. Computer Science, Vol. 60, No. 1, pp. 1846–1855, *http://dx.doi.org/10.1016/j.procs.2015.08.295.*

Singhal, Gaurav; Bansod, Babankumar; and Mathew, Lini (2018), Unmanned Aerial Vehicle Classification, Applications and Challenges: A Review, *http://dx.doi.org/10.20944/preprints201811.0601.v1.*

Sinsley, Gregory; Long, Lyle N.; Niessner, Albert F.; and Horn, Joseph F. (2008), Intelligent Systems Software for Unmanned Air Vehicles, In 46th AIAA Aerospace Sciences Meeting, Reno, NV Jan., AIAA Paper No. 2008-0871.

Skorup, Brent; and Haaland, Connor (2020), How Drones Can Help Fight the Coronavirus, Mercatus Center at George Mason University, Policy Brief, Special Edition, *https://www.mercatus.org/system/files/skorup-drones-coronavirus-mercatus-v1.pdf.*

Sola-Guirado, Rafael R.; Castillo-Ruiz, Francisco J.; Jiménez-Jiménez, Francisco; Blanco-Roldan, Gregorio L.; Castro-Garcia, Sergio; and Gil-Ribes, Jesus A. (2017), Olive Actual "on Year" Yield Forecast Tool Based on the Tree Canopy Geometry Using UAS Imagery, Sensors, MDPI.

Song, Lingyang; Zhang, Rui; Saad, Walid; Han, Zhu; and Schober, Robert (2020), Best Readings in UAV Assisted Wireless Networks, IEEE Communication Society Publications.

Sowmya, M.; and Subramani, Bojan (2018), A Study on Data Mining With Image Processing Via UAV Images, International Conference on Computing Intelligence and Data Science (ICCIDS 2018), IOSR Journal of Engineering (IOSRJEN) ISSN (e): 2250-3021, ISSN (p): 2278–8719, pp 24–30.

Stenger, A.; Fernando, B.; and Heni, M. (2012), Autonomous Mission Planning for UAVs: A Cognitive Approach, Deutscher Luft- und Raumfahrtkongress 2012, Document ID: 281398.

Stewart, J. (2014), Google Tests Drone Deliveries in Project Wing Trials, BBC World Service Radio.

Stoica, A.A. (2018), Emerging Legal Issues Regarding Civilian Drone Usage, Challenges Knowl. Soc., Vol. 2018, No. 12, pp. 692–699.

Stöcker, Claudia; Bennett, Rohan; Nex, Francesco; Gerke, Markus; and Zevenbergen, Jaap (2017), Review of the Current State of UAV Regulations, Remote Sens., Vol. 9, p. 459, DOI: 10.3390/rs9050459.

Sun, Y.; Ng, D.W.K.; Xu, D.; Dai, L.; and Schober, R. (2018), Resource Allocation for Solar Powered UAV Communication Systems, arXiv preprint arXiv:1801.07188.

Šustek, Michal; and Úøedníèek, Zdenìk (2018), The Basics of Quadcopter Anatomy, EDP Sciences, CSCC 2018, MATEC Web of Conferences 210, 01001 (2018), *https://doi.org/10.1051/matecconf/201821001001*.

Sutheerakul, C.; Kronprasert, N.; Kaewmoracharoen, M.; and Pichayapan, P. (2017), Application of Unmanned Aerial Vehicles to Pedestrian Traffic Monitoring and Management for Shopping Streets, Transportation Research Proceed., Vol. 25, pp. 1717–1734.

Suzuki, S. (2018), Recent Researches on Innovative Drone Technologies in Robotics Field, Advanced Robotics, Vol. 32, No. 19, pp. 1008–1022, DOI: 10.1080/01691864.2018.1515660.

Swaminathan, R. (2015), Drones in India, Exploring Policy and Regulatory Challenges Posed By Civilian Unmanned Aerial Vehicles, ORF Occasional Paper #58, Observer Research Foundation, *https://www.orfonline.org*.

Swiss Federal Office of Civil Aviation (2014), Drones and Aircraft Models *http://www.bazl.admin.ch/dienstleistungen/02658/index.html?lang=en Swiss Federal Office of Civil Aviation*.

Sylvester, Gerard (2018), E-agriculture in Action: Drones for Agriculture, Food and Agriculture Organization of the United Nations and International Telecommunication Union, Bangkok.

Taha, Bilal; and Shoufan, Abdulhadi (2019), Machine Learning-Based Drone Detection and Classification: State-of-the-Art in Research, IEEE Access, Sept, Vol. 7.

Tahir, Anam; Boling, Jari; Haghbayan, Mohammad-Hashem; Toivonen, Hannu T.; and Plosila, Juha (2019), Swarms of Unmanned Aerial Vehicles – A Survey, Journal of Industrial Information Integration (16), *https://doi.org/10.1016/j.jii.2019.100106.*

Takacs, Attila; Lin, Xingqin; Hayes, Stephen; and Tejedor, Erika (2018), Drones and networks: Ensuring Safe and Secure Operations, Ericsson White Paper, GFMC-18:000526, November, *https://www.ericsson.com/en/reports-and-papers/white-papers/drones-and-networks-ensuring-safe-and-secure-operations.*

Taladay, Katie (2018), Pros and Cons of Using Unmanned Aircraft Systems (UAS) for Generating Geospatial Data: Case Studies from Hawai'i. DOI: 10.13140/RG.2.2.10224.35847.

Tasevski, Stefan (2018), Drones for Humanitarian Aid, November 27, *https://dronebelow.com/2018/11/27/drones-for-humanitarian-aid/.*

Tatum, Mark C.; and Liu, Junshan (2017), Unmanned Aircraft System Applications, In Construction, Creative Construction Conference 2017, CCC 2017, 19–22 June, Primosten, Croatia.

Terra Times (2019), Drone Regulations, In Other Countries, *https://www.terra-drone.net/en media/2017/10/13/droneregulationsoc/.*

The Economist (2020), To Curb COVID-19, China Is Using Its High-Tech Surveillance Tools, The Economist, Code Red, Chinese Section, February 29, *https://www.economist.com/china/2020/02/29/to-curb-covid-19-china-is-using-its-high-tech-surveillance-tools.*

Tomaštík, Julian; Mokroš, Martin; Surový, Peter; Grznárová, Alžbeta; and Mergani, Jan (2019), UAV RTK/PPK Method—An Optimal Solution for Mapping Inaccessible Forested Areas, Remote Sens, Vol. 11, p. 721, DOI: 10.3390/rs11060721.

Transport Canada (2017), www.tc.gc.ca/eng/civilaviation/menu.htm.

Tsourdos, A.; White, B.; and Shanmugavel, M. (2011), Cooperative Path Planning of Unmanned Aerial Vehicles, John Wiley & Sons, West Sussex, UK.

Tsouros, Dimosthenis C.; Bibi, Stamatia; and Sarigiannidis, Panagiotis G. (2019), A Review on UAV-Based Applications for Precision Agriculture, Information, MDPI, Vol. 10, p. 349, DOI: 10.3390/info10110349.

Turkmen, Zeynep; and Kuloglu, Merve (2018), A New Era for Drug Trafficking: Drones, Forensic Science and Addiction Research, Vol. 2, No. 2, Crimson Publishers.

Turner, D.; Lucieer, A.; and Watson, C. (2012), An Automated Technique for Generating Georectified Mosaics from Ultra-high Resolution Unmanned Aerial Vehicle (UAV) Imagery, Based on Structure from Motion (SFM) Point Clouds. Remote Sensing, Vol. 4, pp. 1392–1410.

UAV Coach (2020), UAV Coach, *https://uavcoach.com/*.

Uddin, Mohammad Ammad Uddin; Mansour, Ali; Jeune, Denis Le; Ayaz, Mohammad; and Aggoune, el-Hadi M. Aggoune (2018), UAV-Assisted Dynamic Clustering of Wireless Sensor Networks for Crop Health Monitoring, Sensors, MDPI, Vol. 18, p. 555, DOI: 10.3390/s18020555.

Uerkwitz, Andrew; Dean, Paul; and Yang, Martin (2016), Drone Industry Report, Oppenheimer & Co Inc., New York, NY, Oppenheimer & Co Inc., New York, NY. *http://pdf.zacks.com/pdf/FA/H4947044.PDF*.

UK Civil Aviation Authority (2012), Unmanned Aircraft System Operations in UK Airspace – Guidance. Report CAP 722.

USAID (2018), UAVs in Global Health: Defining a Collective Path Forward, USAID's Center for Accelerating Innovation and Impact (CII) 2018, www.usaid.gov/cii.

US Department of Transportation (2013), Integration of Civil Unmanned Aircraft Systems (UAS) in the National Airspace System (NAS) Roadmap, Federal Aviation Authority.

Vacanasa, Yiannis; Themistocleousa, Kyriacos; Agapioua, Athos; and Hadjimitsisa, Diofantos (2015), Building Information Modelling (BIM) and Unmanned Aerial Vehicle (UAV) Technologies in Infrastructure Construction Project Management and Delay and Disruption Analysis, Proceedings of SPIE – The International Society for Optical Engineering, June, DOI: 10.1117/12.2192723.

Valvanis, K.; and Vachtsevanos, G. (2015), Future of Unmanned Aviation. In Handbook of Unmanned Aerial Vehicles; Springer: Dordrecht, The Netherlands; pp. 2993–3009.

Vattapparamban, E.; Guvenc, I.; Yurekli, A.I.; Akkaya, K.; and Uluagac, S. (2016), Drones for Smart Cities: Issues in Cybersecurity, Privacy, and Public Safety, in International Wireless Communications and Mobile Computing Conference (IWCMC), IEEE, pp. 216–221.

Veres, S.M., and Molnar, L. (2010), Documents for Intelligent Agents, In: Proceedings of IASTED Conference on Artificial Intelligence Applications, Innsbruck, Austria. IASTED, 2010 on P2P, Parallel, Grid, Cloud and Internet Computing, DOI 10.1109/3PGCIC.2013.18.

Vergouw, B.; Nagel, H.; Bondt, G.; and Custers, B. (2016), Drone Technology: Types, Payloads, Applications, Frequency Spectrum Issues and Future Developments, in the Future of Drone Use. Springer, pp. 21–45.

Vidyadharan, Akash; Philpott, Robert; Kwasa, Benjamin J.; and Bloebaum, Christina L. (2017), Analysis of Autonomous Unmanned Aerial Systems Based on Operational Scenarios Using Value Modelling, Drones, MDPI, Vol. 1, p. 5, DOI: 10.3390/drones1010005 www.mdpi.com/journal/drones.

Viljanen, N.; Honkavaara, E.; Näsi, R.; Hakala, T.; Niemeläinen, O.; and Kaivosoja, J. (2018), A Novel Machine Learning Method for Estimating Biomass of Grass Swards Using a Photogrammetric Canopy Height Model, Images and Vegetation Indices Captured by a Drone. Agriculture, Vol. 8, p. 70.

Visockiene, Jurate Suziedelyte; Brucas, Domantas; and Ragauskas, Ugnius (2014), Comparison of UAV Images Processing Softwares, JVE International Ltd. ISSN Print 2335-2124, ISSN Online 2424-4635, Kaunas, Lithuania.

Wresearch (2016), India UAV Market (2017–2023), 6Wresearch, August, *https://www.6wresearch.com/industry-report/india-unmanned-aerial-vehicle-uav-market-2017-2023-forecast-by-types-by-uav-range-applications-regions-competitive-landscape*.

Wackwitz, Kay; Boedecker, Hendrik; and Froehlich, Andreas (2016), Drone Business Development Guide, Drone Industry Insights, Hamburg, Germany.

Wang, C.; and Ma, Z. (2016), Design of Wireless Power Transfer Device for UAV, IEEE International Conference on Mechatronics and Automation, pp. 2449–2454, *http://ieeexplore.ieee.org/document/7558950/*.

Wang, C.; and Ma, Z. (2018), Design of Wireless Power Transfer Device for UAV, IEEE International Conference on Mechatronics and Automation, pp. 2449–2454, *http://ieeexplore.ieee.org/document/7558950/*.

Wang, D.C.; Zhang, G.L.; Zhao, M.S.; Pan, X.Z.; Zhao, Y.G.; Li, D.C.; and Macmillan, B. (2015), Retrieval and Mapping of Soil Texture Based on Land Surface Diurnal Temperature Range Data from MODIS, PLoS one, Vol. 10, No. 6, p. e0129977.

Wang, H.; Ding, G.; Gao, F.; Chen, J.; Wang, J.; and Wang, L. (2017), Power Control in UAV-supported Ultra Dense Networks: Communications, Caching, and Energy Transfer, arxiv.org/abs/1712.05004.

Wang, Y.; and Wang, Z. (2019), A Survey of Recent Work on Fine-grained Image Classification Techniques. J. Vis. Commun. Image Represent, Vol. 59, pp. 210–214.

Watts, Adam C.; Ambrosia, Vincent G.; and Hinkley, Everett A. (2012), Unmanned Aircraft Systems in Remote Sensing and Scientific Research: Classification and Considerations of Use, Remote Sens, Vol. 4, pp. 1671–1692, DOI: 10.3390/rs4061671.

Westoby, M.J.; Brasington, J.; Glasser, N.F.; Hambrey, M.J.; and Reynolds, J.M. (2012), Structure-From-Motion Photogrammetry: A Low-Cost, Effective Tool for Geoscience Applications. Geomorph., Vol. 179, pp. 300–314.

Wetsman, Nicole (2020), Everything You Need to Know About the Coronavirus Stay Alert and Be Prepared, The Verge, *https://www.theverge.com/2020/1/23/21078457/coronavirus-outbreak-china-wuhan-quarantine-who-sars-cdc-symptoms-risk*.

Whitehead, K.; and Hugenholtz, C.H. (2019), Remote Sensing of the Environment with Small Unmanned Aircraft Systems (UASs), Part 1: A Review of Progress and Challenges, J. Unmanned Vehicle Syst., Vol. 2, No. 3, pp. 69–85, 2014.

White Paper (2018), UAV Workflow Integration, Drone Industry Insights, May. DroneII.com.

Wijeyasinghe, Nilushi; and Ghaffarzadeh, Khasa (2020), LiDAR 2020–2030: Technologies, Players, Markets & Forecasts, IDTechEx, *https://www.idtechex.com/en/research-report/lidar-2020-2030-technologies-players-markets-and-forecasts/694*.

Wingtra (2019), Drone Photogrammetry vs. LiDAR: What Sensor to Choose for a Given Application, *https://wingtra.com/drone-photogrammetry-vs-lidar/*.

Wolfert, S., Ge, L., Verdouw, C.; and Bogaardt, M.J. (2017), Big Data in Smart Farming – A Review, Agricultural Systems, Vol. 153, pp. 69–80.

Wu, B.; Xie, L.; Hu, H.; Zhu, Q.; and Yau, E. (2018), Integration of Aerial Oblique Imagery and Terrestrial Imagery for Optimized 3D Modeling in Urban Areas, ISPRS J. Photogramm. Remote Sens., Vol. 139, pp. 119–132.

Xiao, Z.; Xia, P.; and Xia, X.G. (2016), Enabling UAV Cellular with Millimeterwave Communication: Potentials and Approaches, IEEE Communications Magazine, Vol. 54, No. 5, pp. 66–73.

Xu, Zhiqiang; Yang, Jiansi; Peng, Chaoyong; Wu, Ying; Jiang, Xudong; Li, Rui; Zheng, Yu; Gao, Yu; Liu, Sha; and Tin, Baofeng (2014), Development of an UAS for post-earthquake Disaster Surveying and Its Application in MS7.0 Lushan Earthquake, Sichuan, China, Computers & Geosciences, Vol. 68, pp. 22–30.

Yang, P.; Cao, X.; Yin, C.; Xiao, Z.; Xi, X.; and Wu, D. (2017), Proactive Drone-cell Deployment: Overload Relief for a Cellular Network Under Flash Crowd Traffic, IEEE Transactions on Intelligent Transportation Systems, Vol. 18, No. 10, pp. 2877–2892.

Yazdani, A.M.; Sammut, K.; Yakimenko, O.A.; Lammas, A.; Tang, Y.; and Mahmoud Zadeh, S. (2017), IDVD-Based Trajectory Generator for Autonomous Underwater Docking Operations. Robot. Auton. Syst., Vol. 92, pp. 12–29.

Yu, Seunghee; Heo, Jinyeong; Jeong, Sekyung; and Kwon, Yongjin (2016), Technical Analysis of VTOL UAV. Journal of Computer and Communications, Vol. 04, pp. 92–97, DOI: 10.4236/jcc.2016.415008.

Zadeh, L.A. (1965), Fuzzy Sets, Information and Control, Vol. 8, pp. 338–353.

Zadeh, S. Mahmoud; Powers, D.M.W.; and Sammut, K. (2017), An Autonomous Reactive Architecture for Efficient AUV Mission Time Management in Realistic Dynamic Ocean Environment, Robotics and Autonomous Systems, Vol. 87, pp. 81–103.

Zadeh, S. Mahmoud; Powers, D.M.W.; Sammut, K.; and Yazdani, A.M. (2018), A Novel Versatile Architecture for Autonomous Underwater Vehicle's Motion Planning and Task Assignment, Soft Computing, Vol. 22, No. 5, pp. 1687–1710.

Zafar, W.; and Khan, B.M. (2016), Flying Ad-Hoc Networks: Technological and social Implications, IEEE Technol. Soc. Mag., Vol. 35, No. 2, pp. 67–74, June.

Zeng, Yong; Wu, Qingqing; and Zhang, Rui (2019), Accessing From The Sky: A Tutorial on UAV Communications for 5G and Beyond, IIEE Proceedings, 14 March 2019, arXiv:1903.05289v2 [eess.SP].

Zeng, Young; Zhang, Rui; and Lim, Teng Joon (2016), Wireless Communications with Unmanned Aerial Vehicles: Opportunities and Challenges. IEEE Communications Magazine, Vol. 54, No. 5, pp. 36–42.

Zhang, B.; Liu, C, H.; Tang, J.; Xu, Z.; Ma, J.; and Wang, W. (2017), Learning based Energy-Efficient Data Collection by Unmanned Vehicles in Smart

Cities, IEEE Transactions on Industrial Informatics, Vol. 3203, No. c, pp. 1–1, *http://ieeexplore.ieee.org/document/8207610/*.

Zhang, C.; and Elaksher, A. (2012). An Unmanned Aerial Vehicle-based Imaging System for 3D Measurement of Unpaved road Surface Distresses, Comput.-Aided Civ. Infrastruct. Eng., Vol. 27, No. 2, pp. 118–129, *https://doi.org/10.1111/j.1467-8667.2011.00727.x*.

Zhang, Xiao; and Duan, Lingjie (2019), Fast Deployment of UAV Networks for Optimal Wireless Coverage, IEEE Transactions on Mobile Computing, March, Vol. 18, pp. 588–601, DOI: 10.1109/TMC.2018.2840143.

Zhao, Yi; Ma, Jiale; Li, Xiaohui; and Zhang, Jie (2018), Saliency Detection and Deep Learning-Based Wildfire Identification in UAV Imagery, Sensors, MDPI, Vol. 18, p. 712, DOI: 10.3390/s18030712.

Zhi, Y.; Fu, Z.; Sun, X.; and Yu, Jingnan (2020), Security and Privacy Issues of UAV: A Survey. Mobile Network Application, Vol. 25, pp. 95–101, *https://doi.org/10.1007/s11036-018-1193-x*.

Zhou, X.; Zhao, Z.; Li, R.; Zhou, Y.; Chen, T.; Niu, Z.; and Zhang, H. (2014), Toward 5G: When Explosive Bursts Meet Soft Cloud, IEEE Network, Vol. 28, No. 6, pp. 12–17.

Zollhöfer, M.; Stotko, P.; Görlitz, A.; Theobalt, C.; Nießner, M.; Klein, R.; and Kolb, A. (2018), State of the Art on 3D Reconstruction with RGB-D Cameras, Wiley: Hoboken, NJ, USA.

Zuckerberg, Mark (2016), The Technology Behind Aquila. Facebook, *https://www.facebook.com/notes/mark zuckerberg/ the Technology Behind Aquila/10153916136506634/*.

Zunino, A.; Crocco, M.; Martelli, S.; Trucco, A.; Bue, A.D.; and Murino, V. (2015), Seeing the Sound: A New Multimodal Imaging Device for Computervision, in Proc. IEEE Int. Conf. Comput. Vis. Workshop (ICCVW), December, pp. 693–701.

WEB RESOURCES

https://www.acute3d.com/contextcapture/

https://www.agisoft.com/features/professional-edition/

https://www.agisoft.com/

https://www.airdronecraze.com/new-drones/

References

http://www.applied-geosolutions.github.io/lidar2dems/

http://www.appliedimagery.com/download

https://www.autodesk.com/recap/introducing-recap-photo/

https://www.autodesk.com/solutions/3d-mapping-software

http://www.autodesk.com/products/autocad-map-3d/features/all/gallery-view

http://www.bcal.geology.isu.edu/Envitools.html

http://www.bentley.com/en-US/Promo/Pointools/pointools.htm?skid=CT_PRT_POINTOOLS_B

http://www.bluemarblegeo.com/products/global-mapper-lidar-download.php

http://www.ccwu.me/vsfm/

http://www.cloudcompare.org/

https://www.cloud.sdsc.edu:443/v1/AUTH_opentopography/www/files%2Fp2g%2FGEON-points2grid-Utility-Setup-v1_3.exe

http://www.code.google.com/p/fullanalyze/

http://www.computree.onf.fr/?lang=en

http://www.cs.unc.edu/~isenburg/lastools/

http://www.3dflow.net

https://www.3dflow.net/3df-zephyr-feature-comparison/

https://delair.aero/portfolio/surveying-and-mapping/

http://www.digitalsky.dgca.gov.in.

https://www.dji.com/ground-station-pro

https://www.dronedeploy.com/product/platform/

https://www.dronezon.com/drone-reviews/drone-deploy-app-review-with-tutorials-pricing-contact-details/

https://www.dronedeploy.com

https://www.esri.com/en-us/arcgis/products/drone2map/overview

http://www.eyeondrones.com/a-breakthrough-in-autonomous-drones/

https://www.flylitchi.com/

http://www.forsys.cfr.washington.edu/fusion/fusionlatest.html

http://www.fugroviewer.com/

https://www.grasswiki.osgeo.org/wiki/LiDAR

http://www.greenvalleyintl.com/product/lidar360/

https://www.liblas.org/start.html
http://www.liforest.com/
http://www.lizardtech.com/products/lidar–
https://www.mapsmadeeasy.com/guide
http://www.merrick.com/index.php/geospatial/services-gss/mars-software
https://www.micmac.ensg.eu
http://www.nframes.com
https://www.opendronemap.org/
https://www.pix4d.com/product/pix4dmapper-photogrammetry-software
https://www.pix4d.com/product/pix4dcapture
http://www.pointclouds.org
https://www.precisionhawk.com/precisionmapper http://www.pix4d.com
http://www.rapidlasso.com/LAStools/
http://www.riegl.com/media-events/multimedia/3d-app/
http://www.saga-gis.org
https://www.simactive.com/
http://www.sourceforge.net/projects/mcclidar/
http://www.spdlib.org/doku.php
https://www.support.dronesmadeeasy.com
http://www.terrasolid.fi
https://www.thewiredshopper.com/top-5-latest-technology-drone/
https://www.ugcs.com
http://www.vcg.isti.cnr.it/vcglib/
http://www.vls-inc.com/login/downloads/LIDAR_Analyst_v5_1_2_ArcGIS.zip
http://www.washingtonpost.com/blogs/the-switch/wp/2013/08/15/its-a-bird-its-a-plane-its-a-drone-that-makes-movies/

Index

A

Above Ground Level (AGL), 245, 251
Accelerometer, 21, 61, 72, 75–79, 102, 153
Aerodynamics, 45–48, 55, 58–60, 64
Agisoft PhotoScan, 179, 185, 189, 209–212, 221, 223, 294
Agriculture, 16, 82, 258–260, 280–283, 290–292
Airborne Collision Avoidance System (ACAS), 17
Airborne Laser Scanner (ALS), 88
Aircraft, 1, 17–18, 34–38, 46–99
Air Traffic, 18, 117, 151, 227, 249, 251, 365
Air Traffic Control, 18, 131, 158, 228, 251, 307, 365
Air Traffic Management (ATM), 131, 249, 331
Altitude, 3, 20, 36–39
Applications, 255–325
ArduPilot, 73, 74, 136, 139
Artificial Intelligence (AI), 12, 200, 261, 352
Artificial Neural Network (ANN), 204
Artificial Potential Fields (APF), 133
Augmented Reality (AR), 12
Australia, 233, 234, 240, 242, 318, 356, 357
Automatic, 18, 22, 59, 99–102, 110, 122, 123
Autonomous UAVs, 3, 99–149, 237, 259, 299, 306, 328
Aviation, 123, 131, 141, 225–252, 254, 350
Aviation Regulations, 233, 250

B

Barometer, 7, 34, 75, 78, 79
Base Station (BS), 34, 55, 95, 154–163, 168, 170, 176, 205, 282
Battery Management, 344, 345
Beyond Visual Line-of-Sight (BVLOS), 17, 124, 243
Big Data, 40, 199, 201, 261, 297, 313, 324, 340–342
Biological Sensors, 80, 258
Bluetooth, 125, 137, 229, 308
Bridges, 84, 89, 187, 195, 274, 275, 324
Building Information Modelling (BIM), 212

C

Canada, 234, 241, 243, 356
Chemical Sensors, 74, 76, 80
China, 226, 234, 243, 249, 250, 315
Cloud Computing, 180, 208, 298, 313, 340
Collision Avoidance, 100, 108, 126, 127, 131, 135, 143, 149, 155, 274
Communication, 151–177
Communication System, 136, 152–157, 163, 168, 233, 259, 335
Compound Annual Growth Rate (CAGR), 16, 256, 323, 350, 353–361
Concept of Autonomy, 99, 105
Construction, 6, 16, 189, 256–260, 264–268, 275–276, 357, 358, 362

Convolutional Neural Network (CNN), 206, 207
Cost of UAVs, 147
COVID-19, 315–319
Crop, 174, 260, 280–292, 312
Crop disease, 281, 283, 290
Crop yield, 209, 292, 313, 324
Crowd sourcing, 309–311, 313
Cyber-security, 166–167, 254, 333

D

Dams, 60, 81, 258, 274, 276
Data Mining, 201
Data Processing, 185, 194, 199, 221
Data Products, 180, 181
Decision Support System (DSS), 279
Deep Learning, 202, 206, 207, 304
Digital Elevation Model (DEM), 82, 84, 116, 132, 179, 184–189
Digital Surface Model (DSM), 192–193
Digital Terrain Model (DTM), 193–194
Directorate General of Civil Aviation (DGCA), 247–249
Disaster, 11, 172, 304–311
Distance Sensors, 80
3-D Modelling, 294, 295
Drag, 46–48, 72
Dronecode, 136–139
DroneDeploy, 139, 179, 211, 213
Drone Taxis, 349

E

Electronic Speed Controller (ESC), 69, 70
Emergency, 161, 174, 175, 258, 301, 307, 312–315, 335
Energy sources, 21, 342, 343
Entertainment, 247, 257, 319
Environmental, 24, 75, 80, 82, 106
European Union (EU) Regulations, 236
Extended Visual Line of Sight (EVLOS), 18, 353

F

FAA Regulations, 235–236
Federal Aviation Administration (FAA), 1, 5, 126, 233–236, 320
Fixed-wing UAVs, 10, 19, 35, 59–61
Flight Controller, 34, 69, 72
Flight Planning, 96, 116, 119
Flight Safety, 45, 46, 123, 143
Flying Ad-hoc Networks (FANETs), 170–172
Forestry, 9, 33, 276, 277
FPV approach, 59, 75, 164, 322
Free Space Optical (FSO), 162–164
Fully Autonomous, 100, 103, 108, 110, 360
Future insights, 328
Fuzzy Logic, 133, 202–203

G

Geofencing, 18, 100, 124–126, 129
Geographic Information System (GIS), 184–185
Georeferencing, 25, 33, 184, 185
Germany, 216, 234, 244
Gimbals, 35, 75
Global Navigation Satellite System (GLONASS), 17, 37, 58
Global Navigation Satellite System (GNSS), 22, 24, 25, 37, 327
Global Positioning System (GPS), 1, 17, 18, 37, 58, 80, 334, 351
Ground Control Points (GCPs), 184–186, 192
Ground Control Station (GCS), 2, 18, 26, 95, 242, 305
Gyroscope, 18, 72, 76

H

Health, 99, 177, 256, 313
Hexacopter, 19, 60, 64–66, 111
High Altitude Long Endurance (HALE), 52, 53, 74

High Altitude Platform (HAP), 56–58
Historical developments, 3–6
Hong Kong, 244
Hydromorphological, 294
Hyperspectral sensors, 21, 81, 82, 85

I

Image processing, 84, 127, 180, 200–217
India, 234, 247–254, 313, 315, 356
Inertial Measurement Unit (IMU), 18, 27
Inertial Navigation System (INS), 25, 37, 74, 88
Infrared (IR), 34, 79, 86, 143
Innovative UAVs/Drones, 144
Inspection, 15, 35, 82, 259–276
Insurance, 234, 237, 241, 252, 312
Intelligent Flight, 141, 142, 146, 180
Intelligent Transportation System (ITS), 300, 363
International Civil Aviation Organisation (ICAO), 233, 240, 241
Internet of Things (IoT), 161, 172–177, 261, 268, 282, 297
Irrigation, 209, 280–283, 289
Israel, 5, 8, 238, 244
Issues of concern, 226, 228

J

Japan, 234, 245
Journalism, 12, 16, 42, 320

K

Key challenges, 161, 286, 330

L

Legal issues, 226, 230, 232
LiDAR sensors, 27, 29, 31, 35
LiDAR software, 217–220
LiDAR UAVs, 21, 27–32, 41, 187
Lift, 10, 46, 47

Light Detection and Ranging (LiDAR), 27–33, 75, 80, 81, 88–90, 182
Line of Sight (LOS), 17, 18, 130, 154
Lithium ion (Li-Ion), 232, 342–345
Lithium Polymer (LiPo), 70, 91, 93, 342–347
Logistics, 263, 264, 269, 306–312
Low Altitude Platform (LAP), 55–57

M

Machine Learning, 114, 199–206
Machine to Machine (M2M), 173, 174
Magnetometer, 75, 78, 81
Management, 104, 108, 118
Maneuver, 34–38, 40, 61, 64, 135
Mapping, 22, 31–36, 61, 82, 89, 179, 211, 262, 263
Mean Sea Level (MSL), 246
Media, 180, 257, 258, 319
Medium Altitude Long Endurance (MALE), 52, 53
Micro Unmanned Aerial Vehicle (MAV), 52, 114
Military, 2–13, 16, 42, 53, 255, 300, 320–323, 329, 359–361
Mining, 31, 33, 201, 259, 260, 271
Mini Unmanned Aerial Vehicle (MUAV), 52
Mission planning, 25, 26, 111, 116–120
Mission planning and control station, 120, 121
Mobile Ad-hoc Networks (MANETs), 170–171
Monitoring, 39, 42, 82, 85, 164, 172–175, 181, 259, 270–283, 298–301
Moral and ethical issues, 231
Motors, 49, 50, 60–69
Multispectral, 58, 82–86
Multi-View Stereopsis (MVS), 195–198

N

Nano Air Vehicles (NAV), 52
Near Infrared (NIR), 83, 90, 221, 285
New Zealand, 245, 356

Nickel Cadmium (NiCd), 343
Nickel Metal Hydride (NiMH), 70, 343
Normalised Difference Vegaetation Index (NDVI), 180, 209, 214, 215, 290–292

O

Octocopter, 65
Oil and gas, 258, 268
Open source software, 136, 137, 179, 209–211, 218–220
Orthophotos, 179, 194–195
Overview, 1, 142, 363

P

Paparazzi, 100, 138
Path planning, 108, 110, 113, 116–120, 132–141
Payloads, 20, 41, 60, 62, 74
Photogrammetry software, 37, 179, 209–217
Photogrammetry UAVs, 22–27, 31–33, 187, 193
Pitch, 20, 34, 46–50
Pix4D, 26, 37, 138, 200, 209
Point of Interest (POI), 17, 125
Position Hold (PH), 18
Post-disaster, 37, 306, 311
Power, 28, 33, 68, 87, 91–94, 176, 260, 269
Precision agriculture, 11, 16, 82, 258, 283–286
Privacy issues, 230
Propellers, 20, 47, 61, 69, 70

Q

Quadcopter, 19, 21, 29, 64
Quality of Service (QoS), 295, 296, 364

R

Radio-Frequency Identification (RFID), 125, 229, 268, 362
Ready to Fly (RTF), 143, 288

Real Time Kinematic (RTK), 24, 80, 180
Receiver, 33, 61, 63
Reconnaissance, 255, 256, 311, 321, 329
Red, Green, Blue (RGB), 81–86, 97, 207
Region of Interest, 17
Remotely Piloted Aircraft (RPA), 1, 2, 247, 250, 251
Remotely Piloted Aircraft System (RPAS), 247
Remotely Piloted Vehicle, 108
Remote Sensing, 21, 39, 81, 82, 87, 88, 181, 205, 284, 290, 353
Remote sensing sensors, 81, 82, 181
Response, 41, 73, 80, 142, 175
Return to Home (RTH), 19, 140, 142
Robot, 1, 103, 110, 114, 132
Roll, 46–49
Root Mean Square Error (rmse), 185, 188
Rotary-wing UAVs, 10, 19, 35, 51, 60
Russia, 200, 212, 234, 246

S

Safety issues, 99, 124, 231, 350
Scale-Invariant Feature Transform (SIFT), 198
Search and Rescue (SaR), 304–306
Semi-autonomous, 22, 53, 108, 169
Sensors, 1, 2, 21, 22, 29, 34, 72–90, 101, 104, 120–122
Sheridan scale, 109
Simultaneous Localization and Mapping (SLAM), 140, 295, 362
Smart cities, 161, 173, 296–298
Social media, 54, 309
Soil, 82, 280–286
Solar-powered UAVs, 164, 347, 350
South Africa, 234, 240, 246
Spraying, 60, 225, 280, 291
Structural Health Monitoring (SHM), 106, 272–273
Structure from Motion (SfM), 141, 195–198, 209

Support Vector Machine (SVM), 204
Surveillance, 5, 9, 38, 52, 100, 174, 227, 301
Surveying, 106, 215, 228, 263
Swarm, 152, 300–301, 337–339
Swarm Architecture, 169–170
SWOT, 349, 351
Synthetic Aperture Radar (SAR), 82, 87

T

Tactical Unmanned Aerial Vehicle (TUAV), 53
Thermal Infrared (TIR), 85–87
Thermal Infrared Sensors, 86, 87
Throttle, 51, 63
Thrust, 46, 48
Tilt sensors, 79
Time of Flight (ToF) sensors, 79, 81, 90
Traffic monitoring, 16, 172, 298
Transmitter, 19, 33, 63
Triangular Irregular Network (TIN), 190
Tricopter, 19, 60, 64

U

UAV rating, 110, 250
Ukraine, 247
Unique Identification Number (UIN), 248
United Kingdom (UK), 231–234, 238, 241, 356
United Nations (UN), 225, 240
United States (US), 1, 6, 235, 269
Unmanned Aerial System (UAS), 1, 2, 5, 106, 242

V

Vehicle-to-Vehicle (V2V), 132
Vehicular Ad-hoc Networks (VANETs), 170, 171
Vertical Takeoff and Landing (VTOL), 16, 41, 57, 349
Virtual Reality (VR), 12, 140, 261
Visualization, 39, 184, 187, 199
Visual Line-of-Sight (VLOS), 17, 124, 239
Voice over Internet Protocol (VoIP), 339

W

Waypoints, 17–19, 25, 33
WiFi, 125, 145, 161, 229, 308, 340
Wireless charging, 344, 346, 347
Wireless network, 157–163, 177, 295
Wireless Sensor Network (WSN), 158–162

Y

Yaw, 46, 49–51